Selected Topics on Generalized Integration

editors

Tin Lam Toh
Nanyang Technological University, Singapore

Hemanta Kalita
VIT Bhopal University, India

Anca Croitoru
University "Alexandru Ioan Cuza" of Iasi, Romania

Tomas Perez Becerra
Universidad Tecnológica de la Mixteca, Mexico

Bipan Hazarika
Gauhati University, India

NEW JERSEY • LONDON • SINGAPORE • BEIJING • SHANGHAI • TAIPEI • CHENNAI

Published by

World Scientific Publishing Co. Pte. Ltd.
5 Toh Tuck Link, Singapore 596224
USA office: 27 Warren Street, Suite 401-402, Hackensack, NJ 07601
UK office: 57 Shelton Street, Covent Garden, London WC2H 9HE

Library of Congress Control Number: 2025019880

British Library Cataloguing-in-Publication Data
A catalogue record for this book is available from the British Library.

SELECTED TOPICS ON GENERALIZED INTEGRATION

Copyright © 2025 by World Scientific Publishing Co. Pte. Ltd.

All rights reserved. This book, or parts thereof, may not be reproduced in any form or by any means, electronic or mechanical, including photocopying, recording or any information storage and retrieval system now known or to be invented, without written permission from the publisher.

For photocopying of material in this volume, please pay a copying fee through the Copyright Clearance Center, Inc., 222 Rosewood Drive, Danvers, MA 01923, USA. In this case permission to photocopy is not required from the publisher.

ISBN 978-981-98-1219-6 (hardcover)
ISBN 978-981-98-1220-2 (ebook for institutions)
ISBN 978-981-98-1221-9 (ebook for individuals)

For any available supplementary material, please visit
https://www.worldscientific.com/worldscibooks/10.1142/14283#t=suppl

© 2025 World Scientific Publishing Company
https://doi.org/10.1142/9789819812202_fmatter

Preface

In 1956, Jaroslav Kurzweil studied unusual events in ordinary differential equations. Specifically, he researched ordinary differential equations with rapidly oscillating inputs that were not explained by theories or methods available at the time. He developed a technique to explain the observed results, which in some ways strongly resembled the construction of the Perron integral using minor and major functions. The narrative concluded successfully, and the generalised Perron integral - the tool became a stand-alone, self-contained entity. Jaroslav Kurzweil wrote a detailed description of the generalised Perron integral in his work on the subject and used it in a number of later works on ordinary differential equations because it turned out to be highly intriguing and, due to the need for research, it was described via integral sums of Riemann type. At the same time Ralph Henstock worked on variational approaches to integrals, separately from Kurzweil due to the lack of connection to him. In the early 1960s, it was found that both the Henstock and Kurzweil techniques are equivalent when dealing with real-valued functions. Naturally, this led to the construction of the extremely general non-absolutely convergent integral based on Riemann-type integral sums taking centre stage. Henstock-Kurzweil integrals, which are generalized integrals, are also called non-absolute integral. This is a generalized integral. The abstract concept of measure theory makes Lebesgue integrals complicated. In the late 1960s, McShane defined a Riemann-type integral and proved that it is identical to the Lebesgue integral. Being a Riemann-type integral, it is more user-friendly to work than Lebesgue integral.

In Chapter 1, the authors introduce Henstock-Kurzweil and McShane's gauge integrals on metric measure spaces. For real-valued functions on a compact real interval, these extend the concepts of Riemann, Lebesgue, and improper integrals. Regarding the Henstock-Kurzweil integration on metric measure spaces, the authors fill up several gaps in the existing literature. To ensure the validity of Cousin's lemma's conclusion, the authors pay special attention to the critical task of choosing suitable candidates for "intervals"

v

in metric spaces. The authors go over a few definitions of completeness and compactness in metric spaces using the Cousin's lemma, a fundamental finding in gauge integrals. The authors go into Radon-Nikodym type theorems, measure-theoretic characterisations of gauge integrals, a variant of Hake's theorem for metric measure spaces, links between different integrals on metric spaces, and extensions of absolute continuity. Additionally, the authors present alternative proofs for generalizations of two results in Lebesgue integration. The first proof bypasses the use of the Vitali covering lemma to establish a result on absolute continuity, while the second proof presents a version of the fundamental theorem of calculus within our specific setting.

In Chapter 2, the author investigates the variational Henstock integral of functions defined on $[0, 1]$, taking values in a topological vector space X that is locally convex. It is demonstrated that a variational Henstock integrable function whose values are in a locally convex space is measurable by semi-norm, and that its primitive is continuous and nearly always differentiable by semi-norm. Additionally, the author describes locally convex spaces with the Radon-Nikodym condition using additive interval functions whose Henstock vatiational measures are completely continuous with regard to the Lebesgue measure. The author demonstrates that the Radon-Nikodym property of the locally convex space exists if and only if every X-valued additive interval function has an absolutely continuous variational measure with respect to the Lebesgue measure.

In Chapter 3, the author discusses variational version of Henstock type integral with respect to the basis. An equivalency of variational equivalence and variational Henstock-type integral is established here. In application, the author demonstrates variational integral in Harmonic analysis.

In Chapter 4, the authors report some findings on the Riemann-Lebesgue integral of a vector (real resp.) function. These findings are then extended to the case of interval-valued multifunctions that are Riemann-Lebesgue integrable.

In Chapter 5, the authors present some generalizations for the Sugeno integral of vector mutifunctions with respect to a submeasure. Also, properties for these integrals and relationships between them are established.

In Chapter 6, the authors study the convergence of sequences and series of Riemann integrable functions over Banach spaces on time scales. The authors discuss several classical convergence theorems in the context of Riemann integrable functions over Banach spaces on time scales.

In Chapter 7, the authors present some comparative results concerning set-valued integrals: a Dunford type integral of multifunctions with respect to a multimeasure, a Gould type integral of real functions in relation to a multimeasure, and a Gould type integral of multifunctions relative to a non-negative set function.

In Chapter 8, the heat equation on the real line is solved in the space of distributions. The initial data is taken to be the distributional derivative of an $L^p(\mathbb{R})$ function. The solutions are shown to be smooth functions. Initial conditions are taken on in norm. Sharp estimates of solutions are obtained and a uniqueness theorem is proved.

In Chapter 9, generalized symmetric integral, the symmetric Laplace integral, is defined using the concept of symmetric Laplace derivative which is an integral with continuous primitives and more general than the Henstock integral. After presenting the basic properties of this integral, a useful application to the trigonometric series is provided.

In Chapter 10, a variety of finite and infinite integrals involving a family of incomplete I-functions are investigated.

In Chapter 11, the authors study q-Homotopy Analysis Method, a generalised and more powerful variation of Homotopy Analysis Method to solve the Nonlinear Fredholm Integral Equations of the Second type (NFIES) and test the efficiency and accuracy of the method with some examples. q-Homotopy Analysis Method is a stronger technique as it increases the interval of convergence that exists in Homotopy Analysis Method and the approximation series solutions are more likely to converges.

In Chapter 12, the author presents some properties of operator I_ν in the spaces by considering vector measures on σ-rings.

In Chapter 13, the author provides a treatment of the integration theory that is more intuitive, unifying and technically straightforward. The author uses Riemann-like approach for the vector space valued theory of integral. The approach leads to generalizations of the Dvoretsky-Rogers Theorem, and the Spectral Theorem on Hilbert spaces. By the same token, it can be used to push stochastic integration towards the more general setting of integrable processes taking values in topological vector spaces.

In Chapter 14, the authors propose some unified integral representations for the four-parameter Mittag-Leffler function, and their findings are evaluated in terms of generalized special functions. Additionally, a special case of the four-parameter Mittag-Leffler function has been corollarily presented.

Tin Lam Toh, Nanyang Technological University, Singapore
Hemanta Kalita, VIT Bhopal University, India
Anca Croitoru, University "Alexandru Ioan Cuza" of Iasia, Romania
Tomas Perez Becerra, Mixteca Technological University, Mexico
Bipan Hazarika, Gauhati University, India

© 2025 World Scientific Publishing Company
https://doi.org/10.1142/9789819812202_fmatter

Contents

Preface   v

1. Gauge Integrals on Metric Measure Spaces   1

    S. P. S. Kainth and N. Singh

2. Variational Henstock Integral and its Variational Measure in
    Locally Convex Space   25

    S. Bhatnagar

3. Variational Version of Henstock type Integral and Application
    in Harmonic Analysis   43

    V. Skvortsov

4. A Survey on the Riemann-Lebesgue Integrability in
    Non-additive Setting   59

    A. Croitoru, A. Gavriluţ, A. Iosif and A. R. Sambucini

5. Some Nonlinear Integrals of Vector Multifunctions with
    Respect to a Submeasure   95

    C. Stamate and A. Croitoru

6. Convergence of Riemann Integrable Functions over Banach
    Spaces on Time Scales   123

    H. Bharali, V. Sekhose and H. Kalita

Contents

7. Comparative Results among Different Types of Generalized Integral 143

 H. Kalita and A. Croitoru

8. The Heat Equation with the L^p Primitive Integral 157

 E. Talvila

9. On the Symmetric Laplace Integral and Its Application to Trigonometric Series 175

 S. Mahanta

10. Finite and Infinite Integral Formulae Associated with the Family of Incomplete I-Functions 185

 S. Bhatter, Nishant, S. D. Purohit

11. Homotopy Analysis Method for Solving Nonlinear Fredholm Integral Equations of Second Kind 205

 S. Paul and S. Koley

12. L^1-space of Vector Measures with Density Defined on δ-rings 221

 C. Avalos-Ramos

13. More on Unified Approach to Integration 263

 M. A. Robdera

14. Some Unified Integral Representations of the Four-parameter Mittag-Leffler Functions 291

 A. Pal and K. Kumari

Author Index 305

© 2025 World Scientific Publishing Company
https://doi.org/10.1142/9789819812202_0001

Chapter 1

Gauge Integrals on Metric Measure Spaces

Surinder Pal Singh Kainth

*Department of Mathematics, Panjab University,
Chandigarh, 160014, India
Email: sps@pu.ac.in*

Narinder Singh

*Department of Mathematics, Lovely Professional University,
Phagwara, 144411, India
Email: narinder.30462@lpu.co.in*

This chapter will present the gauge integrals of Henstock-Kurzweil and McShane on metric measure spaces. These generalize the notions of Riemann, Lebesgue, and improper integrals for real-valued functions on a compact real interval.

We address certain gaps in the current literature concerning the Henstock-Kurzweil integration on metric measure spaces. Our focus lies particularly on the crucial aspect of selecting appropriate candidates for 'intervals' in metric spaces, ensuring the validity of the conclusion of Cousin's lemma. We shall discuss some characterizations of compactness and completeness in metric spaces in terms of the Cousin's lemma, a basic result in gauge integrals.

We discuss a version of Hake's theorem for metric measure spaces, measure-theoretic characterizations of gauge integrals, Radon-Nikodym type theorems, relationships among various integrals on metric spaces, and extensions of absolute continuity. Additionally, we present alternative proofs for generalizations of two results in Lebesgue integration. The first proof bypasses the use of the Vitali covering lemma to establish a result on absolute continuity, while the second proof presents a version of the fundamental theorem of calculus within our specific setting.

1. Introduction

In 1957, Jaroslav Kurzweil introduced a modified version of the classical Riemann integral in his paper on differential equations. Subsequently, Ralph Henstock proposed this modified integral as a new approach in 1961, now known as the Henstock-Kurzweil (HK) integral (see [1–3]). It is well known that this integral is non-absolute and generalizes the notions of Riemann, Lebesgue, and improper integrals for real valued functions on compact intervals (see [4]).

Several authors have made efforts to extend the Henstock-Kurzweil integral to metric spaces and some other forms of topological spaces, see e.g. [5–10]. It requires suitable representatives of 'real intervals' to more general spaces so that the corresponding *Cousin's lemma* holds, that is, the existence of the corresponding fine tagged partitions is ensured.

There seems to be an error in the proof of the lemma presented in [6]. Specifically, on lines 4-5 on page 37 of this paper, the authors assume that $x_i \notin B_k; i \neq k$, implies that x_i belongs to the interior of $\bigcup_{k=1}^{i-1}(\overline{B_i} \setminus \overline{B_k})$. A similar issue exists within the proof of the Cousin's lemma on [8, p. 16, line 8]. In this chapter, we fix up this issue by redefining 'intervals' in metric spaces and establishing the corresponding Cousin's lemma.

In Section 3, we focus on the Cousin's lemma and its relationship with topological properties in metric spaces. We demonstrate that a metric space is compact if and only if it satisfies the conclusion of Cousin's lemma. This result allows us to characterize complete metric spaces. Additionally, we offer alternative proofs that show how, in metric spaces, Cousin's lemma implies the Cantor intersection property, total boundedness, and sequential compactness.

In Section 4, we provide some extensions of the Hake's theorem on metric measure spaces in terms of the Henstock variational measure V_F.

In Section 5, we establish the differentiability theorem for our integral and the measurability of Henstock-Kurzweil integrable functions, which is required for the equivalence of the Lebesgue, McShane, and absolute Henstock-Kurzweil integrals. We rectify some errors in similar results from [5], which have been overlooked by the authors (see Remark 2).

In Section 6 of this chapter, we propose short proofs of the descriptive characterizations of the HK-integral, in terms of the variational measure V_F as well as the ACG^Δ functions. We offer clear and streamlined demonstrations of these characterizations, providing a deeper understanding of the HK-integral.

The final section of this chapter offers alternative proofs for certain generalizations of two results in Lebesgue integration. The first proof circumvents the use of the Vitali covering lemma to establish a result on absolute continuity. The second proof presents a version of the fundamental theorem of calculus within our specific setting.

2. Preliminaries

Let (X, d) be a metric space. For any $E \subset X$, let $\overline{E}, \partial E, E^o$ and $\overline{E}^o$, respectively, denote the topological closure, boundary, interior and the interior of closure of E in X. Let $B(x; r)$ denote the open ball in X, centered at $x(\in X)$ with radius $r(> 0)$. We generalize the notion of intervals to metric spaces, as follows.

Let $\mathcal{I}_0$ be the collection of scalloped balls, with sets of the form $B_1 \setminus B_2$, where B_1 and B_2 are open balls in X such that neither $B_1 \subset B_2$ nor $B_2 \subset B_1$. Define $\mathcal{I}$ to be the collection of closures of finite intersections of sets from $\mathcal{I}_0$. In other words,

$$\mathcal{I} := \left\{ \overline{\bigcap_{i=1}^{n} J_i} : J_1, \ldots, J_n \in \mathcal{I}_0, n \in \mathbb{N} \right\}.$$

Sets in $\mathcal{I}$ will be termed as *generalized intervals* of X. If there is no ambiguity, we will simply use the term *intervals* in place of generalized intervals.

In this sense, the generalized intervals in the metric space $X = \mathbb{R}$, equipped with the usual metric, are simply closed, bounded and real intervals.

Example 1. Let $X = \mathbb{R}^2$. For every $x = (x_1, x_2), y = (y_1, y_2) \in X$, define

$$d_1(x, y) := |x_1 - y_1| + |x_2 - y_2|,$$
$$d_2(x, y) := \sqrt{(x_1 - y_1)^2 + (x_2 - y_2)^2}, \text{ and}$$
$$d_\infty(x, y) := \max\{|x_1 - y_1|, |x_2 - y_2|\}.$$

It is well known that both d_1 and d_∞ are metrics on X with open balls as open squares with edges having slope ± 1 and having only vertical or horizontal edges, respectively. Hence, our intervals in these spaces are finite unions of polygons having such edges.

Further, the open balls in (X, d_2) are open disks and thus intervals in this case are finite unions of sets in the $\mathbb{R}^2$ plane with piecewise circular boundaries.

Two intervals I and J of a metric space (X, d) will be called *non-overlapping* if they have disjoint interiors, that is, $I^o \cap J^o = \emptyset$. If $E \subset X$, any function $\delta : E \longrightarrow (0, \infty)$ will be termed as a *gauge* on E.

Let $I \in \mathcal{I}$ be arbitrary. A finite collection $\mathcal{P} := \{(I_j, t_j) : j = 1, \ldots, n\}$ of interval-point pairs will be called a *partial partition* in I if I_j's are pairwise non-overlapping intervals such that $\cup_{j=1}^n I_j \subset I$ and $t_j \in I_j$ for all $j = 1, \ldots, n$. Further, this partial partition $\mathcal{P}$ will be called:

- a *tagged partition* of I, if $\cup_{j=1}^n I_j = I$,
- *δ-fine,* if $I_j \subset B(t_j; \delta(t_j))$ for all $j = 1, \ldots, n$, and
- *E-anchored,* if $t_j \in E$ for all $j = 1, \ldots, n$.

Let μ be a non-atomic Radon measure on a sigma algebra containing the family of intervals $\mathcal{I}$ of X such that

$$\mu(\partial I) = 0 \text{ for all } I \in \mathcal{I}.$$

In fact, any non-atomic probability measure is enough for our results of this chapter, as every probability measure on a metric space is regular (see [11, Proposition 19.13]). In the beginning of Section 5, we shall establish the existence of such measures on compact metric spaces.

The *Riemann sum* of a function $f : X \longrightarrow \mathbb{R}$ w.r.t. a partial partition $\mathcal{P} = \{(I_j, t_j) : j = 1, \ldots, n\}$ is defined as

$$S(\mathcal{P}, f) := \sum_{j=1}^n f(t_j)\mu(I_j).$$

The function $f : X \longrightarrow \mathbb{R}$ is said to be *Henstock-Kurzweil integrable* (or simply HK-integrable) if there exists some $\lambda \in \mathbb{R}$ such that for every $\epsilon > 0$ there is a gauge δ on X such that

$$\left| S(\mathcal{P}, f) - \lambda \right| < \epsilon,$$

for every δ-fine tagged partition $\mathcal{P}$ of X. The real number λ is known as the value of Henstock-Kurzweil integral (HK-integral) of f over X which is denoted by $(HK) \int_X f d\mu$.

By keeping the tags independent of their associated subintervals, the *McShane integral* is defined. It is pertinent to mention that the generalized intervals in the HK-integral cannot be replaced with measurable sets or closed sets, as in that case the integral will be reduced to the McShane integral even for functions on $\mathbb{R}^m$, see [12] for more details.

Note that the HK-integral is well-defined only if for each gauge δ on X there exists at least one δ-fine tagged partition of X. This result on closed bounded real intervals is known as *Cousin's lemma*, given by Pierre Cousin in [13] for his study of complex functions of n-variables. Its standard proof is a direct application of the nested intervals property of reals (see [4, p. 11]).

Remark 1. In [5], pairwise non-overlapping 'cells' (a class of closed sets satisfying some particular conditions) were considered, while pairwise disjoint scalloped balls in the tagged partitions were considered in [6]. In [10], we had chosen the domain to be a compact metric space, while the same is done over locally compact metric spaces in [8].

It must be noted that in [5], no explicit construction of 'intervals' in general metric spaces is given. Instead, the authors assume the existence of a family of closed subsets of metric spaces satisfying some 'nice' conditions.

In [14], we establish the Cousin's lemma for our choice of intervals on compact metric spaces. The same will be provided in Theorem 1.

2.1. *Metric Outer Measures*

In the sequel, let X be a compact metric space and $\mathcal{I}$ be the collection of intervals in X. In this section, we first ensure that there exists a metric outer measure satisfying our required properties, before establishing the measurability of HK-integrable functions on X.

Let $\mathcal{C}$ be any collection of subsets of X containing the collection $\{\partial I : I \in \mathcal{I}\} \cup \{\emptyset, X\}$ and $\nu : \mathcal{C} \longrightarrow [0, \infty)$ be any function such that

$$\nu(\emptyset) := 0, \nu(X) < \infty, \text{ and } \nu(\partial I) := 0 \text{ for all } I \in \mathcal{I}.$$

We outline *Method II outer measure on X induced by ν* (see [15, p. 159]).

For $\epsilon > 0$, let $\mathcal{C}_\epsilon := \{C \in \mathcal{C} : diam(C) \le \epsilon\}$. For $A \subset X$, define

$$\mu_\epsilon^*(A) := \inf \left\{ \sum_{n=1}^\infty \nu(C_n) : C_n \in \mathcal{C}_\epsilon, \bigcup_{n=1}^\infty C_n \supset A \right\}.$$

Here, we adopt the standard convention that infimum over an empty set is $+\infty$. If $0 < \epsilon' < \epsilon$, note that $\mathcal{C}_{\epsilon'} \subset \mathcal{C}_\epsilon$ and hence $\mu_\epsilon^* \le \mu_{\epsilon'}^*$. The *Method II outer measure induced by ν* is defined as

$$\mu^*(A) := \lim_{\epsilon \longrightarrow 0} \mu_\epsilon^*(A) = \sup_{\epsilon > 0} \mu_\epsilon^*(A) \text{ for all } A \subset X.$$

By [15, p. 159, Theorem 5.4.4], μ^* is a metric outer measure on X. Let $\mathcal{M}$ denote the collection of μ^*-measurable subsets of X, that is, $\mathcal{M}$ is the collection of sets $E \subset X$ such that

$$\mu^*(A) \ge \mu^*(A \cap E) + \mu^*(A \setminus E) \text{ for all } A \subset X.$$

It is immediate that every μ^*-null set is μ^*-measurable. Further, $\mathcal{M}$ is a sigma algebra and μ^* is countably additive on $\mathcal{M}$ (see [15, p. 151, Theorem 5.2.5]). Write $\mu := \mu^*|_{\mathcal{M}}$.

Since μ^* is a metric outer measure on X, every Borel subset of X is μ^*-measurable (see [15, p. 157, Theorem 5.4.2]).

From the definition of μ^*, it follows that $\mu^*(A) \leq \nu(A)$ for all $A \in \mathcal{C}$. Hence $\mu(X) = \mu^*(X) \leq \nu(X) < \infty$. By [15, p. 157, Theorem 5.4.3], μ is regular on every Borel subset of X.

Next, it is pertinent to note that the HK-integral on X with respect to μ satisfies the basic properties such as linearity, additivity, monotonicity, the Saks-Henstock lemma and the monotone convergence theorem. Their proofs are analogous to the case of functions on compact real intervals (see [4, 16, 17]).

In particular, if f is HK-integrable over X, then so is $f.\chi_I$ for all $I \in \mathcal{I}$. Further, $(HK) \int_X f.\chi_I d\mu$ is known as the HK-integral of f over I and is also denoted by $(HK) \int_I f d\mu$. In this case, a set function $F : \mathcal{I} \longrightarrow \mathbb{R}$ is called the *primitive* of f if

$$F(I) = (HK) \int_I f d\mu \text{ for all } I \in \mathcal{I}.$$

3. The Cousin's Lemma and its Topological Equivalents

We say that a metric space X has *Cousin's property* if for every gauge δ on X, there exists some δ-fine tagged partition of X.

In this section, we establish that a metric space is compact if and only if it satisfies Cousin's property. As a consequence, we conclude a characterization of complete metric spaces. In [6] and [8, p. 11], the authors present some sufficient conditions for the existence of fine tagged partitions. Some arguments have been overlooked in the proofs therein. It has been intrinsically assumed that if $\overline{A}$ has a δ-fine tagged partition, then so does the set A. We fix these issues here.

Recall that a metric space is compact if and only if it is complete and totally bounded (see [18, p. 132, Theorem 5.27]).

Theorem 1. *Every compact metric space satisfies Cousin's property.*

Proof. Let (X, d) be a compact metric space and δ be a gauge on X. Assume that there exists no δ-fine tagged partition of X.

Since X is compact, it is totally bounded. So there are finitely many balls $B_1, \ldots, B_n$, having diameter < 1, such that $\bigcup_{i=1}^{n} B_i = X$. Further, by dropping some B_i, we assume that $B_i \not\subset B_j$, for all $i \neq j$. Let $A_i :=$ $B_i \setminus \bigcup_{k<i} B_k$, for every i. We *claim* that $A_1, \ldots, A_n$ are non-overlapping intervals.

Suppose not. Then there exist $i < j$ and some $x \in (A_i)^o \cap (A_j)^o$. Hence there exists an open ball B such that $x \in B \subset A_i \cap A_j$. Therefore

$$x \in B \subset \overline{\left(B_i \setminus \bigcup_{k<i} B_k\right)} \cap \overline{\left(B_j \setminus \bigcup_{k<j} B_k\right)}$$

$$\subset \overline{B_i} \cap \overline{(B_j \setminus B_i)} \subset \overline{B_i} \cap \overline{(\overline{B_j} \setminus B_i)} = \overline{B_i} \cap (\overline{B_j} \setminus B_i).$$

This enforces that $B \cap B_i = \emptyset$ and x is an adherent point of B_i, a contradiction. This establishes our claim.

Since X has no δ-fine tagged partition, there exists some j such that A_j is non-empty and has no δ-fine tagged partition. Otherwise, the union of a δ-fine tagged partition of each A_i will produce a δ-fine tagged partition of X. Let $X_1 := A_j$. Note that $diam(X_1) = diam(A_j) \leq diam(\overline{B_j}) \leq 1$, here $diam(E)$ denotes the diameter of $E(\subset X)$ defined as the supremum of the set $\{d(x,y) : x, y \in E\}$.

Note that X_1, being a subset of the totally bounded metric space X, is also totally bounded. As earlier, we choose a non-empty closed interval $X_2 \subset X_1$ such that $diam(X_2) \leq 1/2$ and X_2 has no δ-fine tagged partition. Inducting like this, we obtain a nested decreasing sequence of non-empty closed intervals $\{X_n\}$ of X such that $diam(X_n) \longrightarrow 0$.

Since X is compact, it is complete. Therefore X satisfies the Cantor intersection property (see [18, p. 96, Corollary 4.15]). Hence $\bigcap_{n=1}^{\infty} X_n = \{x_0\}$, for some $x_0 \in X$.

Since $\delta(x_0) > 0$, we have $1/m < \delta(x_0)$, for some $m \in \mathbb{N}$. Therefore $\{(X_m, x_0)\}$ is a δ-fine tagged partition of X_m, a contradiction. This proves the result. $\qquad\square$

The converse is much easier.

Theorem 2. *Let X be a metric space satisfying Cousin's property. Then X is compact.*

Proof. Let $\Omega = \{O_\alpha : \alpha \in \wedge\}$ be any open cover of X. Then $X = \bigcup_{\alpha \in \wedge} O_\alpha$. For each $x \in X$, choose $\alpha_x \in \wedge$ such that $x \in O_{\alpha_x}$. Since O_{α_x} is an open set containing x, there exists some $\delta(x) > 0$ such that $B(x; \delta(x)) \subset O_{\alpha_x}$. This defines a gauge δ on X.

Applying Cousin's property of X, let $P := \{(I_j, t_j) : j = 1, \ldots, n\}$ be any δ-fine tagged partition of X. Then $I_j \subset B(t_j; \delta(t_j))$ for all j and

$\bigcup_{j=1}^{n} I_j = X$. Hence

$$X = \bigcup_{j=1}^{n} I_j \subset \bigcup_{j=1}^{n} B(t_j; \delta(t_j)) \subset \bigcup_{j=1}^{n} O_{\alpha_{t_j}}.$$

Therefore $\{O_{\alpha_{t_1}}, \ldots, O_{\alpha_{t_n}}\}$ is a finite subcover of X from the collection Ω. Hence X is compact. $\qquad\square$

Finally, we present a characterization of complete metric spaces.

Theorem 3. *A metric space X is complete if and only if every totally bounded closed subspace of X satisfies Cousin's property.*

Proof. Let E be a totally bounded closed subset of X. If X is complete, then so is E. Hence E is compact. By Theorem 1, E satisfies Cousin's property.

To establish the converse, let $\{x_n\}$ be a Cauchy sequence in X and $E := \{x_n : n \in \mathbb{N}\}$. Note that for each $\epsilon > 0$, there exists some $N \in \mathbb{N}$ such that $d(x_n, x_m) < \epsilon$ for all $n > m \geq N$ which implies $E \subset \bigcup_{i=1}^{N} B(x_i; \epsilon)$. Hence E is totally bounded. Consequently, $\overline{E}$ is also totally bounded. By hypothesis, the subspace $\overline{E}$ satisfies Cousin's property.

Applying Theorem 2, $\overline{E}$ is compact and hence complete. Since $\{x_n\}$ is a Cauchy sequence in E, there exists some $x \in \overline{E} \subset X$ such that $x_n \longrightarrow x$. Therefore X is complete. $\qquad\square$

If X is a metric space satisfying Cousin's property, then it is compact and hence complete, totally bounded, and sequentially compact. Now we provide some direct alternative proofs of these results. First the quickest one.

Theorem 4. *Let X be a metric space satisfying Cousin's property. Then X is totally bounded.*

Proof. Let $\epsilon > 0$ be given. Define $\delta : X \longrightarrow (0, \infty)$ as the constant gauge $\delta(x) := \epsilon$ for all $x \in X$. By Cousin's property, let $\{(X_i, t_i) : i = 1, \ldots, n\}$ be a δ-fine tagged partition of X. Then $X_i \subset B(t_i; \delta(t_i))$ for each i. Hence we conclude that $X = \bigcup_{i=1}^{n} X_i \subset \bigcup_{i=1}^{n} B(t_i; \epsilon)$. $\qquad\square$

The completeness of a metric space is equivalent to the Cantor intersection property (see [18, p. 96, Corollary 4.15]). Therefore the following is another proof for the fact that every metric space satisfying Cousin's property is complete.

Theorem 5. *Let (X, d) be a metric space satisfying Cousin's property. Then X satisfies the Cantor intersection property.*

Proof. Assume that X does not satisfy the Cantor intersection property. Then there is a nested decreasing sequence $\{F_n\}$ of non-empty closed subsets of X such that the intersection $\bigcap_{n=1}^{\infty} F_n$ is empty. For each $n \in \mathbb{N}$, pick any $a_n \in F_n$. Define a gauge δ on X as follows:

Since $\bigcap_{n=1}^{\infty} F_n = \emptyset$, for each $t \in X$ there exists some positive integer n such that $t \notin F_n$. Let $n(t) := \min\{n : t \notin F_n\}$ and

$$\delta(t) := \frac{1}{2} dist(t, F_{n(t)}) = \frac{1}{2} \inf\{d(t, x) : x \in F_{n(t)}\}.$$

This defines a gauge δ on X. By Cousin's property, let $\{(J_k, t_k) : 1 \leq k \leq m\}$ be a δ-fine tagged partition of X. Then for every k, we have $J_k \subset B(t_k; \delta(t_k))$, which implies that $J_k \cap F_{n(t_k)} = \emptyset$. Since the sequence of sets $\{F_n\}$ is nested and decreasing, we obtain $a_n \notin J_k$, for all $n \geq n(t_k)$. Let

$$n_0 := \max\{n(t_1), \ldots, n(t_m)\}.$$

Then $a_{n_0} \notin J_k$ for all $1 \leq k \leq m$. Therefore $a_{n_0} \notin \bigcup_{k=1}^{m} J_k = X$, a contradiction, as $a_{n_0} \in F_{n_0} \subset X$. Hence the result. $\square$

Recall that a metric space X is *sequentially compact* if every sequence in X has a convergent subsequence (see [19, p. 179]).

Theorem 6. *Every metric space satisfying Cousin's property is sequentially compact.*

Proof. Let X be a metric space having Cousin's property. Assume that X has a sequence $\{x_n\}$ which has no convergent subsequence. Therefore $\{x_n\}$ has no constant subsequence and hence it contains infinitely many distinct terms. By throwing away repeating terms, without loss of generality, we assume that $\{x_n\}$ is a sequence with distinct terms. Let $E := \{x_n : n \in \mathbb{N}\}$.

Since $\{x_n\}$ has no convergent subsequence, the set E has no limit points. Therefore for every $t \in X$, there exists some $\epsilon_t > 0$ such that $B(t; \epsilon_t) \cap E \subset \{t\}$. Define $\delta(t) := \epsilon_t$ for each $t \in X$.

This defines a gauge δ on X. By Cousin's property of X, there exists a δ-fine tagged partition $\{(J_k, t_k) : k = 1, \ldots, m\}$ of X. Then for every $k =$

$1, \ldots, m$, we have $J_k \subset B(t_k; \epsilon_{t_k})$ which implies $J_k \cap E \subset \{t_k\}$. Therefore

$$E = X \cap E = \Big(\bigcup_{k=1}^{m} J_k \Big) \cap E = \bigcup_{k=1}^{m} (J_k \cap E) \subset \{t_1, \ldots, t_m\}$$

is a finite set, a contradiction. $\qquad \square$

4. Hake's Theorem on Metric Measure Spaces

The Hake's Theorem for real functions is as follows (see [16, Theorem 9.21]).

Theorem 7 (Hake). *A function $f : [0,1] \longrightarrow \mathbb{R}$ is Henstock-Kurzweil integrable if and only if f is Henstock-Kurzweil integrable over each subinterval $[c,1]$ with $0 < c < 1$ and the following limit exists*

$$\lim_{c \longrightarrow 0} \int_c^1 f.$$

It can be restated as follows:

Theorem 8. *Let f and F be real valued functions over $[0,1]$ such that for each interval $[c,1]$ with $0 < c < 1$, f is HK-integrable over $[c,1]$ with $(\mathcal{HK}) \int_c^1 f = F(1) - F(c)$.*

Then f is HK-integrable over $[0,1]$ if and only if F is continuous at 0. Moreover, in that case we have, $\int_0^1 f = F(1) - F(0)$.

In this section, we generalize this version of the Hake's theorem over metric spaces as given in [10], which also extends our previous results on Hake-type theorems as in [20, Theorem 5.2, Theorem 5.4]. For some other versions of this theorem on $\mathbb{R}^m$, we refer to [20–22].

First we define the Henstock variational measure. Unless specified, let $\mathcal{F}$ denote the family of all subintervals of X and $F : \mathcal{F} \longrightarrow \mathbb{R}$ be a finitely additive set function.

Definition 1. Given $E \subset X$ and a gauge δ on E, we define

$$V(F, E, \delta) := \sup_{\mathcal{P}} \sum_{i=1}^{n} |F(Q_i)|,$$

where the supremum is taken over all δ-fine E-anchored partial partitions $\mathcal{P} := \{(Q_i, x_i) : 1 \leq i \leq n\}$ in X. The *Henstock variational measure* corresponding to F is defined as

$$V_F(E) := \inf\{V(F, E, \delta) : \delta \text{ is a gauge on } E\}.$$

Gauge Integrals on Metric Measure Spaces

It is known that if F is finitely additive, then V_F is a metric outer measure on X (see [17, 23]). A variant of V_F appeared in [24].

We observe that the following partial result, as a particular case of [25, Proposition 2], is valid even on metric measure spaces. For this section, let E be any compact interval in the metric space (X, d) and by a figure in X, we refer to a finite union of intervals.

Theorem 9. *Let $f : E \longrightarrow \mathbb{R}$ be an HK-integrable function with primitive F. Then V_F is absolutely continuous with respect to μ.*

Next we present some extensions of Theorem 8 in our setting.

Theorem 10. *Let $I = \overline{B}(x, r)$ be a closed ball in E and its boundary be $\partial I := \{y \in X : d(x, y) = r\}$. Assume that for each compact interval $J \subset I$ with $J \cap \partial I = \emptyset$, f is HK-integrable over J, with $(\mathcal{HK}) \int_J f d\mu = F(J)$.*

Then f is HK-integrable over I if and only if $V_F(\partial I) = 0$. Moreover, in that case we have, $(\mathcal{HK}) \int_I f d\mu = F(I)$.

Proof. If f is HK-integrable over I then by Theorem 9, V_F is absolutely continuous with respect to μ. Thus $V_F(\partial I) = 0$.

For the converse, assume that $V_F(\partial I) = 0$ and let $\epsilon > 0$ be given. We choose an increasing sequence of closed balls $A_n = \overline{B}\left(x, r - \frac{1}{n}\right)$ inside I such that $(\cup_{n=1}^{\infty} A_n) \cup \partial I = I$.

By our hypothesis, f is HK-integrable over A_n, for each $n \in \mathbb{N}$. Using Saks-Henstock Lemma, we choose a gauge $\delta_n : A_n \longrightarrow (0, \infty)$ so that the inequality

$$\sum_{i=1}^{p} |f(t_i)\mu(J_i) - F(J_i)| \leq \frac{\epsilon}{2^{n+1}}$$

is satisfied for any δ_n−fine partial division $\{(J_i, t_i) : 1 \leq i \leq p\}$ of A_n.

Now we divide the proof in two cases. First we consider the case when $f(t) = 0$, for all $t \in \partial I \cup (\cup_n \partial A_n)$. Set $B := \partial I \cup (\cup_n \partial A_n)$. Since f is HK-integrable over each A_n, Theorem 9 implies that $V_F(\partial A_n) = 0$, for all $n \in \mathbb{N}$. Now since V_F is a metric outer measure, we have

$$V_F(B) = V_F(\partial I \cup (\cup_n \partial A_n)) \leq V_F(\partial I) + \sum_n V_F(\partial A_n) = 0.$$

Since $V_F(B) = 0$, we can choose a gauge $\delta_0 : B \longrightarrow (0, \infty)$ such that for every δ_0-fine partial division $\{(J_i, t_i) : 1 \leq i \leq p\}$ anchored in B, the following inequality is satisfied

$$\sum_{i=1}^{p} |F(J_i)| < \frac{\epsilon}{2}.$$

Now, we define a gauge $\delta : I \longrightarrow (0, \infty)$ as follows:

$$\delta(t) = \begin{cases} \delta_0(t), & \text{for } t \in B, \\ \min\{\delta_n(t), \frac{1}{2}\text{dist}(t, \partial A_n \cup \partial A_{n-1})\}, & \text{for } t \in (A_n \setminus A_{n-1})^\circ. \end{cases}$$

For any given δ-fine division $P = \{(I_i, t_i) : 1 \leq i \leq p\}$ of I, we have

$$\begin{aligned} |\sum_{i=1}^{p} f(t_i)\mu(I_i) - F(I)| &\leq \sum_{t_i \in B} |f(t_i)\mu(I_i) - F(I_i)| + \sum_{t_i \notin B} |f(t_i)\mu(I_i) - F(I_i)| \\ &\leq \sum_{t_i \in B} |F(I_i)| + \sum_{n} \sum_{t_i \in (A_n \setminus A_{n-1})^\circ} |f(t_i)\mu(I_i) - F(I_i)| \\ &< \frac{\epsilon}{2} + \sum_{n} \frac{\epsilon}{2^{n+1}} < \epsilon. \end{aligned}$$

This proves that f is HK-integrable over I with primitive F, when $f(t) = 0$ for all $t \in B$.

For the general case, we define a function $g : I \longrightarrow \mathbb{R}$ as $g = f - f.\chi_B$, where χ_B denotes the characteristic function of the set B. Then $g(t) = 0$ for all $t \in B$. Note that for a compact interval $J \subset I \setminus \partial I$, g is HK-integrable over J with integral $F(J)$, as $g(t) = f(t)$ for almost all $t \in I$. As earlier, we get $V_F(B) = 0$.

Thence, using the previous case, we conclude that g is HK-integrable over I with integral $F(I)$. Since $f(t) = g(t)$ for almost all $t \in I$, we see that f is HK-integrable over I with $(\mathcal{HK}) \int_I f d\mu = F(I)$, as the desired conclusion. $\qquad\square$

On the similar lines one can also prove the above theorem when I is a generalized interval. Now we present an alternative proof of [20, Theorem 5.4] for V_F. We observe that the following version of [23, Lemma 3.7], is true for V_F, even in our setting.

Lemma 1. *Let $f : E \longrightarrow \mathbb{R}$ be HK-integrable with $(HK) \int_J f = F(J)$, for every interval $J \subset E$. Then for every $M \subset E$*

$$V_F(M) \leq \mu(E). \sup\{|f(t)| : t \in M\}.$$

Theorem 11. *Let $A \subset E$ be a closed set such that*

(1) f is HK-integrable over A.
(2) For each compact interval $J \subset E \setminus A$, f is HK-integrable over J, with integral $F(J)$.

Then $V_F(A) = 0$ if and only if f is HK-integrable over E with

$$(\mathcal{HK}) \int_E f d\mu = F(E) + (\mathcal{HK}) \int_A f d\mu. \tag{1}$$

Proof. Since A is a closed subset of E, the set $E \setminus A$ can be written as a union of balls, open in the metric space (E, d). Being a compact metric space, (E, d) is Lindeloff and thus there exists a countable subfamily of those balls, say $\{B_n : n \in \mathbb{N}\}$, which covers $E \setminus A$. For each $n \in \mathbb{N}$, define a figure U_n as $U_n := B_n \setminus \cup_{m < n} B_m$.

As in the previous theorem, we first take the case when $f(t) = 0$, for all $t \in A \cup (\cup_n \partial U_n)$. Set $B := A \cup (\cup_n \partial U_n)$. If f is HK-integrable over E, Lemma 1 gives us $V_F(A) = 0$. For the converse, we assume that $V_F(A) = 0$. For any $n \in \mathbb{N}$, we write

$$\partial U_n = (\partial U_n \cap A) \cup (\partial U_n \cap (E \setminus A)).$$

We find a compact figure $J \subset (E \setminus A)$ such that $\partial U_n \cap (E \setminus A) \subset J$. Using our hypothesis, f is HK-integrable over J. Now by Theorem 9, we have $V_F \ll \mu$ on J and thence

$$V_F(\partial U_n \cap (E \setminus A)) = 0.$$

Since V_F is an outer measure, we have

$$V_F(\partial U_n) \le V_F(\partial U_n \cap A) + V_F(\partial U_n \cap (E \setminus A)) \le V_F(A) = 0.$$

Again the outer measurability of V_F implies

$$V_F(B) \le V_F(A) + \sum_n V_F(\partial U_n) = 0.$$

Let $\epsilon > 0$ be given. We can choose a gauge $\delta_0 : B \longrightarrow (0, \infty)$ such that for each δ_0-fine partial division $P := \{(J_i, t_i) : 1 \le i \le p\}$ anchored in B, the following inequality is satisfied

$$\sum_{i=1}^{p} |F(J_i)| < \frac{\epsilon}{2}.$$

Since $V_F(\partial(U_n)) = 0$, f is HK-integrable over each $\overline{U}_n$. Using Saks-Henstock Lemma, we choose a gauge $\delta_n : \overline{U}_n \longrightarrow (0, \infty)$ such that the inequality

$$\sum_{i=1}^{p} |f(t_i)\mu(J_i) - F(J_i)| \le \frac{\epsilon}{2^{n+1}}$$

is satisfied for any δ_n-fine partial division $\{(J_i, t_i) : 1 \le i \le p\}$ of $\overline{U}_n$. Next, we define a gauge $\delta : E \longrightarrow (0, \infty)$ as follows:

$$\delta(t) = \begin{cases} \delta_0(t), & \text{for } t \in B, \\ \min\{\delta_n(t), \frac{1}{2}\text{dist}(t, \partial U_n)\}, & \text{for } t \in (U_n)^o. \end{cases}$$

Now for any δ-fine division $P := \{(I_i, t_i) : 1 \le i \le p\}$ of E, the following assertions hold true, due to our choice of δ.

$$\left| \sum_{i=1}^{p} f(t_i)\mu(I_i) - F(E) \right| \le \sum_{t_i \in B} |f(t_i)\mu(I_i) - F(I_i)| + \sum_{t_i \notin B} |f(t_i)\mu(I_i) - F(I_i)|$$

$$\le \sum_{t_i \in B} |F(I_i)| + \sum_{n} \sum_{t_i \in (U_n)^o} |f(t_i)\mu(I_i) - F(I_i)|$$

$$< \frac{\epsilon}{2} + \sum_{n} \frac{\epsilon}{2^{n+1}} = \epsilon.$$

Thus f is HK-integrable over E with $(\mathcal{HK}) \int_E f d\mu = F(E)$. Hence we have proved our result for the case when $f(t) = 0$ for all $t \in B$.

For the general case, define a function $g : E \longrightarrow \mathbb{R}$ as $g = f - f.\chi_B$, where χ_B is the characteristic function of B. Then $g(t) = 0$ for all $t \in B$ and $g(t) = f(t)$ for all $t \in E \setminus B$.

Note that for any compact interval $J \subset (E \setminus B) \subset (E \setminus A)$, since f is HK-integrable over J with integral $F(J)$ and $f(t) = g(t)$ for almost all $t \in J$, g is HK-integrable over J with integral $F(J)$.

Now as above, we have $V_F(A) = 0$ if and only if g is HK-integrable over E with $(\mathcal{HK}) \int_E g d\mu = F(E)$, that is, if and only if $f - f.\chi_A$ is HK-integrable over E with $(\mathcal{HK}) \int_E (f - f.\chi_A) d\mu = F(E)$.

Since f is given to be integrable over A we observe that $V_F(A) = 0$ if and only if f is HK-integrable over E with $(\mathcal{HK}) \int_E (f - f.\chi_A) d\mu = F(E)$, that is,

$$(\mathcal{HK}) \int_E f d\mu = F(E) + (\mathcal{HK}) \int_A f d\mu. \qquad \square$$

5. Differentiation and Integration in Metric Spaces

In this section, we discuss a differentiation theorem for real valued HK integrable functions on metric spaces, which leads to the measurability of such functions and various results like the equivalence of McShane, Lebesgue and the absolute HK integrals.

Recall that the Vitali covering lemma plays an important role for such a differentiation theorem for real valued functions on compact real intervals. We adopt the following notion of fine covers and Vitali family from [5].

Let A be any non-empty subset of a metric space X. A subcollection $\mathcal{G}$ of intervals $\mathcal{I}$ will be called a *fine-cover* of A if

$$\inf\{diam(Q) : x \in Q \in \mathcal{G}\} = 0 \text{ for all } x \in A.$$

Further, a collection $\mathcal{V} \subset \mathcal{I}$ will be called a μ-*Vitali family* if it satisfies the following Vitali covering condition:

For every $A \subset X$, if $\mathcal{C} \subset \mathcal{V}$ is a fine-cover of A, then there exists a countable collection of non-overlapping sets $\{Q_n\} \subset \mathcal{C}$ satisfying

$$\mu^*\Big(A \setminus \bigcup_{n=1}^{\infty} Q_n\Big) = 0.$$

From now onwards, $\mathcal{F}$ will denote an arbitrary μ-Vitali family of (X, d). Elements of $\mathcal{F}$ will also be called *cells*.

A function $F : \mathcal{F} \longrightarrow \mathbb{R}$ will be called a *cell function* if it is *finitely additive*, that is,

$$F(I \cup J) = F(I) + F(J),$$

for all non-overlapping intervals I and J from $\mathcal{F}$ such that $I \cup J \in \mathcal{F}$.

Let $F : \mathcal{F} \longrightarrow \mathbb{R}$ be a cell function and $x \in X$. We define the *upper derivative* of F at x

$$\overline{D}F(x) := \limsup_{B \longrightarrow x} \frac{F(B)}{\mu(B)},$$

where $B \longrightarrow x$ means $x \in B \in \mathcal{F}, diam(B) \longrightarrow 0$ and $\mu(B) \neq 0$. Analogously, the *lower derivative* of F at x is defined by taking corresponding lim inf and is denoted by $\underline{D}F(x)$. Whenever $\overline{D}F(x) = \underline{D}F(x) \neq \infty$, we say that F is differentiable at x and with this common value as its *derivative*, denoted by $F'(x)$.

Remark 2. We observe that in [5], Theorems 5.1 and 5.3 are proved for non-negative functions, but are used for real valued functions in some results of that chapter, for instance see the proof of [5, Theorem 6.4]. Also there is an error in the proof of [5, Theorem 5.3], while using the definition of derivative therein. Most importantly, the presentation in [5] uses measure on all subsets of X, which we settle by using the metric outer measure μ^* instead. In this section, we rectify such errors and also generalize Theorems 5.1 and 5.3 from [5] for real valued functions.

Theorem 12. *If $f : X \longrightarrow \mathbb{R}$ is HK-integrable with primitive F, then F is differentiable μ-almost everywhere on X with derivative f.*

Proof. Note that it is enough to show that $\overline{D}F \leq f \leq \underline{D}F$, μ-almost everywhere on X. Let

$$A := \{x \in X : f(x) < \overline{D}F(x)\} = \bigcup_{p,q \in \mathbb{Q}} \{x \in X : f(x) < p < q < \overline{D}F(x)\}.$$

Write $A_{p,q} := \{x \in X : f(x) < p < q < \overline{D}F(x)\}$, for rationals $p < q$. It is enough to prove that $\mu(A_{p,q}) = 0$, for all $p, q \in \mathbb{Q}$, as that will imply $\mu(A) = 0$. That is, $\overline{D}F \leq f$ μ-almost everywhere on X. Similarly, we can prove that $\underline{D}F \geq f$ μ-almost everywhere on X. This will establish the result.

Let p and q be rationals such that $p < q$. If $A_{p,q} = \emptyset$, we are done. Suppose $A_{p,q} \neq \emptyset$. Let $\epsilon > 0$ be given. Choose a gauge δ on X, by Saks-Henstock lemma. That is, choose a gauge δ on X such that for every δ-fine partial partition $\{(Q_i, t_i) : 1 \leq i \leq n\}$ in X, we have

$$\left| \sum_{j=1}^{n} [F(Q_j) - f(t_j)\mu(Q_j)] \right| < \epsilon.$$

Consider the collection of subcells of X given by

$$\Omega := \left\{ B \in \mathcal{F} : \frac{F(B)}{\mu(B)} \geq q \text{ and there exists } x \in B \cap A_{p,q} \text{ with } diam(B) < \delta(x) \right\}.$$

Note that $\Omega \neq \emptyset$, as $\overline{D}F(x) > f(x)$, for every $x \in A_{p,q}$. It is also easy to see that Ω is a fine-cover of $A_{p,q}$.

Since Ω is a μ-Vitali family, there are non-overlapping $B_1, \ldots, B_m \in \Omega$ such that

$$\mu^*(A_{p,q}) \leq \sum_{j=1}^{m} \mu(B_i) + \epsilon. \tag{2}$$

For every j, let $x_j \in B_j \cap A_{p,q}$ be such that $diam(B_j) < \delta(x_j)$. This is possible as each $B_j \in \Omega$.

Then $\{(B_j, x_j) : 1 \le j \le m\}$ is a δ-fine partial partition in X and therefore

$$q \sum_{j=1}^{m} \mu(B_i) \le \sum_{j=1}^{m} F(B_i)$$

$$= \sum_{j=1}^{m} [F(B_i) - f(x_j)\mu(B_j)] + \sum_{j=1}^{m} f(x_j)\mu(B_j)$$

$$\le \left| \sum_{j=1}^{m} [F(B_i) - f(x_j)\mu(B_j)] \right| + p \sum_{j=1}^{m} \mu(B_j)$$

$$< \epsilon + p \sum_{j=1}^{m} \mu(B_j).$$

Consequently, we obtain $\sum_{j=1}^{m} \mu(B_j) < \epsilon/(q-p)$. Since $\epsilon > 0$ was arbitrary, from (2), we obtain $\mu^*(A_{p,q}) = 0$. Hence the result. $\qquad \square$

Theorem 13. *If $f : X \longrightarrow \mathbb{R}$ is an HK-integrable function, then f is a μ-measurable function.*

Proof. Let F denote the primitive of f on X. For each $k \in \mathbb{N}$, let $\mathcal{P}_k$ be a $\frac{1}{k}$-fine partition of X, say

$$\mathcal{P}_k := \{(B_{k_i}, x_{k_i}) : 1 \le i \le n_k\} \text{ and } Q_{k_i} := B_{k_i} \setminus \cup_{j<i} B_{k_j}, \text{ for each } k_i.$$

If $x \in X = \cup_i Q_{k_i}$, then for every $k \in \mathbb{N}$, there exists a unique k_x such that $x \in Q_{k_x}$. Define $f_k(x) := F(B_{k_x})/\mu(B_{k_x})$. That is

$$f_k := \sum_i \frac{F(B_{k_i})}{\mu(B_{k_i})} \chi_{Q_{k_i}}.$$

It is enough to prove that $f_k(x) \longrightarrow f(x)$, μ-almost everywhere on X. Write

$$D := \{x \in X : F'(x) \text{ does not exist or } F'(x) \ne f(x)\}.$$

By Theorem 12, the set D is μ-null. We now claim that $f_k(x) \longrightarrow f(x)$, for all $x \in X \setminus D$, which will conclude the result.

Pick any $x \in X \setminus D$. For each $k \in \mathbb{N}$, there is Q_{k_x} such that $x \in Q_{k_x} \subset B_{k_x}$. Since $f_k(x) = \frac{F(B_{k_x})}{\mu(B_{k_x})}$ and $diam(Q_{k_x}) \le diam(B_{k_x}) < 1/k$, we obtain

$$\lim_{k \longrightarrow \infty} f_k(x) := \lim_{k \longrightarrow \infty} \frac{F(B_{k_x})}{\mu(B_{k_x})} = F'(x) = f(x).$$

This proves our claim. Hence the result. $\qquad \square$

As a consequence of the above theorem, one can establish the following results. The proofs of these are analogous to the case of real valued functions on compact real intervals. Interested readers can find these in [16, p. 163, Theorem 10.3] and [5, Theorem 5.4], respectively.

Theorem 14. *Let* $f : X \longrightarrow \mathbb{R}$. *Then* f *is Lebesgue integrable if and only if* f *is McShane integrable. In this case, the value of two integrals coincide.*

Theorem 15. *If* $f : X \longrightarrow \mathbb{R}$ *is an absolutely Henstock-Kurzweil integrable function, then* f *is Lebesgue integrable. In this case, the values of Henstock and Lebesgue integrals of* f *are the same.*

6. Descriptive Characterizations of the Henstock-Kurzweil integral

In this section, we provide some measure theoretic characterizations of the Henstock-Kurzweil integral in terms of the Henstock variational measure V_F and ACG^{Δ} functions. We define the latter notion below.

Definition 2. Let $E \subset X$. A cell function $F : \mathcal{F} \longrightarrow \mathbb{R}$ is said to be

(i) AC^{Δ} on E if for every $\epsilon > 0$ there exists a gauge δ on E and a real $\eta > 0$ such that $\sum_{j=1}^{n} |F(Q_j)| < \epsilon$, for every δ-fine E-anchored partial partition $\{(Q_i, t_i) : 1 \leq i \leq n\}$ in X satisfying $\sum_{j=1}^{n} \mu(Q_j) < \eta$.

(ii) ACG^{Δ} on X, if $X = \cup_{n=1}^{\infty} E_n$, for a sequence of closed sets $\{E_n\}$ such that F is AC^{Δ} on E_n, for all $n \in \mathbb{N}$.

Recall that a measure μ is said to be *absolutely continuous* with respect to another measure ν ($\mu \ll \nu$) if $\mu(E) = 0$, whenever $\nu(E) = 0$. Next, we state some results from section 6 of [5].

Theorem 16. *Let* $F : \mathcal{F} \longrightarrow \mathbb{R}$ *be differentiable* μ-*almost everywhere such that* $V_F \ll \mu$. *Then* F' *is* HK-*integrable with primitive* F.

Theorem 17. *Let* $f : X \longrightarrow \mathbb{R}$ *be an* HK-*integrable function with primitive* F. *Then* F *is* ACG^{Δ} *on* X.

Theorem 18. *If* $F : \mathcal{F} \longrightarrow \mathbb{R}$ *is an* ACG^{Δ} *on* X, *then* $V_F \ll \mu$.

The above three results can be found in Theorems 6.2, 6.6 and 6.7 of [5]. After that, the authors of [5] provide a few more results to prove their main results (see [5, Lemma 6.8 to Theorem 6.11]). We restate [5, Lemma 6.9] which appeared suspicious in the very first reading.

Gauge Integrals on Metric Measure Spaces

Lemma 2. *Let $F : \mathcal{F} \longrightarrow \mathbb{R}$ be AC^{Δ} on a closed subset A of a cell I of X. Then $F' = 0$, μ almost everywhere on A.*

This is not true, even for various elementary functions on $X = [0, 1]$ with μ as the Lebesgue measure on X. For example, take $f(x) := x$ for all $x \in [0, 1]$ and let $F(J)$ denote the Lebesgue integral of f over J, for every interval J. Then $F'(x) = f(x)$ for every $x \in X$. Note that for $A = I = [0, 1]$, the set function F is absolutely continuous and hence AC^{Δ} on A. However $F'(x) \neq 0$ for all $x \in [0, 1]$.

Hence, we completely avoid this lemma, and the subsequent results of [5]. We still obtain the same characterization theorems. The following is a much shorter approach.

Theorem 19. *Let $f : X \longrightarrow \mathbb{R}$. The following are equivalent:*

(i) *f is HK-integrable on X.*
(ii) *There exists a cell function $F : \mathcal{F} \longrightarrow \mathbb{R}$ such that $F' = f$, μ-almost everywhere on X and F is ACG^{Δ} on X.*
(iii) *There exists a cell function $F : \mathcal{F} \longrightarrow \mathbb{R}$ such that $F' = f$, μ-almost everywhere on X and $V_F \ll \mu$ on X.*

Proof. First assume that f is HK-integrable on X with primitive F. Then (ii) holds by Theorem 12 and Theorem 17.

Further (ii) $\Rightarrow$ (iii) follows from Theorem 18, and (iii) $\Rightarrow$ (i) follows from Theorem 16. Hence the result. $\qquad\square$

Analogous to [17, p. 110, Theorem 4.2.3], the following characterization of Lebesgue integrable functions can also be established.

Theorem 20. *Let $f : X \longrightarrow \mathbb{R}$ be an HK-integrable function with primitive F and E be a measurable subset of X. Then f is Lebesgue integrable on E if and only if $V_F(E) < \infty$. Moreover in this case, we have*

$$V_F(E) = (L) \int_E |f| d\mu.$$

Remark 3. In [16, p. 146, Definition 9.14], ACG_{δ} functions are defined as a variant of the notion of ACG^{Δ}-functions. Then in [16, p. 147, Theorem 9.17], it is established that a function f is HK-integrable if and only if there exists an ACG_{δ} function F such that $F' = f$ almost everywhere. That proof also uses the measurability of HK-integrable functions. It can be verified that an analogous proof is valid even on metric spaces.

7. Some Alternative Proofs in Lebesgue Integration

Now we provide alternative proofs of two results in the theory of Lebesgue integration on metric measure spaces. The first one even bypasses the Vitali covering lemma.

For this section, let F be any real valued finitely additive set function on the collection of intervals $\mathcal{I}$ of a compact metric space (X, d). Then F will be called *absolutely continuous* (w.r.t. μ) on X if for every $\epsilon > 0$, there exists some $\eta > 0$ such that

$$\left| \sum_{j=1}^{m} F(I_j) \right| < \epsilon$$

for any finitely many non-overlapping intervals $I_1, \ldots, I_m$ from $\mathcal{I}$ such that $\mu(\bigcup_{j=1}^{m} I_j) < \eta$.

In contrast with the absolute continuity of measures, the same for finitely additive set functions splits up into three different versions (see [26, p. 159]).

In standard measure theory courses, the Vitali covering lemma is often used to establish that every absolutely continuous real valued function on a compact real interval is a constant (see [27, p. 103]). We provide an alternative proof of this result to even the general setting of real valued functions on metric spaces, without involving the Vitali lemma. It is inspired by [28, Theorem 12].

Theorem 21. *Let $F : \mathcal{I} \longrightarrow \mathbb{R}$ be an absolutely continuous function on X such that $F'(t) = 0$, for almost all $t \in X$. Then $F \equiv 0$, on $\mathcal{I}$.*

Proof. Let $\epsilon > 0$ be given. By hypothesis $\mu(E) = 0$, where the set E is defined as

$$E := \{ t \in X : \text{ either } F'(t) \text{ does not exist or } F'(t) \neq 0 \}.$$

Let $\epsilon > 0$ be given and choose some $\eta > 0$ as per the definition of absolute continuity of F, corresponding to $\epsilon > 0$. Pick any open set O such that $O \supset E$ and $\mu(O) < \eta$. Let $t \in X$. If $t \in E$, choose $\delta(t) > 0$ such that $B(t; \delta(t)) \subset O$. If $t \notin E$, choose $\delta(t) > 0$ such that

$$|F(I)| \le \epsilon \mu(I), \text{ whenever } t \in I \in \mathcal{I} \text{ and } diam(I) < \delta(t).$$

This defines a gauge δ on X. Let $J \in \mathcal{I}$ be arbitrary. Let $\{(I_j, t_j) : j = 1, \ldots, n\}$ be a δ-fine tagged partition of J. Therefore we obtain

$$|F(J)| \leq \sum_{j=1}^{n} |F(I_j)| = \sum_{t_j \in E} |F(I_j)| + \sum_{t_j \notin E} |F(I_j)|$$

$$\leq \epsilon + \epsilon \sum_{t_j \notin E} \mu(I_j) \leq \epsilon(1 + \mu(X)).$$

Since $\epsilon > 0$ is arbitrary, we have $F(J) = 0$ for all $J \in \mathcal{I}$. Hence the result.

$\square$

Lemma 3. *If $\epsilon > 0$ and $I \in \mathcal{I}$, then there are finitely many open balls $B_1, \ldots, B_m$ of X such that*

$$I \subset \bigcup_{i=1}^{m} B_i \text{ and } \mu(\overline{B_i}) < \epsilon \text{ for all } i = 1, \ldots, m.$$

Proof. Fix any $\epsilon > 0$ and $I \in \mathcal{I}$. Pick any $x \in I$. Note that $\{B(x; 1/n)\}$ is a nested decreasing sequence of measurable sets such that $\bigcap_{n=1}^{\infty} B(x; 1/n) = \{x\}$. Since μ is non-atomic Radon measure, we obtain

$$0 = \mu(\{x\}) = \lim_{n \to \infty} \mu(B(x; 1/n)).$$

Thus, there exists some $n_x \in \mathbb{N}$ such that $\mu(B(x; 1/n_x)) < \epsilon$. Recall that I is a closed subset of the compact metric space X and hence it is also compact.

Note that $\{B(x; 1/n_x) : x \in I\}$ is an open cover of the compact space I. Hence there are finitely many $x_1, \ldots, x_m \in I$ such that $I \subset \bigcup_{i=1}^{m} B(x_i; 1/n_{x_i})$. Moreover

$$\mu(\overline{B(x_i; 1/n_{x_i})}) = \mu(B(x_i; 1/n_{x_i})) < \epsilon \text{ for all } i = 1, \ldots, m. \qquad \square$$

Theorem 22. *Let $f : X \longrightarrow \mathbb{R}$ be an HK-integrable function with primitive F. If F is absolutely continuous, then f is Lebesgue integrable.*

Proof. Applying Theorem 20, it is enough to prove that $V_F(X) < \infty$.

Since F is absolutely continuous, there exists some $\eta > 0$ such that

$$|F(\bigcup_{j=1}^{m} I_j)| = |\sum_{j=1}^{m} F(I_j)| < 1$$

for any finitely many non-overlapping intervals $I_1, \ldots, I_m$ from $\mathcal{I}$ such that $\mu(\bigcup_{j=1}^{m} I_j) < \eta$. By Lemma 3, there are finitely many open balls $B_1, \ldots, B_n$

of X such that

$$\bigcup_{i=1}^{n} B_i = X \text{ and } \mu(\overline{B_i}) < \eta \text{ for all } i = 1, \ldots, n.$$

Let $t \in X = \bigcup_{i=1}^{n} B_i$. Let i_t denote the least $i \in \{1, \ldots, n\}$ such that $t \in B_{i_t}$. Since B_{i_t} is open, there exists some $\delta(t) > 0$ such that $B(t, \delta(t)) \subset B_{i_t}$. This defines a gauge δ on X.

Let $\{(J_k, t_k) : k = 1, \ldots, p\}$ be any δ-fine tagged partition of X. For every $i = 1, \ldots, n$, write

$$A_i^+ := \{k : J_k \subset B_i, F(J_k) \geq 0\} \text{ and}$$
$$A_i^- := \{k : J_k \subset B_i, F(J_k) < 0\}.$$

Hence, we obtain

$$\sum_{k=1}^{p} |F(J_k)| = \sum_{i=1}^{n} \sum_{k \in A_i^+} |F(J_k)| + \sum_{i=1}^{n} \sum_{k \in A_i^-} |F(J_k)|$$

$$= \sum_{i=1}^{n} \sum_{k \in A_i^+} F(J_k) - \sum_{i=1}^{n} \sum_{k \in A_i^-} F(J_k)$$

$$= \sum_{i=1}^{n} \left| F\left(\bigcup_{k \in A_i^+} J_k \right) \right| + \sum_{i=1}^{n} \left| F\left(\bigcup_{k \in A_i^-} J_k \right) \right| < 2n.$$

This ensures that $V_F(X) < \infty$ and hence the result. $\square$

Finally, we provide an alternative proof of a version of the fundamental theorem of calculus for the Lebesgue integral of real valued functions on metric measure spaces. Also see [27, p. 110].

Theorem 23. *Let $F : \mathcal{I} \longrightarrow \mathbb{R}$ be an absolutely continuous function on X such that F is differentiable μ almost everywhere on X. Then there exists a Lebesgue integrable function on X with primitive F.*

Proof. Applying Theorem 22, it is enough to show that there exists an HK-integrable function on X with primitive F. Let $\epsilon > 0$ be given. Write

$$E := \{t \in X : F'(t) \text{ does not exist}\}.$$

It is given that $\mu(E) = 0$. Since F is absolutely continuous, there exists some $\eta > 0$ such that $|F(\bigcup_{j=1}^{m} I_j)| < \epsilon$ for every finite collection $\{I_1, \ldots, I_m\}$ of non-overlapping cells from $\mathcal{I}$ such that $\mu(\bigcup_{j=1}^{m} I_j) < \eta$.

Using the outer regularity of μ ensures that there exists some open set O containing E such that $\mu(O) < \eta$. Define $f : X \longrightarrow \mathbb{R}$ as

$$f(t) := \begin{cases} F'(t) & , t \in X \setminus E, \\ 0 & , t \in E. \end{cases}$$

Let $t \in X$. If $t \in E$, choose $\delta(t) > 0$ such that $B(t; \delta(t)) \subset O$. If $t \in X \setminus E$, choose $\delta(t) > 0$ such that

$$|F(I) - f(t)\mu(I)| \leq \epsilon\mu(I), \text{ whenever } t \in I \in \mathcal{I} \text{ and } diam(I) < \delta(t).$$

This defines a gauge δ on X. Let $J \in \mathcal{I}$ be arbitrary and $\{(J_i, t_i) : i = 1, \ldots, n\}$ be a δ-fine tagged partition of J. Then we obtain

$$\left| \sum_{i=1}^{n} f(t_i)\mu(J_i) - F(J) \right| \leq \left| \sum_{t_i \in J \cap E} F(J_i) \right| + \sum_{t_i \in J \setminus E} |f(t_i)\mu(J_i) - F(J_i)|$$

$$\leq \left| F\left(\bigcup_{t_i \in J \cap E} J_i \right) \right| + \sum_{t_i \in J \setminus E} \epsilon\mu(J_i)$$

$$\leq \epsilon + \epsilon\mu(J) \leq \epsilon(1 + \mu(X)).$$

Since $\epsilon > 0$ is arbitrary, we conclude that f is HK-integrable over J with integral $F(J)$ for all $J \in \mathcal{I}$. Hence the result. $\square$

References

[1] R. Henstock, *Definitions of Riemann Type of the Variational Integrals*, Proc. London Math. Soc. **11(3)**, (1961), 402-418.

[2] R. Henstock, *Theory of Integration*, Butterworths, London, (1963).

[3] J. Kurzweil, *Generalized Ordinary Differential Equations and Continuous Dependence on a Parameter. (Russian)*, Czechoslovak Math. J. **82(7)**, (1957), 418-449.

[4] R. G. Bartle, *A Modern Theory of Integration*, American Mathematical Society, Providence, RI, (2001).

[5] D. Bongiorno and G. Corrao, *An Integral on a Complete Metric Measure Space*, Real Analysis Exchange **40(1)**, (2014/15), 157-178.

[6] W. Ng Leng and P. Y. Lee, *Nonabsolute Integral on Measure Spaces*, Bulletin of the London Mathematical Society **32(1)** (2000), 34-38.

[7] W. Ng Leng, *A Nonabsolute Integral on Measure Spaces that Includes the Davies-McShane Integral*, New Zealand J. Math. **30(2)** (2001), 147-155.

[8] W. Ng Leng, *Nonabsolute Integration on Measure Spaces*, Series in Real Analysis, Vol. 14, World Scientific, Singapore, (2017).

[9] A. Boccuto, V. A. Skvortsov and F. Tulone, *A Hake-type theorem for integrals with respect to abstract derivation basis in the Riesz space setting*, Mathematica Slovaca, **65**, (2015), 1319-1336.

[10] S. P. Singh and I. K. Rana, *The Hake's Theorem on Metric Measure Spaces*, Real Analysis Exchange **39(2)**, (2013/14), 447-458.

[11] K. R. Parthasarathy, *Introduction to Probability and Measure*, Springer-Verlag New York Inc., New York and London, (1978).

[12] S. P. Singh and I. K. Rana, *Some Alternative Approaches to the McShane Integral*, Real Anal. Exchange 35(1) (2010) 229-233.

[13] P. Cousin, *Sur les fonctions de n variables complexes. (French)*, Acta Mathematica **19(1)**, (1895), 1-61.

[14] S. P. S. Kainth and N. Singh, *Henstock Integral on Metric Spaces Revisited*, Real Anal. Exchange 47(2) (2022), 377-396.

[15] G. Edgar, *Measure, Topology, and Fractal Geometry*, Springer (2008).

[16] R. A. Gordon, *The Integrals of Lebesgue, Denjoy, Perron and Henstock*, American Mathematical Society, (1994).

[17] Lee Tuo-Yeong, *Henstock-Kurzweil Integration on Euclidean spaces*, Ser. Real Anal., vol. 12, World Scientific Publishing Co., Singapore, (2011).

[18] S. P. S. Kainth, *A Comprehensive Textbook on Metric Spaces*, Springer-Nature, Singapore, 2023.

[19] J. R. Munkres, *Topology (Second Edition)*, Pearson New International, Edition, (2014).

[20] S. P. Singh and I. K. Rana, *The Hake's theorem and variational measures*, Real Analysis Exchange 37(2) (2011/2012), 477-488.

[21] C. A. Faure and J. Mawhin, *The Hake's property for some integrals over multidimensional intervals*, Real Anal. Exchange 20(2) (1994-95), 622-630.

[22] P. Muldowney and V. A. Skvortsov, *Improper Riemann integral and the Henstock integral in $\mathbb{R}^n$*, Math. Notes 78, no. 1-2, (2005), 228-233.

[23] S. Schwabik, *Variational Measures and the Kurzweil-Henstock Integral*, Math. Slovaca **59(6)** (2009), 731-752.

[24] N. Singh, S. P. S. Kainth, *Variational Measure with respect to Measurable Gauges*, Real Analysis Exchange **46(1)**, (2021), 247-260.

[25] L. Di Piazza *Variational measures in the theory of the integration in $\mathbb{R}^m$*, Czechoslovak Math. J. 51(126) (2001), no. 1, 95-110.

[26] K. P. S. B. Rao and M. B. Rao, *Theory of Charges; A Study of Finitely Additive Measures (second edition)*, Cambridge University Press, (1970).

[27] H. L. Royden, *Real Analysis (third edition)*, Macmillian Publishing Company, (1987).

[28] R. A. Gordon, *The Use of Tagged Partitions in Elementary Real Analysis*, American Mathematical Monthly **105(2)**, (1998), 107-117.

© 2025 World Scientific Publishing Company

https://doi.org/10.1142/9789819812202_0002

Chapter 2

Variational Henstock Integral and its Variational Measure in Locally Convex Space

Savita Bhatnagar

Centre for Advanced Study in Mathematics, Panjab University, Chandigarh, India

Email: bhsavita@pu.ac.in

We study variational Henstock integral of functions defined on $[0, 1]$ and taking values in a locally convex topological vector space X. We show that a variational Henstock integrable function with values in a locally convex space is measurable by semi-norm, and that its primitive is continuous and differentiable by semi-norm almost everywhere. Further, we characterize locally convex spaces possessing the Radon-Nikodym property in terms of additive interval functions whose Henstock variational measures are absolutely continuous with respect to the Lebesgue measure. We show that the locally convex space has Radon-Nikodym property if and only if each X-valued additive interval function possessing absolutely continuous variational measure with respect to Lebesgue measure for each continuous semi-norm of the separating family of semi-norms is differentiable by semi-norm, F' is variationally Henstock integrable and $F(I) = (vH) \int_I F'(t)dt$ for each compact subinterval I of $[0, 1]$. We also study locally convex spaces possessing the weak Radon-Nikodym property in terms of additive interval functions whose Henstock variational measures are absolutely continuous with respect to the Lebesgue measure.

1. Introduction

In the papers [9] and [10], Henstock-Kurzweil and McShane integrals of functions defined on $[0, 1]$ and taking values in a locally convex space are defined and their properties are studied. In this chapter, we take this study further and define variational Henstock integral, its variational Henstock measure for each continuous semi-norm and characterize locally convex spaces possessing the Radon-Nikodym property in terms of additive interval functions whose Henstock variational measure for each continuous semi-norm is absolutely continuous with respect to the Lebesgue measure. In [3], a characterization of Banach spaces possessing the Radon-Nikodym property is given in terms of additive interval functions whose Henstock variational measures are absolutely continuous with respect to the Lebesgue measure. For an application of this result to vector valued multipliers of Banach algebra valued variationally Henstock integrable functions, see [1].

It is known that for Banach valued functions, Saks-Henstock Lemma fails to be true and holds if and only if the Banach space is finite dimensional. An exception is the case when the integrator is a stochastic process and the Banach space has stochastic properties, see [5, 17, 18]. Therefore, for Banach valued functions, we can define variational Henstock integral. Analogously, in Fréchet space X, Henstock lemma holds iff X is nuclear (see [9, Theorem 4]). We show that a variational Henstock integrable function with values in a locally convex space is measurable by semi-norm, its primitive is continuous and differentiable by semi-norm almost everywhere. Finally, we show that X satisfies Radon-Nikodym property if and only if each X-valued additive interval function possessing absolutely continuous variational measure with respect to Lebesgue measure for each continuous semi-norm of the separating family of semi-norms is a variational Henstock integral of an X-valued function.

2. Preliminaries

This section contains the preliminary material from which we shall draw throughout the rest of the paper.

Let $[0, 1]$ be a compact real interval, $\mathcal{I}$ be the family of compact subintervals of $[0, 1]$, $\mathcal{L}$ be the σ-algebra of all Lebesgue measurable subsets of $[0, 1]$, m stand for Lebesgue measure on $[0, 1]$, m^* stand for Lebesgue outer measure and X be a complete Hausdorff locally convex topological vector space (briefly, locally convex space) with topology τ and topological dual X^*. $\mathcal{P}$ denotes a separating family of τ-continuous semi-norms on X so that the Hausdorff topology on X is generated by $\mathcal{P}$. For any $p \in \mathcal{P}$, we denote by X_p the quotient space $X/p^{-1}(0)$, by $\pi_p : X \to X_p$ the canonical quotient map. Note that X_p is a Banach space with norm $\tilde{p}(\pi_p(x)) = p(x)$, $x \in X$ (see [13, Chapter 1, page 31]) and $X^* = \{x_p^* \circ \pi_p : p \in \mathcal{P}, x_p^* \in X_p^*\}$. If $x^* \in X^*$, then $x^* = x_p^* \circ \pi_p$ for some $p \in \mathcal{P}$ and $x_p^* \in X_p^*$. So there exists a constant $C_p = \|x_p^*\|$ such that $|x^*(x)| \leq C_p \ p(x)$ for all $x \in X$. For every $p, q \in \mathcal{P}$ such that $p \leq q$, we denote by $g_{pq} : X_q \to X_p$ the continuous linear map defined by $g_{pq}(\pi_q(x)) = \pi_p(x)$, $x \in X$.

If Y is the subspace of $\prod_{p \in \mathcal{P}} X_p$ whose elements $x = (x_p)$ satisfy the relation $x_p = g_{pq}(x_q)$ whenever $p \le q$, then Y is the projective limit of the family $\{X_p, p \in \mathcal{P}\}$ with respect to the mappings $\{g_{pq}, p, q \in \mathcal{P}, p \le q\}$. By [14, II.5.4] X is isomorphic to Y.

The above result is used several times in the sequel. So we state [14, II.5.4] for the sake of completeness:

Every complete locally convex space E is isomorphic to a projective limit of a family of Banach spaces; this family can be so chosen that its cardinality equals the cardinality of a given 0-neighbourhood base in E.

A set function $F : \mathcal{I} \to X$ is said to be *additive* if $F(J \cup K) = F(J) + F(K)$, for all non-overlapping intervals $J, K \in \mathcal{I}$ such that $J \cup K \in \mathcal{I}$.

A collection $\{(t_i, J_i); i = 1, \ldots, k\}$ of point-interval pairs is called a *tagged partition* of (respectively, in) the interval $[0, 1]$ if each $t_i \in J_i$ and $\{J_i : i = 1, \ldots, k\}$ are pairwise non-overlapping compact subintervals of $[0, 1]$ with $[0, 1] = \bigcup_{i=1}^{k} J_i$ (respectively $\bigcup_{i=1}^{k} J_i \subseteq [0, 1]$).

Any positive function $\delta : [0, 1] \to (0, \infty)$ is called a *gauge* on $[0, 1]$ and the above tagged partition is said to be δ-*fine* if $J_i \subset (t_i - \delta(t_i), t_i + \delta(t_i))$, for every $i = 1, \ldots, k$.

We recall the following definitions (see [2, Definition 2.4, for 1-4]).

Definition 1. A function $f : [0, 1] \to X$ is said to be *strongly (Bochner) integrable* if there exists a sequence $\{f_n\}$ of simple functions such that

(1) $f_n(t) \to f(t)$ a.e. So f is strongly measurable;
(2) $p(f(t) - f_n(t)) \in L^1([0, 1])$ for each $n \in \mathbf{N}$ and for all $p \in \mathcal{P}$,
$$\lim_{n \to \infty} \int_0^1 p(f(t) - f_n(t))dt = 0;$$
(3) $\int_E f_n$ converges in X for each measurable subset E of $[0, 1]$.

In this case, we put $(B) \int_E f = \lim_{n \to \infty} \int_E f_n$.

Definition 2. A function $f : [0, 1] \to X$ is said to be *integrable by semi-norm* if for any $p \in \mathcal{P}$ there exists a sequence $\{f_n^p\}$ of simple functions and a subset $Z_p \subseteq [0, 1]$ with $m(Z_p) = 0$ such that

(1) $\lim_{n \to \infty} p(f_n^p(t) - f(t)) = 0$ for all $t \in [0, 1] \setminus Z_p$ (i.e., f is measurable by semi-norm);

(2) $p(f_n^p(t) - f(t)) \in L^1([0,1])$ for each $n \in \mathbf{N}$ and for all $p \in \mathcal{P}$,
$$\lim_{n\to\infty} \int_0^1 p(f_n^p(t) - f(t))dt = 0;$$
(3) For each measurable subset E of $[0,1]$ there exists an element $y_E \in X$ such that $\lim_{n\to\infty} p(\int_E f_n^p(t)dt - y_E) = 0$ for every $p \in \mathcal{P}$.

In this case, we put $\int_E f = y_E$.

Clearly a Bochner integrable function is integrable by semi-norm, and the two definitions coincide in a Banach space.

Definition 3. A function $f : [0,1] \to X$ is said to be *Pettis integrable* if $x^* f$ is Lebesgue integrable on $[0,1]$ for every $x^* \in X^*$ and if for every measurable set E in $[0,1]$ there is a vector $\nu(E) \in X$ such that $x^*(\nu(E)) = \int_E x^* f(t)dt$ for all $x^* \in X^*$. We write $\nu(E) = (P) \int_E f$.

The set function $\nu : \mathcal{L} \to X$ is called the indefinite Pettis integral of f. It is known that ν is a countably additive vector measure, absolutely continuous with respect to m (in the sense $m(E) = 0$ implies $\nu(E) = 0$ for $E \in \mathcal{L}$). We write $\nu \ll m$ to denote absolute continuity of ν with respect to m.

Definition 4. An interval function $F : \mathcal{I} \to X$ is said to be *differentiable by semi-norm* if there is a function $f : [0,1] \to X$ satisfying the following property: Given $p \in \mathcal{P}$, there exists a subset $Z_p \subseteq [0,1]$ with $m(Z_p) = 0$ such that
$$\lim_{h\to 0} p\left(\frac{F \prec t, t+h \succ}{|h|} - f(t)\right) = 0$$
for all $t \in [0,1] \setminus Z_p$, where $\prec a, b \succ = [\min\{a,b\}, \max\{a,b\}]$.

In a similar manner, we can define differentiability by semi-norm for point functions.

Definition 5. An interval function $F : \mathcal{I} \to X$ is said to be *pseudo-differentiable* if there is a function $f : [0,1] \to X$ satisfying the following property: Given $x^* \in X^*$, there exists a subset $Z_{x^*} \subseteq [0,1]$ with $m(Z_{x^*}) = 0$ such that
$$\lim_{h\to 0} x^*\left(\frac{F \prec t, t+h \succ}{|h|} - f(t)\right) = 0$$
for all $t \in [0,1] \setminus Z_{x^*}$.

Clearly, if F is differentiable by semi-norm, then it is pseudo-differentiable, while pseudo-differentiability of F implies differentiability by semi-norm if X is representable by semi-norm, i.e. $p(x) = \sup\limits_{|x^*|\leq p} |x^*(x)|$ for all $p \in \mathcal{P}$ and $x \in [0,1]$. If X is separable by semi-norm, then it is representable by semi-norm (see [6, page 185]).

Definition 6. A function $f : [0,1] \to X$ is said to be *Henstock-Kurzweil (HK) integrable* on $[0,1]$ if there is a $w \in X$ such that for every $\epsilon > 0$ and $p \in \mathcal{P}$ there exists a gauge δ_p on $[0,1]$, such that the inequality

$$p\left(\sum_{i=1}^{k} f(t_i)\, m(J_i) - w\right) < \epsilon,$$

is satisfied, for every δ_p-fine tagged partition $P = \{(t_i, J_i) : i = 1, \ldots, k\}$ of $[0,1]$.
We write $w = (HK) \int_0^1 f$.

A function $f : [0,1] \to X$ is said to be *variational Henstock (vH) integrable* on $[0,1]$ if there is an additive function $F : \mathcal{I} \to X$ such that for every $\epsilon > 0$ and $p \in \mathcal{P}$ there exists a gauge δ_p on $[0,1]$, such that the inequality

$$\sum_{i=1}^{k} p(f(t_i)\, m(J_i) - F(J_i)) < \epsilon \tag{1}$$

is satisfied, for every δ_p-fine tagged partition $P = \{(t_i, J_i) : i = 1, \ldots, k\}$ of $[0,1]$.
We write $F(I) = (vH) \int_I f$ for all $I \in \mathcal{I}$.

Clearly $vH \subseteq HK$, equality holds in case of a Fréchet (complete, metrizable locally convex space) space if and only if X is nuclear (see [9, Theorem 4]).

If the inequality (1) is satisfied by Henstock-Kurzweil integrable functions for partition P in $[0,1]$, f is said to satisfy Henstock Lemma. For real-valued functions, Henstock Lemma holds for all Henstock-Kurzweil integrable functions.

A function $f : [0,1] \to X$ is said to be *scalarly Henstock-Kurzweil integrable* on $[0,1]$ if for all $x^* \in X^*$ the function $x^*(f)$ is Henstock-Kurzweil integrable. A scalarly Henstock-Kurzweil integrable function is *Henstock-Kurzweil-Pettis (HKP)integrable* if for all $I \in \mathcal{I}$ there exists $w_I \in X$ such that

$$< x^*, w_I > = (HK) \int_I x^*(f) dm.$$

We call w_I the Henstock-Kurzweil-Pettis integral of f over I and we write $w_I = (HKP) \int_I f dm$.

We shall use the point function $\tilde{F} : [0,1] \to X$ and the interval function $F : \mathcal{I} \to X$ related by $\tilde{F}(v) - \tilde{F}(u) = F([u,v])$ interchangeably.

The function $F : \mathcal{I} \to X$ is said to be of *strong bounded variation* (BV^*) if, for each $p \in \mathcal{P}$, the function $\pi_p \circ F : \mathcal{I} \to X_p$ is BV^*.

The function $F : \mathcal{I} \to X$ is said to be of *strong generalized bounded variation* (BVG^*) if, for each $p \in \mathcal{P}$, the function $\pi_p \circ F : \mathcal{I} \to X_p$ is BVG^*.

For the definitions of BV^* and BVG^* in Banach spaces, see [[15], p. 194 and p. 200].

Given $\tilde{F} : [0,1] \to X$, $p \in \mathcal{P}$, a subset $E \subseteq [0,1]$ and a gauge δ on E, we define

$$V_{p,F}(E, \delta) = \sup \sum_{(t,I) \in P} p(F(I))$$

where the supremum is taken over all E-tagged, δ-fine, tagged partitions P in $[0,1]$. Then we set

$$V_{p,F}(E) = \inf\{V_{p,F}(E, \delta) : \delta \quad \text{is a gauge on } E\}.$$

The set function $V_{p,F}(.)$ is said to be the *Henstock variational measure* generated by F with respect to semi-norm p. As in [16], it follows that the set function $V_{p,F}(.)$ is a Borel outer measure on $[0,1]$.

We say that X has *Radon-Nikodym property* (RNP) (respectively *weak Radon-Nikodym property* (WRNP)) (see [11]) if, for each countably additive absolutely continuous vector measure $\nu : \mathcal{L} \to X$ of bounded variation (respectively σ-finite variation), there is an integrable by semi-norm (respectively Pettis integrable) function $f : [0,1] \to X$ such that $\nu(E) = \int_E f \, dm$ for all $E \in \mathcal{L}$.

3. Variational Henstock integral

Theorem 1. *Let X be a locally convex space and $f : [0,1] \to X$ be a variationally Henstock integrable function with $\tilde{F}$ as its variational primitive. Then $\tilde{F}$ is continuous and F is differentiable by semi-norm almost everywhere.*

Proof. We first show that $\tilde{F}$ is continuous. For $p \in \mathcal{P}$, $\pi_p \circ \tilde{F} : [0,1] \to X_p$ is defined by

$$\pi_p \circ \tilde{F}(t) = \pi_p\Big(\int_0^t f(s)ds \Big) = \int_0^t \pi_p \circ f(s)ds.$$

As $\pi_p \circ f$ is variationally Henstock integrable with values in a Banach space, $\pi_p \circ \tilde{F}$ is continuous by [15, Theorem 7.4.1]. Since X is isomorphic to projective limit [14, II.5.4] of the Banach spaces X_p, we get that $\tilde{F}$ is continuous (see [14, II.5.2]) ($\tilde{F}$ is continuous iff $\pi_p \circ \tilde{F}$ is continuous for each $p \in \mathcal{P}$.)

By [15, Theorem 7.4.2], we get that the primitive $\pi_p \circ \tilde{F}$ of variationally integrable function $\pi_p \circ f$ is differentiable almost everywhere and $(\pi_p \circ \tilde{F})'(t) = \pi_p \circ f(t)$ a.e. for all $p \in \mathcal{P}$. Thus F is differentiable by semi-norm almost everywhere with $F' = f$ almost everywhere.

$\square$

Theorem 2. *If the function $f : [0,1] \to X$ is variationally Henstock integrable on $[0,1]$, then f is measurable by semi-norm.*

Proof. As $f : [0,1] \to X$ is variationally Henstock integrable on $[0,1]$ there is an additive function $F : \mathcal{I} \to X$ such that for every $\epsilon > 0$ and $p \in \mathcal{P}$ there exists a gauge δ_p on $[0,1]$, such that the inequality

$$\sum_{i=1}^k p(f(t_i) \, m(J_i) - F(J_i)) < \epsilon, \tag{2}$$

is satisfied, for every δ_p-fine tagged partition $P = \{(t_i, J_i) : i = 1, \ldots, k\}$ of $[0,1]$. If $x^* \in X^*$ then there exists $p \in \mathcal{P}$, a constant C_p such that $|x^*(x)| \leq C_p \, p(x)$ for all $x \in X$. So

$$\left| \sum_{i=1}^{k} x^*(f(t_i)) \; m(J_i) - x^*(F(J_i)) \right|$$

$$= \left| x^* \left(\sum_{i=1}^{k} \big(f(t_i) \; m(J_i) - F(J_i) \big) \right) \right|$$

$$\leq C_p \; p \left(\sum_{i=1}^{k} \big(f(t_i) \; m(J_i) - F(J_i) \big) \right)$$

$$< C_p \; \epsilon$$

for every δ_p-fine tagged partition $P = \{(t_i, J_i) : i = 1, \ldots, k\}$ of $[0,1]$ by (2)

So the real function $x^* f$ is Henstock integrable and therefore measurable (see [8]) for each $x^* \in X^*$, i.e. f is weakly measurable.

The primitive $\tilde{F}(t) = \int_0^t f$ is continuous, so the set $\{\tilde{F}(t); t \in [0,1]\} \subseteq X$ is compact and therefore totally bounded. Therefore the closed linear span of $\tilde{F}([0,1])$ is separable by semi-norm. Also, by Theorem 1, F is differentiable by semi-norm almost everywhere and $F' = f$ a.e. on $[0,1]$. If Y_p is the closed linear span of $\pi_p \circ \tilde{F}(t); t \in [0,1]$, then Y_p is separable in X_p and contains the set $\{\pi_p \circ f(t) : (\pi_p \circ \tilde{F})'(t) = \pi_p \circ f(t)\}$. Hence $\pi_p \circ f$ is essentially separably valued. By Pettis measurability theorem (see [2, Theorem 2.2] and [7, page 248]), it follows that f is measurable by semi-norm.

$$\square$$

Remark. If X is a locally convex space with countable family of semi-norms, in particular a Fréchet space, then a measurable by semi-norm function is measurable (see [7, p. 247]).

Some special measurable by semi-norm functions are variationally Henstock integrable as shown in the next result. For proof of $(1) \Rightarrow (2)$, we follow the corresponding result in Banach space from [12, Proposition 4.1]. See also, [15, Propositions 5.4.1, 5.4.2].

Proposition 1. *Let $\{a_n\}$ be a decreasing sequence converging to zero such that $a_1 = 1$. For $\{x_n\} \subset X$, define $f : [0,1] \to X$ by $f = \sum_{n=1}^{\infty} x_n \chi_{E_n}$ where each $E_n \subset [a_{n+1}, a_n)$ is Lebesgue measurable. Then the following are equivalent:*

(1) The series $\sum_{n=1}^{\infty} x_n m(E_n)$ is convergent in X;

(2) f is variationally Henstock integrable;

(3) f is Henstock-Kurzwil integrable.

In each case,

$$(vH) \int_I f = \sum_{n=1}^{\infty} x_n m(E_n \cap I) \tag{3}$$

for every $I \in \mathcal{I}$.

Proof. $(1) \Rightarrow (2)$: Observe that if $\sum_{n=1}^{\infty} x_n m(E_n)$ is convergent (for definition of convergence, see [14, Chapter III, p. 120]), then so is $\sum_{n=1}^{\infty} x_n m(E_n \cap I)$ for each $I \in \mathcal{I}$. Let $F(I) = \sum_{n=1}^{\infty} x_n m(E_n \cap I)$. To prove (2), we assume without loss of generality that $f(0) = 0$.

By convergence of the series $\sum_{n=1}^{\infty} x_n m(E_n)$ in X we have: Given $\epsilon > 0$ and $p \in \mathcal{P}$, there exists a natural number K_p such that for $s \geq n \geq K_p$, we have

$$p\left(\sum_{k=n}^{s} x_k m(E_k) \right) < \epsilon/4.$$

Also, for each natural number n, let $\delta_{p,n} : [a_{n+1}, a_n] \to (0, \infty)$ be a gauge such that if $P = \{(t_i, J_i) : i = 1, \ldots, k_n\}$ is a $\delta_{p,n}$-fine tagged partition of $[a_{n+1}, a_n]$, then

$$\sum_{i=1}^{k_n} p\big(f(t_i)\, m(J_i) - F(J_i)\big) < \frac{\epsilon}{2^{n+1}}.$$

We may assume that $\delta_{p,n}(a_n) = \delta_{p,n-1}(a_n)$. Define δ_p on $[0,1]$ as follows:

$$\delta_p(t) = \begin{cases} \delta_{p,n}(t), \ t \in (a_{n+1}, a_n], \\ a_{K_p}, \quad t = 0. \end{cases}$$

Now consider a δ_p-fine partition $P = \{(t_i, J_i) : i = 1, \ldots, k\}$ of $[0,1]$ and the corresponding sum $\sum_{i=1}^{k} p(f(t_i)\, m(J_i) - F(J_i))$. If $q \geq K_p$ is the largest integer such that $J_1 \subset [0, a_q)$, then

$$p(f(t_1)\, m(J_1) - F(J_1)) = p\left(\sum_{n=1}^{\infty} x_n m(E_n \cap J_1) \right)$$

$$= p\left(\sum_{n=q}^{\infty} x_n m(E_n \cap J_1) \right)$$

$$= p\left(x_q m(E_q \cap J_1) + \sum_{n=q+1}^{\infty} x_n m(E_n \cap J_1) \right)$$

$$\leq p(x_q) m(E_q \cap J_1) + p\left(\sum_{n=q+1}^{\infty} x_n m(E_n) \right)$$

$$< \frac{\epsilon}{4} + \frac{\epsilon}{4} = \frac{\epsilon}{2}.$$

Therefore,

$$\sum_{i=1}^{k} p(f(t_i)\, m(J_i) - F(J_i)) = p(f(t_1)\, m(J_1) - F(J_1))$$

$$+ \sum_{n=1}^{\infty} \sum_{t_i \in (a_{n+1}, a_n]} p(f(t_i)\, m(J_i \cap [a_{n+1}, a_n]) - F(J_i \cap [a_{n+1}, a_n]))$$

$$\leq \frac{\epsilon}{2} + \sum_{n=1}^{\infty} \frac{\epsilon}{2^{n+1}} = \epsilon.$$

Hence, f is variationally Henstock integrable and equation (3) holds for $I = [0, 1]$. In the same way, if we consider $F_I(t) = (vH) \int_a^t f$ if $t \in I = [a, b] \subset [0, 1]$, we get (3).

$(2) \Rightarrow (3)$: Trivial.

$(3) \Rightarrow (1)$: If f is Henstock-Kurzweil integrable, then $\pi_p \circ f$ is Henstock-Kurzweil integrable for each $p \in \mathcal{P}$. By $(3) \Rightarrow (1)$ of [12, Proposition 4.1], we get that $\sum_{n=1}^{\infty} \pi_p(x_n) m(E_n)$ is convergent to say y_p in X_p for each $p \in \mathcal{P}$. For every $p, q \in \mathcal{P}$ such that $p \leq q$, $g_{pq}(\pi_q(x_n)) = \pi_p(x_n)$. Since g_{pq} and π_p are continuous and linear, we get $g_{pq}(y_q) = y_p$. By [14, Chapter II.5.4], there exists $y \in X$ such that $\pi_p(y) = y_p$ for every $p \in \mathcal{P}$.

It is easy to check that $\sum_{n=1}^{\infty} x_n m(E_n)$ is convergent to y in X.

$\square$

Variational Henstock Integral and its Variational Measure 35

Proposition 2. *Let $f : [0,1] \to X$ be a variationally Henstock integrable function with $\tilde{F}$ as its primitive. Then, $V_{p,F} \ll m$ for all $p \in \mathcal{P}$.*

Proof. Since $V_{p,F} = V_{\pi_p \circ F}$, and $\pi_p \circ \tilde{F} : [0,1] \to X_p$ is primitive of the Banach space valued variationally Henstock integrable function $\pi_p \circ f$, the result is a consequence of [15, Theorem 7.5.1].

$\square$

The converse of above theorem holds under the condition that X satisfies RNP (see Theorem 4 ahead). To prove this result, we need the following:

Definition 7. A function $f : [0,1] \to X$ is said to be *Lipschitz* at the point $t \in [0,1]$ if for each $p \in \mathcal{P}$ there exists two positive constants M_p and δ_p such that

$$p(f(t+h) - f(t)) \leq M_p |h|$$

for all real h with $|h| < \delta_p$.

Proposition 3. *Let X be a locally convex space having RNP and $f : [0,1] \to X$ be an arbitrary function. Denote by G the set of all points $t \in [0,1]$ at which f is Lipschitz. Then f is differentiable by semi-norm almost everywhere in G.*

Proof. Since $\tilde{p}(\pi_p(f(t+h) - f(t))) = p(f(t+h) - f(t))$, we have that $\pi_p \circ f$ is X_p-valued Lipchitz function in G. By [3, Lemma 3.5], $\pi_p \circ f$ is differentiable a.e. in G with derivative k_p, i.e. $(\pi_p \circ f)' = k_p$ a.e. in G.

Suppose there are two continuous semi-norms p and q such that $p \leq q$. Since g_{pq} is a continuous linear map satisfying $g_{pq}(\pi_q(\frac{f(t+h)-f(t)}{h})) = \pi_p(\frac{f(t+h)-f(t)}{h})$, $t \in [0,1]$, and π_p is continuous for each $p \in \mathcal{P}$, we get, on taking limit as h approaches 0 that $g_{pq}(k_q(t)) = k_p(t)$, a.e., $t \in G$. By [14, II.5.4] there exists a unique function k on G such that $\pi_p \circ k(t) = k_p(t)$, $t \in G$ for every $p \in \mathcal{P}$. Therefore, $(\pi_p \circ f)' = \pi_p \circ k$ a.e. in G for all $p \in \mathcal{P}$.

Since $\mathcal{P}$ is a separating family of semi-norms, we get that $f' = k$ a.e. in G. Therefore, f is differentiable by semi-norm almost everywhere in G. $\square$

Continuing, we shall need the following result.

Theorem 3. *[2, Theorem 2.7] Let X be a locally convex space and $f : [0,1] \to X$ be a Pettis integrable function which is measurable by semi-norm. Then the induced vector measure ν has finite variation if and only if f is integrable by semi-norm.*

The following result, when X is a Banach space has been proved in [3, Theorem 3.6].

Theorem 4. *Let X be a locally convex space. Then the following are equivalent:*

(1) X has RNP;

(2) If $F : \mathcal{I} \to X$ is BVG^* on $[0,1]$, then F is differentiable by semi-norm a.e. on $[0,1]$;

(3) If $V_{p,F}$ is σ-finite for all $p \in \mathcal{P}$, then F is differentiable by semi-norm a.e.;

(4) If $V_{p,F} \ll m$ for all $p \in \mathcal{P}$, then F is differentiable by semi-norm a.e., $F' \in vH([0,1], X)$ and

$$F(I) = (vH) \int_I F'(t)dt, \ I \in \mathcal{I};$$

(5) If $V_{p,F} \ll m$ for all $p \in \mathcal{P}$ then there exists $f \in vH([0,1], X)$ such that

$$F(I) = (vH) \int_I f(t)dt, \ I \in \mathcal{I}.$$

Proof. $(1) \Rightarrow (2)$: Let $[0,1] = \bigcup_{n=1}^{\infty} E_n$ be a decomposition of $[0,1]$ such that F is BV^* on each E_n. Define $f(t) = F[0,t] = \tilde{F}(t) - \tilde{F}(0)$ for $t \in [0,1]$, and for $p \in \mathcal{P}$, set $G_{p,n} = \{t \in [0,1] : p(f(s) - f(t)) < n|s - t|, \text{for all } s \in [0,1] \text{ with } |s - t| < 1/n\}$, and $G_p = [0,1] \setminus \bigcup_n G_{p,n}$. We show that $m(G_p) = 0$ for each $p \in \mathcal{P}$.

If $m^*(G_p) > 0$ for some $p \in \mathcal{P}$, then there exists a natural number k such that $m^*(E_k \cap G_p) > 0$. By definition of G_p we have: For each $t \in G_p$ and each natural number n there exists $s \in [0,1]$ with $|s - t| < 1/n$ and $p(f(s) - f(t)) \geq n|s - t|$. Given $M > 0$, choose n such that $n \, m^*(E_k \cap G_p) > 2M$ and let

$$\mathbf{A_p} = \{[t, s] : t \in E_k \cap G_p : |s - t| < 1/n, p(f(s) - f(t)) \geq n|s - t|\}.$$

Then $\mathbf{A_p}$ is a Vitali covering of $E_k \cap G_p$. So there exist finitely many disjoint intervals $\{[t_i, s_i]\}$ such that $[t_i, s_i] \in \mathbf{A_p}$ and $m^*(E_k \cap G_p) < 2 \sum_i |s_i - t_i|$.

Hence

$$\sum_i p(F[t_i, s_i]) = \sum_i p(f(s_i) - f(t_i)) \geq n \sum_i |s_i - t_i| > n \frac{m^*(E_k \cap G_p)}{2} > M.$$

As M is arbitrary, we get F is not BV^* on E_k, which contradicts the hypothesis. Therefore, $m^*(G_p) = 0$ for each $p \in \mathcal{P}$. Thus f is Lipschitz almost everywhere. By Proposition 3, f is differentiable by semi-norm almost everywhere.

$(2) \Rightarrow (3)$: If $V_{p,F}$ is σ-finite for all $p \in \mathcal{P}$, then using the fact that $V_{p,F} = V_{\pi_p \circ F}$ and [3, Theorem 2.5], we get that $\pi_p \circ F$ is BVG^* on $[0,1]$. Therefore, F is BVG^* on $[0,1]$. By (2), we get F is differentiable by semi-norm almost everywhere on $[0,1]$. So (3) holds.

$(3) \Rightarrow (4)$: If $V_{p,F} \ll m$ for all $p \in \mathcal{P}$, we get that $V_{p,F} = V_{\pi_p \circ F}$ is σ-finite for all $p \in \mathcal{P}$, using [3, Corollary 2.3]. By (3), F is differentiable by semi-norm almost everywhere on $[0,1]$. Thus, the function F' satisfies the property:

Given $p \in \mathcal{P}$, there exists a subset $Z_p \subseteq [0,1]$ with $m(Z_p) = 0$ such that

$$\lim_{h \to 0} p\left(\frac{F \prec t, t+h \succ}{|h|} - F'(t)\right) = 0 \quad \text{for all } t \in [0,1] \setminus Z_p.$$

Equivalently,

$$\lim_{h \to 0} \tilde{p}\left(\pi_p\left(\frac{F \prec t, t+h \succ}{|h|} - F'(t)\right)\right) = 0 \quad \text{for all } t \in [0,1] \setminus Z_p.$$

By $(4) \Rightarrow (5)$ of [3, Theorem 3.6] $\pi_p \circ F'$ is in $vH([0,1], X_p)$ and

$$\pi_p \circ F(I) = (vH)\int_I \pi_p \circ F'(t)dt \quad \text{for every } I \in \mathcal{I}.$$

Since $\sum_{i=1}^k p(F'(t_i)\, m(J_i) - F(J_i)) = \sum_{i=1}^k \tilde{p}(\pi_p \circ F'(t_i)\, m(J_i) - \pi_p \circ F(J_i))$ for every δ_p-fine tagged partition $P = \{(t_i, J_i) : i = 1, \ldots, k\}$ of $[0,1]$, we get $F' \in vH([0,1], X)$ and

$$\pi_p\left(F(I) - (vH)\int_I F'(t)dt\right) = 0$$

for every $I \in \mathcal{I}$ and every $p \in \mathcal{P}$. As $\mathcal{P}$ is a separating family of semi-norms, we get the result.

$(4) \Rightarrow (5)$: Trivial.

$(5) \Rightarrow (1)$: Let $\nu : \mathcal{L} \to X$ be a measure of bounded variation absolutely continuous with respect to the Lebesgue measure on $[0, 1]$. Define $F : \mathcal{I} \to X$ by $F(I) = \nu(I)$. Then $V_{p,F} = V_{\pi_p \circ F} \ll m$ for each $p \in \mathcal{P}$ using [3, Lemma 3.3]. By hypothesis, there exists a function $f \in vH([0, 1], X)$ such that

$$F(I) = (vH) \int_I f(t) dm \quad \text{for every } I \in \mathcal{I}.$$

Consequently, for $x^* \in X^*$ we have

$$x^*(\nu(I)) = x^*(F(I)) = (vH) \int_I x^*(f(t)) dm \tag{4}$$

for every $I \in \mathcal{I}$ and $V_{x^* \circ F} \ll m$.

Moreover, for every $x^* \in X^*$ the measure $x^* \circ \nu$ has finite variation. Therefore, $V_{x^* \circ F}([0, 1]) \leq |x^* \circ \nu|([0, 1]) < \infty$. As $V_{x^* \circ F}(E) = \int_E |x^* f| dm$, we get that $x^* f \in L^1[0, 1]$ for each $x^* \in X^*$.

Fix $x^* \in B(X^*)$ and let $\mathcal{A}$ be the algebra generated by all the intervals $(a, b] \subset [0, 1]$. It then follows from (4) and countable additivity of both sides that

$$x^*(\nu(A)) = \int_A x^*(f(t)) dm \quad \text{for every } A \in \mathcal{A}.$$

But both sides of the above equality have unique extension to a measure on Borel σ-algebra $\mathcal{B}$. Thus

$$x^*(\nu(E)) = \int_E x^*(f(t)) dm \quad \text{for every } E \in \mathcal{B}.$$

Since both sides have unique extension to $\mathcal{L}$, the above equality holds for all $E \in \mathcal{L}$. Thus ν has a Pettis integrable Radon-Nikodym density f. As f is variationally Henstock integrable, f is measurable by semi-norm with variation of ν finite. By Theorem 4, f is integrable by semi-norm. As X^* separates points of X, we have

$$\nu(E) = \int_E f(t) dm \quad \text{for every } E \in \mathcal{L}.$$

So X has RNP.

$\square$

Remark. If X is a Fréchet space, then Bochner integrability and integrability by semi-norm coincide (see [2, Theorem 2.12]). Thus, in case of Fréchet space, the derivative F' obtained in the above theorem is Bochner integrable.

We now study the analogous results when X satisfies WRNP. For this, we first prove the following result.

Proposition 4. *Let X be a locally convex space having WRNP and $f : [0,1] \to X$ be an arbitrary function. Denote by G the set of all points $t \in [0,1]$ at which f is Lipschitz. Then f is pseudo-differentiable in G.*

Proof. Since $\tilde{p}\big(\pi_p(f(t+h) - f(t))\big) = p(f(t+h) - f(t))$, we have that $\pi_p \circ f$ is X_p-valued Lipchitz function in G. By [4, Lemma 4.2], $\pi_p \circ f$ is pseudo-differentiable in G with pseudo-derivative k_p.

Suppose there are two continuous semi-norms p and q such that $p \leq q$. Since g_{pq} is a continuous linear map satisfying $g_{pq}\big(\pi_q\big(\frac{f(t+h)-f(t)}{h}\big)\big) = \pi_p\big(\frac{f(t+h)-f(t)}{h}\big)$, $t \in [0,1]$, and π_p is continuous for each $p \in \mathcal{P}$, we get, on taking limit as h approaches 0 that

$$x_p^*(g_{pq}(k_q(t))) = x_p^*(k_p(t)), t \in G, \quad x_p^* \in X_p^*.$$

Using the fact that X_p^* separates points of X_p and [14, Chapter II.5.4], there exists a unique function k on G such that $\pi_p \circ k(t) = k_p(t)$, $t \in G$ for every $p \in \mathcal{P}$. Therefore, the pseudo-derivative of $\pi_p \circ f$ is $\pi_p \circ k$ a.e. in G for all $p \in \mathcal{P}$. In other words, $x_p^*(\pi_p \circ f)' = x_p^*(\pi_p \circ k)$ for all $x_p^* \in X_p^*$.

If $x^* \in X^*$, then there exists $p \in \mathcal{P}$ such that $x^* = x_p \circ \pi_p$, so that we have $x^*(f)' = x^*(k)$. Hence f is pseudo-differentiable in G with pseudo-derivative k. $\qquad\square$

The following result, when X is a Banach space, has been proved in [4].

Theorem 5. *Let X be a locally convex space. Then the following are equivalent:*

(1) X has WRNP;

(2) If $F : \mathcal{I} \to X$ is BVG^* on $[0,1]$, then F is pseudo-differentiable a.e. on $[0,1]$;

(3) If $V_{p,F}$ is σ-finite for all $p \in \mathcal{P}$, then F is pseudo-differentiable a.e.;

(4) If $V_{p,F} \ll m$ for all $p \in \mathcal{P}$, then F is pseudo-differentiable with pseudo-derivative F_P'; $F_P' \in HKP([0,1],X)$ and

$$F(I) = (HKP) \int_I F_P'(t)dt, \ I \in \mathcal{I};$$

(5) If $V_{p,F} \ll m$, for all $p \subset \mathcal{P}$ then there exists $f \in HKP([0,1], X)$ such that

$$F(I) = (HKP) \int_I f(t)dt, \ I \in \mathcal{I}.$$

Proof. (1) $\Rightarrow$ (2): As in the proof of (1) $\Rightarrow$ (2) of Theorem 4 we get that F is Lipschitz almost everywhere. By Proposition 4, F is pseudo-differentiable a.e. on $[0,1]$.

(2) $\Rightarrow$ (3): If $V_{p,F}$ is σ-finite for all $p \in \mathcal{P}$, then using the fact that $V_{p,F} = V_{\pi_p \circ F}$ and [4, Proposition 3.4] we get that $\pi_p \circ F$ is BVG^* on $[0,1]$. Therefore, F is BVG^* on $[0,1]$. By (2), we get F is pseudo-differentiable on $[0,1]$. So (3) holds.

(3) $\Rightarrow$ (4): If $V_{p,F} \ll m$ for all $p \in \mathcal{P}$, we get that $V_{p,F} = V_{\pi_p \circ F}$ is σ-finite for all $p \in \mathcal{P}$, using [4, Proposition 3.3]. By (3), F is pseudo-differentiable on $[0,1]$ with pseudo-derivative F'_P. Thus the function F'_P satisfies the property:

Given $x^* \in X^*$, there exists a subset $Z_{x^*} \subseteq [0,1]$ with $m(Z_{x^*}) = 0$ such that

$$\lim_{h \to 0} x^* \left(\frac{F \prec t, t+h \succ}{|h|} - F'_P(t) \right) = 0 \quad \text{for all} \quad t \in [0,1] \setminus Z_{x^*}.$$

Equivalently,

$$\lim_{h \to 0} x_p^* \left(\pi_p \left(\frac{F \prec t, t+h \succ}{|h|} - F'_P(t) \right) \right) = 0 \text{ for all } t \in [0,1] \setminus Z_{x^*}, \ p \in \mathcal{P}.$$

Since $V_{p,F} = V_{\pi_p \circ F} \ll m$, by $(v) \Rightarrow (vi)$ of [4, Theorem 4.5], we have that $\pi_p \circ F'_P$ is in $HKP([0,1], X_p)$ and

$$\pi_p \circ F(I) = (HKP) \int_I \pi_p \circ F'_P(t)dt \quad \text{for every } I \in \mathcal{I}.$$

So

$$x_p^*(\pi_p \circ F(I)) = (HK) \int_I x_p^*(\pi_p \circ F'_P(t))dt \quad \text{for every } I \in \mathcal{I}, \ x_p^* \in X_p^*.$$

Or

$$x^*(F(I)) = (HK) \int_I x^*(F'_P(t))dt \quad \text{for every } I \in \mathcal{I}, \ x^* \in X^*.$$

Therefore (4) holds.

$(4) \Rightarrow (5)$: Trivial.

$(5) \Rightarrow (1)$: Let $\nu : \mathcal{L} \to X$ be a measure of finite variation absolutely continuous with respect to the Lebesgue measure on $[0, 1]$. Define $F : \mathcal{I} \to X$ by $F(I) = \nu(I)$. Then $V_{p,F} = V_{\pi_p \circ F} \ll m$ for each $p \in \mathcal{P}$ using [4, Lemma 4.1]. By hypothesis, there exists a function $f \in HKP([0, 1], X)$ such that

$$F(I) = (HKP) \int_I f(t)dm \quad \text{for every } I \in \mathcal{I}.$$

Consequently, for $x^* \in X^*$ we have

$$x^*(\nu(I)) = x^*(F(I)) = (HK) \int_I x^*(f(t))dm \tag{5}$$

for every $I \in \mathcal{I}$ and $V_{x^* \circ F} \ll m$.

Moreover, for every $x^* \in X^*$ the measure $x^* \circ \nu$ has finite variation. Therefore, $V_{x^* \circ F}([0, 1]) \le |x^* \circ \nu|([0, 1]) < \infty$. As $V_{x^* \circ F}(E) = \int_E |x^* f|dm$, we get that $x^* f \in L^1[0, 1]$ for each $x^* \in X^*$.

Fix $x^* \in B(X^*)$ and let $\mathcal{A}$ be the algebra generated by all the intervals $(a, b] \subset [0, 1]$. It then follows from (5) and countable additivity of both sides that

$$x^*(\nu(A)) = \int_A x^*(f(t))dm \quad \text{for every } A \in \mathcal{A}.$$

But both sides of the above equality have unique extension to a measure on Borel σ-algebra $\mathcal{B}$. Thus,

$$x^*(\nu(E)) = \int_E x^*(f(t))dm \quad \text{for every } E \in \mathcal{B}.$$

Since both sides have unique extension to $\mathcal{L}$, the above equality holds for all $E \in \mathcal{L}$. Thus ν has a Pettis integrable Radon-Nikodym density f. So X has WRNP. $\qquad\square$

References

[1] S. Bhatnagar, *The Radon Nikodym property and Multipliers for the Class of Strongly $\mathcal{HK}$-integrable Functions*, Real Analysis Exchange, 44(2), 2019, 391-402.

[2] C. Blondia, *Integration in locally convex spaces*, Simon Stevin, A Quaterly Journal of Pure and applied Math., No. 3, 55(1981), 81-102.

[3] B. Bongiorno, L. Di Piazza and K. Musial, *A variational Henstock integral characterization of the Radon-Nikodym property*, Ill. J. of Math. 53(2009) No. 1, 87-99.

[4] B. Bongiorno, L. Di Piazza and K. Musial, *A characterization of the weak Radon-Nikodym property by finitely additive interval functions*, Bull. Aus. Math. Soc. 80(2009), 476-485.

[5] T. S. Chew, J. Y. Tay and T. L. Toh, *The non-uniform Riemann approach to Itô's Integral*, Real Anal. Exch., 27(2001-2), 495-514.

[6] H. G. Garnir, M.De Wilde and J. Schmets, *Analyse Fonctionnelle*, T.I., Théorie générale, Birkhauser Verlag, Basel, 1968.

[7] H. G. Garnir, M.De Wilde and J. Schmets, *Analyse Fonctionnelle*, T.II., Measure et intégration dans l'espace Euclidean E_n, Birkhauser Verlag, Basel, 1972.

[8] R.A. Gordon, *The integrals of Lebesgue, Denjoy, Perron and Henstock*, GSM Vol. 4, AMS(1994).

[9] V. Marraffa, *Riemann type integral for functions taking values in a locally convex space*, Czech. Math. J., 56(2006), 475-489.

[10] V. Marraffa, *Non absolutely convergent integrals of functions taking values in a locally convex space*, Rocky Mountain J. Math., 36(2006), 1577-1593.

[11] K. Musial, *Topics in the theory of Pettis integration*, Rend. Instit. Mat. Univ. Trieste 23(1991), 177-262.

[12] L. Di Piazza, V. Marraffa and K. Musial, *Variational Henstock integrability of Banach space valued functions*, Math. Bohemica. 141(2016), No. 2, 287-296.

[13] W. Rudin, *Functional Analysis*, Tata McGraw Hill, New Delhi (1982).

[14] H. H. Schaefer with M. P. Wolff *Topological Vector Spaces*, Second edition, GTM Vol. 3, Springer Verlag, New York (1999).

[15] Stefań Schwabik and Ye Guoju, *Topics in Banach Space Integration*, Series in Real Analysis, Vol. 10, World Scientific, Singapore (2005).

[16] B. Thomson, *Derivatives of interval functions*, Mem. Amer. Math. Soc., 452(1991).

[17] T. L. Toh and T. S. Chew, *A Variational Approach to Itô's Integral*, In Proceedings of SAP's 98, 133-147(1998),

[18] T. L. Toh and T. S. Chew, *On Henstock-Fubini Theorem for Multiple Stochastic Integrals*, Real Anal. Exch., 35(2010), 375-390.

© 2025 World Scientific Publishing Company
https://doi.org/10.1142/9789819812202_0003

Chapter 3

Variational Version of Henstock type Integral and Application in Harmonic Analysis

V. Skvortsov

Lomonosov Moscow State University, Moscow Center of Fundamental and Applied Mathematics Email: vaskvor2000@yahoo.com

A variational version of a Henstock type integral with respect to an abstract derivation basis in a topological measure space is defined for the case of Banach space-valued integrands. It is shown that this integral recovers a primitive from its derivative which is defined with respect to the same basis.

As an example of an application of this theory in harmonic analysis, a derivation bases and the respective Henstock type integrals on a zero-dimensional group are considered. It is shown that the variational integral on such a group solves the problem of recovering, by generalized Fourier formulas, the Banach space-valued coefficients of a series with respect to characters of this group.

Keywords and phrases: Henstock type integrals, variational Henstock integral, zero-dimensional group.

AMS Subject Classification No(2020): 58A30, 26A39, 46B04.

1. Introduction

A Riemann type integral, which solves the problem of recovering a primitive from its derivative and covers the Lebesgue integral, was introduced by Jaroslav Kurzweil in the late 1950s and independently by Ralph Henstock in the early 60s. For a good introduction to the theory of this integral and the history of its creation, including the reason why this integral is usually referred to as Henstock integral, see [3].

One of the ways to generalize the construction of this integral and to apply it in various fields of analysis, is to consider Riemann sums in the definition of the integral with respect to derivation bases more general than a basis constituted by all intervals on the real line. In Subsection 2.1 of this chapter, we define an abstract derivation basis $\mathcal{B}$ in a topological measure space, generalizing some partial cases known in the theory of Henstock integral (see [6], and [13] – [16]). We consider a Henstock type integral with respect to this bases, $H_{\mathcal{B}}$-integral, which integrates Banach space-valued functions. In this Banach space-valued case $H_{\mathcal{B}}$-integral being defined with respect to the usual interval basis, is a generalization of Bochner integral.

An important role in this theory is played by a notion of variational equivalence. Using this notion, we can obtain another, so called variational version of Henstock type integral with respect to the basis $\mathcal{B}$. Subsection 2.2 is devoted to this issue. For such an integral in the case of the full interval basis on the real line in the Banach space-valued case see [12]. While those two definitions are equivalent in the real-valued case, variational integral is strictly included into $H_{\mathcal{B}}$-integral for the Banach space-valued functions.

The advantage of the variational version of an integral with respect to any basis is that it reveals a direct connection of the concept of an integral with a derivative with respect to the same basis. In Subsection 2.3 a theorem on recovering a function from its derivative by the variational integral is proved. As for the problem of differentiability almost everywhere of the indefinite integral, this problem in the classical real-valued case had been solved in fact for the variational version of the integral. In the Banach space-valued case, this theorem on differentiability almost everywhere of the indefinite variational integral holds true for a wide class of bases, including a basis considered in Section 3.

In Section 3 we consider an application of this theory in harmonic analysis. We define derivation bases and the respective integrals on a zero-dimensional group. Typical examples of such a group are the Cantor Dyadic group and the group of p-adic integers.

We consider a problem of recovering, by generalized Fourier formulas, the coefficients of a series with respect to characters of such a group, and show that this problem can be reduced to the one of recovering the primitive from its derivative with respect to a special basis defined on the group. The problem of recovering the primitive in turn can be solved by the variational integral with respect to the same basis (see Subsection 3.2).

As we consider Banach space-valued coefficients of a series, we need to mention cases of strong and weak convergence. In the last case, a Pettis type variation integral is used to solve the coefficient problem.

2. Henstock type integrals with respect to a basis

2.1. *Derivational basis and Henstock integral with respect to the basis*

A *derivation basis* (or simply, *basis*) $\mathcal{B}$ in a measure space $(X, \mathcal{M}, \mu)$ is a non-empty family of non-empty subsets β of the product space $\mathcal{I} \times X$, where $\mathcal{I}$ is a family of measurable subsets of X of positive measure μ called *generalized intervals* or $\mathcal{B}$-*intervals* where

(a) *For every* $\beta_1, \beta_2 \in \mathcal{B}$ *there exists* $\beta \in \mathcal{B}$ *such that* $\beta \subset \beta_1 \cap \beta_2$.

So each basis is a directed set with the order given by "reverse" inclusion. We shall refer to the elements $\beta \in \mathcal{B}$ as *basis sets*. In this paper we shall suppose that all the pairs (I, x) constituting each β are such that $x \in I$, although it is not the case in the general theory (see [12], [6]).

For a set $E \subset X$ and $\beta \in \mathcal{B}$ we write

$$\beta(E) := \{(I, x) \in \beta : I \subset E\} \quad \text{and} \quad \beta[E] := \{(I, x) \in \beta : x \in E\}.$$

We suppose that the basis $\mathcal{B}$ *ignores no point*, i.e., $\beta[\{x\}] \neq \emptyset$ for any point $x \in X$ and for any $\beta \in \mathcal{B}$. We assume also that the basis $\mathcal{B}$ has a *local character* by which we mean that for any family of basis sets $\{\beta_\tau\}$, $\beta_\tau \in \mathcal{B}$ and for any pairwise disjoint sets E_τ there exists $\beta \in \mathcal{B}$ such that $\beta[\bigcup_\tau E_\tau] \subset \bigcup_\tau \beta_\tau[E_\tau]$.

Assuming that X is a topological space, we shall suppose that $\mathcal{B}$ is a *Vitali basis* by which we mean that for any x and for any neighborhood $U(x)$ of x there exists $\beta_x \in \mathcal{B}$ such that $I \subset U(x)$ for each pair $(I, x) \in \beta_x$.

A β-*partition* is a finite collection π of elements of β, where the distinct elements (I', x') and (I'', x'') in π have I' and I'' nonoverlapping, i.e., $\mu(I' \cap I'') = 0$. Let $L \in \mathcal{I}$. If $\pi \subset \beta(L)$ then π is called β-*partition in* L, if $\bigcup_{(I, x) \in \pi} I = L$ then π is called β-*partition of* L and is denoted by $\pi(L)$.

We say that a basis $\mathcal{B}$ has the *partitioning property* if the following conditions hold:

(i) For each finite collection $I_0, I_1, ..., I_n$ of $\mathcal{B}$-intervals with $I_1, ... I_n \subset I_0$ the difference $I_0 \setminus \bigcup_{i=1}^n I_i$ can be expressed as a finite union of pairwise non-overlapping $\mathcal{B}$-intervals;

(ii) For each $\mathcal{B}$-interval I and for any $\beta \in \mathcal{B}$ there exists a β-partition of I.

Let $P_\beta(L)$ denote a set of β-partitions of a fixed $\mathcal{B}$-interval L. Using the partitioning property and property (a) of basis $\mathcal{B}$, it is not difficult to see that *family* $\{P_\beta(L)\}_{\beta \in \mathcal{B}}$ *is a filter base*. Then for functions defined on β-partitions $\pi \in P_\beta(L)$ with values in some metric space we can consider *a limit with respect to this filter base*, and denote it $\lim_{\mathcal{B}} F(\pi)$. In this term Henstock integral with respect to basis $\mathcal{B}$ of a function $\Phi : \mathcal{I} \times X \to Y$, where Y is a Banach space, is defined as follows:

Definition 1. A function $\Phi : \mathcal{I} \times X \to Y$, is said to be *Kurzweil-Henstock integrable with respect to basis $\mathcal{B}$* (or *$H_{\mathcal{B}}$-integrable*) on $L \in \mathcal{I}$, with *$H_{\mathcal{B}}$-integral value $A \in Y$*, if there exists limit

$$\lim_{\mathcal{B}} \sum_{(I,x)\in\pi(L)} \Phi(I,x) = A.$$

We denote the integral value A by $(H_{\mathcal{B}}) \int_L \Phi$.

It is easy to check that the set of all $H_{\mathcal{B}}$-integrable functions on a fixed $\mathcal{B}$-interval constitutes a linear space.

We note that if Φ is $H_{\mathcal{B}}$-integrable on L then it is $H_{\mathcal{B}}$-integrable also on any $\mathcal{B}$-interval $I \subset L$. It can be easily proved that the $\mathcal{B}$-interval function $F : I \mapsto (H_{\mathcal{B}}) \int_I \Phi$ is additive on $\mathcal{I}$ and so we can call it the *indefinite $H_{\mathcal{B}}$-integral* of Φ.

In particular case $\Phi(I, x) = f(x)\mu(I)$, where $f : L \to Y$, we obtain *$H_{\mathcal{B}}$-integral of a function f on L with respect to measure μ*. In this case, the $H_{\mathcal{B}}$-integral is a generalization of Bochner integral. This can be checked in the same way as it is done in the classical case of the basis constituted by usual intervals on the real line (see [8]).

We consider also a Pettis type version of $H_{\mathcal{B}}$-integral. It is natural to denote it by $HP_{\mathcal{B}}$-integral.

Definition 2. A function $f : L \to Y$ is *Henstock-Pettis integrable with respect to basis $\mathcal{B}$* (or *$HP_{\mathcal{B}}$-integrable*) on $L \in \mathcal{I}$ if $y^*(f)$ is $H_{\mathcal{B}}$-integrable on each $\mathcal{B}$-interval $I \subset L$, for each functional $y^* \in Y^*$, and there exists $A_I \in Y$ such that

$$y^*(A_I) = (H_{\mathcal{B}}) \int_I y^*(f)$$

for each y^*. A_I is the value of the *indefinite $HP_{\mathcal{B}}$ integral* on I and we write

$$A_I = (HP_{\mathcal{B}}) \int_I f.$$

2.2. *Variational equivalence and variational Henstock type integral*

In the same terms of the limit with respect to the filter base which was introduced above, we can define a notion of variational equivalence which is in fact an analogue of Kolmogorov notion of differential equivalence (see [4]).

Definition 3. (see [6]) Two functions $\Phi_1 : \mathcal{I} \times X \to Y$ and $\Phi_2 : \mathcal{I} \times X \to Y$ are said to be *variationally equivalent on a $\mathcal{B}$-interval L* if

$$\lim_{\mathcal{B}} \sum_{(I,x) \in \pi(L)} \left\| \Phi_1(I,x) - \Phi_2(I,x) \right\| = 0$$

or, equivalently,

$$(H_{\mathcal{B}}) \int_L \left\| \Phi_1(I,x) - \Phi_2(I,x) \right\| = 0.$$

Variational equivalence can be connected also with another notion, which plays an important role in the Henstock integration theory. Namely, with the notion of variational measure. A standard definition of variational measure with respect to basis $\mathcal{B}$, generated by a function $\Phi : \mathcal{I} \times X \to Y$, on a fixed $\mathcal{B}$-interval L is this. First we define a β-variation on a set $E \subset L$:

$$Var(E, \Phi, \beta) := \sup_{\pi \subset \beta[E]} \sum \|\Phi(I,x)\|.$$

Then variational measure of a set $E \subset L$ is defined by

$$V_\Phi(E) = V(E, \Phi, \mathcal{B}) := \inf_{\beta \in \mathcal{B}} Var(E, \Phi, \beta).$$

Following the proof given in [15] for the interval bases in $\mathbb{R}$ it is possible to show that the extended real-valued set function $V_\Phi(\cdot)$ is an outer measure and a metric outer measure in the case of a metric space X (in the last case the definition of Vitali basis should be used).

Variational measure, generated by a function Φ, of a set $E \subset L$ can be also defined as $H_{\mathcal{B}}$-integral over L of a function defined by

$$\Phi_1(I,x) := \begin{cases} \|\Phi(I,x)\|, & x \in E, \\ 0, & x \in L \setminus E. \end{cases}$$

With this definition, we have to admit that $H_{\mathcal{B}}$-integrals of nonnegative functions can have an infinite value.

Using the notion of variational equivalence, we can obtain another so called variational version of Henstock type integral with respect to basis $\mathcal{B}$. For the full interval basis on the real line such an integral was defined in [12].

Definition 4. A function $f : L \to Y$, where $L \in \mathcal{I}$ and Y is a Banach space, is said to be *$VH_{\mathcal{B}}$-integrable on L*, if there exists an *additive $\mathcal{B}$-interval function $F : \mathcal{I} \to Y$* such that the function $\Phi_1(I,x) = F(I)$ for all $x \in I$, is variationally equivalent to the function $\Phi_2(I,x) = f(x)\mu(I)$. In this case F is *the indefinite $VH_{\mathcal{B}}$-integral* of f, in particular $(VH_{\mathcal{B}}) \int_L f = F(L)$.

It is easy to prove that $VH_\mathcal{B}$-integrable function on L is also $H_\mathcal{B}$-integrable and the integral values coincide. Indeed, let F be the indefinite $VH_\mathcal{B}$-integral. Using Definitions 3, 4 and the definition of the limit with respect to basis $\mathcal{B}$ we can state, that for any $\varepsilon > 0$ there exits $\beta \in \mathcal{B}$ such that for any β-partition $\pi(L)$ we have

$$\sum_{(I,x)\in\pi(L)} \left\| f(x)\mu(I) - F(I) \right\| < \varepsilon.$$

Using additivity of $VH_\mathcal{B}$-integral we obtain from the above estimate that for any β-partition $\pi(L)$ where $\beta \in \mathcal{B}$ is chosen above, we have

$$\left\| \sum_{(I,x)\in\pi} f(x)\mu(I) - F(L) \right\| = \left\| \sum_{(I,x)\in\pi} f(x)\mu(I) - \sum_{(I,x)\in\pi} F(I) \right\|$$
$$\leq \sum_{(I,x)\in\pi} \left\| f(x)\mu(I) - F(I) \right\| < \varepsilon.$$

This means that

$$\lim_{\mathcal{B}} \sum_{(I,x)\in\pi(L)} f(x)|I| = F(L),$$

i.e., f is $H_\mathcal{B}$-integrable on L with $F(L)$ being its $H_\mathcal{B}$-integral.

A result in the opposite direction is known, for a real valued functions and for usual interval basis on an interval of the real line, and is called Saks-Henstock lemma (see [3]). It can easily be generalized for the basis considered here. Note that the version of this Lemma was used and proved by Kolmogorov a long time ago in his paper [4]. So it would be fair to call this result as Kolmogorov-Henstock lemma.

Hence in the real valued case, the $H_\mathcal{B}$-integral and the $VH_\mathcal{B}$-integral are equivalent. Many properties of the $H_\mathcal{B}$-integral are based on this equivalence, i.e., they are proved in fact for the variational version of integral (see [3]).

However, this equivalence of two integrals fails to be true in the Banach valued case. It is proved in [12] (see also [5]) that in the case of basis of usual intervals on the real line the $VH_\mathcal{B}$-integral is equivalent to the $H_\mathcal{B}$-integral if and only if the range space is of finite dimension. It is likely that this result can be extended to the case of our abstract basis.

If we define a variational version of Henstock-Pettis integral then, due to the above result on the real valued case, it would be equivalent to Henstock-Pettis integral (see Definition 2).

It is easy to check that a function which is equal to zero almost everywhere on $L \in \mathcal{I}$, is $VH_{\mathcal{B}}$-integrable (and also $H_{\mathcal{B}}$-integrable) on L with integral value zero. This implies that $H_{\mathcal{B}}$-integrability of a function and the value of the $H_{\mathcal{B}}$-integral does not depend on values of the function on a set of measure zero. This justifies the following extension of Definitions 1 and 4 to the case of functions defined only almost everywhere on L.

Definition 5. A function f defined almost everywhere on $L \in \mathcal{I}$ is said to be $H_{\mathcal{B}}$-*integrable* (or $VH_{\mathcal{B}}$-*integrable*) on L, with integral value A, if the function

$$f_1(g) := \begin{cases} f(g), & \text{where } f \text{ is defined}, \\ 0, & \text{otherwise} \end{cases}$$

is $H_{\mathcal{B}}$-integrable (respectively, $VH_{\mathcal{B}}$-integrable) on L to A in the sense of Definition 1 (or 4).

2.3. *Recovering the primitive and problem of differentiation*

The advantage of the variational version of an integral is that it reveals a direct connection of the concept of an integral with a derivative.

$\mathcal{B}$-*derivative* of a Banach valued function $F : \mathcal{I} \to Y$ at a point x is defined as a limit

$$D_{\mathcal{B}}F(x) := \lim_{\mathcal{B}} \frac{F(I)}{\mu(I)},$$

if the limit exists. In other words, $A \in Y$ is a value of $\mathcal{B}$-derivative $D_{\mathcal{B}}F(x)$ if for any $\varepsilon > 0$ there exists β such that for all $(I, x) \in \beta[\{x\}]$

$$\left\| \frac{F(I)}{\mu(I)} - A \right\| < \varepsilon.$$

We define also a *weak $\mathcal{B}$-derivative* of F at x as an element $wD_{\mathcal{B}}F(x) \in Y$ such that for any $y^* \in Y^*$,

$$\lim_{\mathcal{B}} \frac{y^*(F(I))}{\mu(I)} = y^*(wD_{\mathcal{B}}F(x)).$$

In this case, we say that F is *weakly $\mathcal{B}$-differentiable* at x.

The following statement on recovering a function from its $\mathcal{B}$-derivative holds.

Theorem 1. *Let an additive function $F : \mathcal{I} \to Y$ be $\mathcal{B}$-differentiable everywhere on $L \in \mathcal{I}$, outside a set $E \subset L$ such that $V_F(E) = 0$. Then the function*

$$f(x) := \begin{cases} D_{\mathcal{B}}F(x), & \text{if it exists}, \\ 0, & \text{if } x \in E, \end{cases}$$

is $VH_{\mathcal{B}}$-integrable on L and F is its indefinite $VH_{\mathcal{B}}$-integral.

Proof. Fix $\varepsilon > 0$ and according to definition of variational measure find β such that for any $\beta[E]$-partition π_1 we have $\sum_{\pi_1} \|F(I)\| < \frac{\varepsilon}{2}$.

For each point x at which F is $\mathcal{B}$-differentiable find β_x such that for $(I, x) \in \beta_x[\{x\}]$

$$\|F(I) - f(x)\mu(I)\| < \varepsilon \frac{\mu(I)}{2\mu(L)}.$$

In this way according to property (1) we define β on L. Then for any β-partition $\pi(L)$ of L we get

$$\sum_{(I,x)\in\pi} \|F(I)) - f(x)\mu(I))\| \leq \sum_{(I,x)\in\pi,\, x\notin E} \|f(x)\mu(I) - F(I)\| +$$

$$+ \sum_{(I,x)\in\pi_1,} \|f(x)\mu(I) - F(I)\| \leq \frac{\varepsilon}{2\mu(L)} \sum_{(I,x)\in\pi,\, x\notin E} \mu(I) + \frac{\varepsilon}{2} \leq \varepsilon.$$

Thus F is the indefinite $VH_\mathcal{B}$-integral of f. In particular

$$F(L) = (VH_\mathcal{B}) \int_L f.$$

$\square$

Note that the condition $V_F(E) = 0$ put on the exceptional set E is in fact necessary in the case $\mu(E) = 0$. Moreover, the following descriptive characterization of the indefinite $VH_\mathcal{B}$-integral holds.

Theorem 2. *An additive $\mathcal{B}$-differentiable almost everywhere on $L \in \mathcal{I}$ function $F : \mathcal{I} \to X$ is the indefinite $VH_\mathcal{B}$-integral of its derivative if and only if the variational measure, generated by F, with respect to basis $\mathcal{B}$, is absolutely continuous with respect to μ.*

Proof. The sufficiency follows from Theorem 1. Conversely, let F be the indefinite $VH_\mathcal{B}$-integral of its derivative f. Take any set $E \subset L$, $\mu(E) = 0$. As f is integrable we can assume that $f(x) = 0$ on E. According to Definitions 4 and 3,

$$(H_\mathcal{B}) \int_L \left\| f(x)|I| - F(I) \right\| = 0.$$

Especially, $(H_\mathcal{B}) \int_E \|f(x)|I| - F(I)\| = (H_\mathcal{B}) \int_E \|F(I)\| = 0$. This means that $V_F(E) = 0$ and so we obtain the absolute continuity of V_F. $\square$

We formulate also a weak version of Theorem 1:

Theorem 3. *Let an additive function $F : \mathcal{I} \to X$ be $w\mathcal{B}$-differentiable on L outside a set E such that $V_{y^*F}(E) = 0$ for any $y^* \in Y^*$. Then the function*

$$f(x) := \begin{cases} wD_{\mathcal{B}}F(x), & \text{if it exists}, \\ 0, & \text{if } x \in E, \end{cases}$$

is $HP_{\mathcal{B}}$-integrable on L and F is its indefinite $HP_{\mathcal{B}}$-integral.

The problem of differentiability almost everywhere of the indefinite $H_{\mathcal{B}}$-integral in classical real valued case is also proved in fact for the variational version of integral. In the Banach space-valued case theorem on differentiability almost everywhere of the indefinite $VH_{\mathcal{B}}$-integral holds true for a wide class of bases (see Subsection 3.1 below). It is possible that validity of this result in the case of an abstract basis, we consider here, could depend on some additional requirements for the basis

As for the indefinite $H_{\mathcal{B}}$-integral, we shall see in the next section that, at least in the case of many particular bases, classical interval basis on the real line including, for any Banach space of infinite dimension it is possible to construct an example of an $H_{\mathcal{B}}$-integrable function with the indefinite $H_{\mathcal{B}}$-integral $\mathcal{B}$-differentiable nowhere.

3. Variational Integral in Harmonic Analisys

Henstock type integrals with respect to various bases are especially useful in the problem of recovering, by generalized Fourier formulas, vector-valued coefficients of orthogonal series from their sums. A choice of a particular basis depends on the orthogonal system we are dealing with.

In classical harmonic analysis (i.e., on one-dimensional torus) various kind of symmetric bases are useful and an integral with respect to the so-called approximate symmetric basis (see [16]) solves the problem of recovering the coefficients in this case.

In the case of Haar and Walsh systems (see [2], [14]), considered on the interval $[0,1]$, the Dyadic basis is used where a family $\mathcal{I}$ of $\mathcal{B}$-intervals is constituted by the family of dyadic intervals

$$\left[\frac{j}{2^n}, \frac{j+1}{2^n}\right], \quad 0 \leq j \leq 2^n - 1, \quad n = 0, 1, 2, \dots .$$

3.1. *Henstock type integrals on a zero-dimensional group*

As a very important particular example of application of the above theory in harmonic analysis we consider here in more detail a derivation bases and respective derivatives and integrals defined on a zero-dimensional group with second countability axiom. Typical examples of such a group are Cantor Dyadic group and the group of p-adic integers. We consider here the case of abelian group although in the part related to the construction of basis and integral this is not essential. Non-abelian case, the problem of recovering the coefficients including, is considered in [11].

It is known (see [1]) that with our assumption a topology in such a group can be given by a chain of subgroups

$$G = G_0 \supset G_1 \supset G_2 \cdots \supset G_n \supset \cdots \tag{1}$$

with $G = \bigcup_{n=0}^{+\infty} G_n$ and $\{0\} = \bigcap_{n=0}^{+\infty} G_n$. The subgroups G_n are clopen sets with respect to this topology. As G is compact, the factor group G_0/G_n for each n is finite. Let its order be m_n. We denote by K_n any coset of the subgroup G_n. For any $g \in G$ we denote by $K_n(g)$ the coset of the subgroup G_n which contains the element g, i.e.,

$$K_n(g) = g + G_n. \tag{2}$$

For each $g \in G$, the sequence $\{K_n(g)\}$ is decreasing and $\{g\} = \bigcap_n K_n(g)$.

We denote by λ the normalized Haar measure on the group G. We can make this measure to be complete by including all the subsets of the sets of measure zero into the class of measurable sets.

Since $\lambda(G_0) = 1$ and λ is translation invariant then

$$\lambda(G_n) = \lambda(K_n) = \frac{1}{m_n} \tag{3}$$

for all cosets K_n, $n \geq 0$.

The family of all K_n for all $n \in \mathbb{N}$ is a semiring of sets which forms the family $\mathcal{I}$ of $\mathcal{B}$-intervals of a basis $\mathcal{B}$ on G. For any function $\nu : G \to \mathbb{N}$, we define the basis set

$$\beta_\nu := \{(I, g) : g \in G, I = K_n(g), n \geq \nu(g)\}.$$

Then our derivation basis $\mathcal{B}$ is the family $\{\beta_\nu\}_\nu$ where ν runs over the set of all natural-valued functions on G. This basis has all the properties described in Section 1. But in this case, they are not postulated but easily checked.

If integrals $H_\mathcal{B}$, $VH_\mathcal{B}$ and $HP_\mathcal{B}$ are defined with respect to the basis on the group G described above, we denote them as H_G-integral, VH_G-integral and HP_G-integral, respectively. Definition of the $\mathcal{B}$-derivative is reduced in this basis to the ordinary limit

$$D_G F(x) = \lim_{n \to \infty} \frac{F(K_n(g))}{\lambda(K_n(g))}.$$

A particular case of Theorem 1, in which condition $V_F(E) = 0$ are ensured by requirement that difference ratio is bounded, is formulated in the following way:

Theorem 4. *Let an additive function $F : \mathcal{I} \to Y$ be $\mathcal{B}$-differentiable everywhere on G outside of a set E with $\mu(E) = 0$, and*

$$\overline{\lim}_{n\to\infty} \frac{||F(K_n(g))||}{\mu(K_n(g))} < \infty \tag{4}$$

everywhere on E. Then the function

$$f(x) := \begin{cases} D_{\mathcal{B}}F(x), & \text{if it exists,} \\ 0, & \text{if } x \in E, \end{cases}$$

is $VH_{\mathcal{B}}$-integrable on G and F is its indefinite $VH_{\mathcal{B}}$-integral.

In the weak version of Theorem 4 we need not use variational type integral by the reason mentioned in the previous Section.

Theorem 5. *Let an additive function $F : \mathcal{I} \to X$ be weakly $\mathcal{B}$-differentiable everywhere on G outside of a set E with $\mu(E) = 0$, and for any $x^* \in X^*$*

$$\overline{\lim}_{n\to\infty} \frac{|x^* F(K_n(g))||}{\mu(K_n(g))} < \infty$$

everywhere on E. Then the function

$$f(x) := \begin{cases} wD_{\mathcal{B}}F(x), & \text{if it exists,} \\ 0, & \text{if } x \in E, \end{cases}$$

is $HP_{\mathcal{B}}$-integrable on G and F is its indefinite $HP_{\mathcal{B}}$-integral.

It was proved in [13] that in the scalar-valued case the indefinite H_G-integral of any H_G-integrable function is G-differentiable everywhere on G and $D_G F(g) = f(g)$ almost everywhere. In a similar way, this property can be proved in a case when a range of a function is of finite dimension, and for VH_G-integral it is true in the case of any Banach space:

Theorem 6. *If a function $f : G \to X$ is VH_G-integrable on G then the indefinite VH_G-integral $F(K) = (VH)_G \int_K f$ as an additive function on the set of all $\mathcal{B}$-intervals is G-differentiable almost everywhere on G and*

$$D_G F(g) = f(g) \quad \text{a.e. on} \quad L. \tag{5}$$

The proof follows the line of the argument in [13, Theorem 3.1] for the scalar case provided one replaces a reference to Kolmogorov-Henstock lemma by a reference to variational equivalence of functions $F(I)$ and $f(g)\lambda(I)$ (see Definitions 4 and 3).

The next theorem, proved in [10], shows that this result can not be extended to the case of $H_{\mathcal{B}}$-integral.

Theorem 7. *For any infinite-dimensional Banach space Y, there exists a $H_{\mathcal{B}}$-integrable on G function $f : G \to Y$ with the indefinite H_G-integral which is G-differentiable nowhere on G.*

3.2. Application to harmonic analysis on the group G

Let Γ denotes the dual group of G, i.e., the group of characters of the group G. It is known (see [1]) that under assumption imposed on G the group Γ is a discrete abelian group (with respect to the point-wise multiplication of characters) and it can be represented as a sum of increasing chain of finite subgroups

$$\Gamma_0 \subset \Gamma_1 \subset \Gamma_2 \subset \cdots \subset \Gamma_n \subset \cdots \tag{6}$$

where $\Gamma_0 = \{\gamma_0\}$ with $\gamma_0(g) = 1$ for all $g \in G$. For each $n \in \mathbb{N}$ the group Γ_n is the annulator of G_n, i.e.,

$$\Gamma_n := \{\gamma \in \Gamma : \gamma(g) = 1 \text{ for all } g \in G_n\}.$$

The factor groups Γ_{n+1}/Γ_n and G_n/G_{n+1} are isomorphic (see [1]) and so they are of the same finite order for each $n \in \mathbb{N}$.

It is easy to check that if $\gamma \in \Gamma_n$ then γ is constant on each coset K_n of G_n, and if $\gamma \in \Gamma \setminus \Gamma_n$ then $\int_{K_n} \gamma d\mu = 0$ for each coset K_n.

This implies that the characters γ constitute a countable orthonormal system on G with respect to normalized measure λ, and we can consider a series

$$\sum_{\gamma \in \Gamma} a_\gamma \gamma \tag{7}$$

with respect to this system. We define the convergence of this series at a point g as the convergence of its partial sums

$$S_n(g) := \sum_{\gamma \in \Gamma_n} a_\gamma \gamma(g) \tag{8}$$

when n tends to infinity. If coefficients a_γ are Banach-valued, we can consider strong and weak convergence of this series.

We associate with the series (7) a function F defined on each coset K_n by

$$F(K_n) := \int_{K_n} S_n(g) d\mu \qquad (9)$$

where S_n are partial sums given by (8). Similar to the scalar case (see [13]) it is easy to check that F is an additive function on the family $\mathcal{I}$ of all $\mathcal{B}$-intervals. As it was in the case of Haar and Walsh series (see [9] and [7]) we call this function a *quasi-measure* associated with the series (7).

The properties of the characters, described above, imply that the sum S_n, defined by (8), is constant on each K_n. Then by (4) we have

$$S_n(g) = \frac{F(K_n(g))}{\mu(K_n(g))}. \qquad (10)$$

It follows directly from this equality that if the series (7) converges at some point $g \in G$ to a value $f(g)$ then the associated quasi-measure F is $\mathcal{B}$-differentiable at g and $D_{\mathcal{B}}F(g) = f(g)$. The same is true for the weak convergence.

The following statement is essential for establishing that a given series with respect to characters is the Fourier series in the sense of some general integral.

Theorem 8. *Let some integration process $\mathcal{A}$ be given which produces an integral additive on $\mathcal{I}$. Let a function F defined on $\mathcal{I}$ be the quasi-measure associated with the series (7). Then this series is the Fourier series of an $\mathcal{A}$-integrable function f if and only if $F(K) = (\mathcal{A}) \int_K f$ for any $K \in \mathcal{I}$.*

In view of (10) and Theorem 8, in order to solve the coefficient problem, it is enough to show that the quasi-measure associated with the series (7) is the indefinite integral of its derivative (strong or weak, respectively).

By this we reduce the problem of recovering the coefficients to the one of recovering the primitive and we can use a corresponding theorem on primitives in Subsection 2.1.

Theorem 9. *Suppose that the partial sums (8) of the series (7) converge to a function f everywhere on G outside of a set E with $\mu(E) = 0$, and*

$$\overline{\lim}_{n \to \infty} ||S_n(g)|| < \infty$$

everywhere on E. Then f is VH_G-integrable and (7) is the VH_G-Fourier series of f.

In the same way, using Theorem 8 for the case of the $HP_\mathcal{B}$-integral, we get

Theorem 10. *Suppose that the partial sums* (8) *of the series* (7) *converge weakly to a function f everywhere on G outside of a set E with $\mu(E) = 0$, and for any $y^* \in Y^*$*

$$\overline{\lim}_{n \to \infty} |y^* S_n(g)\| < \infty$$

everywhere on E. Then f is HP_G-integrable and (7) *is the HP_G-Fourier series of f.*

Note that the above theorem covers, in particular, the case of the convergence of the series (7) everywhere on G.

It is remarkable that in Theorems 9 and 10 there is no need to suppose that the sum of the series is integrable in the respective sense. The integrability is an implication of the convergence. However, if we assume appriori that the sum is integrable in the sense of Lebesgue (or Bochner in the Banach space-valued case) then we obtain an analogue of the classical du Bois Reymond–Vallee Poissin theorem.

Theorem 11. *Suppose that the partial sums* (8) *of the series* (7) *converge (converge weakly) to a Lebesgue (Bochner) integrable function f everywhere on G except a countable set. Then the series* (7) *is the Lebesgue–Fourier series (resp. Bochner–Fourier series) of f.*

A proof is reduced to checking that any Lebesgue (Bochner) integrable function is also VH_G-integrable and the integrals coincide what can be done following the line of the proof in the case of the usual interval basis on the real line.

Now we consider the problem of convergence of a Fourier series in the sense of the VH_G-integral and the $H_\mathcal{B}$-integral. The partial sums $S_n(f, g)$ of Fourier series, with respect to the system Γ, of a function $f : G \to Y$ integrable in the sense of these integrals can be represented, according to Theorem 8 and formula (10), as

$$S_n(f, g) = \frac{1}{\mu(K_n(g))} \int_{K_n(g)} f. \tag{11}$$

From this equality together with differentiability property of the indefinite $VH_\mathcal{B}$-integral (see Theorem 6) follows

Theorem 12. *The partial sums $S_n(f, g)$ of the $VH_\mathcal{B}$-Fourier series of a $VH_\mathcal{B}$-integrable on G function f are convergent to f almost everywhere on G.*

At the same time such a theorem fails to be true for $H_{\mathcal{B}}$-Fourier series. Indeed, we get from Theorem 7:

Theorem 13. *For any infinite-dimensional Banach space Y there exists H_G-integrable function $f : G \to Y$ such that partial sums of its H_G-Fourier series with respect to the system Γ diverge everywhere.*

It is interesting to note that the rate of growth of these partial sums can not be made arbitrary large for the whole class of infinite-dimensional Banach spaces. For some particular spaces, this rate has some restriction. For example, it can be deduced from [9] that for some class of infinite-dimensional Banach spaces for a Pettis-integrable function f taking values in such a space, sums of its Fourier series with respect to the Walsh system, which is the systems of characters of a particular case of a zero-dimensional group, satisfy the relation $||S_n(f,g)|| = o(2^{\frac{1}{2}n})$.

References

[1] *Agaev G. N., Vilenkin N. Ya., Dzhafarli G. M., Rubinstein A. I.* Multiplicative system of functions and harmonic analysis on zero-dimensional groups, Elm, Baku, 1981 (in Russian).

[2] *Golubov B., Efimov A., Skvortsov V.* Walsh series and transforms. Theory and applications, Kluwer Academic Publishers Group, Dordrecht, 1991.

[3] *Gordon R. A.* The Integrals of Lebesgue, Denjoy, Perron, and Henstock, Grad. Stud. Math 4, Amer. Math. Soc., 1994.

[4] *Kolmogoroff A.* Untersuchungen uber Integralbegriff, Math. Ann. **103** (1930), 654–696.

[5] *Lukashenko T. P., Skvortsov V. A., Solodov A. P.* Generalized integrals, LIBROKOM (URSS group), Moscow, 2011 (in Russian).

[6] *Ostaszewski K. M.* Henstock integration in the plane, Mem. AMS., **63**:353, 1986.

[7] *Plotnikov M. G.* Quasi-measures, Hausdorff p-measures and Walsh and Haar series, Izvestiya: Mathematics, **74**:4 (2010), 819–848.

[8] *Schwabik S., Ye Guoju* Topics in Banach space integration, Series in Real Analysis, Vol. 10, Hackensack, NJ: World Scientific, 2005.

[9] *Skvortsov, V. A.* Integration of Banach-valued functions and Haar series with Banach-valued coefficients, Moscow Univ. Math. Bull., **72**:1 (2017), 24–30.

[10] *Skvortsov, V. A.* Recovering Banach-valued coefficients of series with respect to characters of zero-dimensional groups, Annales Univ. Sci. Budapest, Sect. Comp., **49** (2019), 379–397.

[11] *Skvortsov V. A.* Reconstruction of a Generalized Fourier Series from Its Sum on a Compact Zero-Dimensional Group in the Non-Abelian Case, Math. Notes, **109**:4, (2021), 630–637.

[12] *Skvortsov V. A., Solodov A. P.* A variational integral for Banach-valued functions, Real Analysis Exchange, **24**:2 (1998), 799–805.

[13] *Skvortsov V., Tulone F.* Kurzweil-Henstock type integral on zero-dimensional group and some of its applications, Czech. Math. J., **58** (2008), 1167–1183.

[14] *Skvortsov V., Tulone F.* Multidimensional dyadic Kurzweil-Henstock- and Perron-type integrals in the theory of Haar and Walsh series, J. Math. Anal. Appl., **421**:2 (2015), 1502–1518.

[15] *Thomson B. S.* Derivation bases on the real line, Real Anal. Exchange, **8**:1,2 (1982/83), 67–207, 278–442.

[16] *Thomson B. S.* Symmetric properties of real functions, Monographs and Textbooks in Pure and Appl. Math., 183, Marcel Dekker, New York, 1994.

© 2025 World Scientific Publishing Company
https://doi.org/10.1142/9789819812202_0004

Chapter 4

A Survey on the Riemann-Lebesgue Integrability in Non-additive Setting

Anca Croitoru

University "Alexandru Ioan Cuza", Faculty of Mathematics, Bd. Carol I, No. 11, Iaşi, 700506, Romania
Email: croitoru@uaic.ro, Orcid iD: 0000-0001-8180-3590

Alina Gavriluţ

University "Alexandru Ioan Cuza", Faculty of Mathematics, Bd. Carol I, No. 11, Iaşi, 700506, Romania
Email: gavrilut@uaic.ro, Orcid iD: 0000-0002-6113-0963

Alina Iosif

Petroleum-Gas University of Ploieşti, Department of Computer Science, Information Technology, Mathematics and Physics, Bd. Bucureşti, No. 39, Ploieşti 100680, Romania
Email: emilia.iosif@upg-ploiesti.ro, Orcid iD: 0000-0003-0144-8811

Anna Rita Sambucini*

University of Perugia, Department of Mathematics and Computer Sciences, 1, Via Vanvitelli N. 1, Perugia, 06123, Italy
Email: anna.sambucini@unipg.it, Orcid iD: 0000-0003-0161-8729

We present some results regarding the Riemann-Lebesgue integral of a vector (real resp.) function relative to an arbitrary non-additive set function. Then these results are generalized to the case of Riemann–Lebesgue integrable interval-valued multifunctions.

---

*Finanziato dall'Unione europea - Next Generation EU, Missione 4, Componente 2, CUP J53D23015950001, Codice del Progetto P20229SH29

1. Introduction

The theory of non-additive set functions and nonlinear integrals has become an important tool in many domains such as: potential theory, subjective evaluation, optimization, economics, decision making, data mining, artificial intelligence, and accident rates estimations (e.g. see [20, 21, 38, 44, 53, 57, 61, 64, 76, 80–82, 85]). In the literature several methods of integration for (multi)functions based on extensions of the Riemann and Lebesgue integrals have been introduced and studied (e.g. see [2–9, 13–15, 17, 18, 25, 29–34, 36, 37, 39–43, 54–56, 59, 69]). In this context, Kadets and Tseytlin [48] have introduced the absolute Riemann-Lebesgue $|RL|$ and unconditional Riemann-Lebesgue RL integrability, for Banach valued functions with respect to countably additive measures. According to [48], in finite measure spaces, the Bochner integrability implies $|RL|$ integrability, which is stronger than RL integrability, that implies Pettis integrability. Contributions in this area are given in [10, 16, 17, 22, 26, 27, 33, 47, 65–68, 70–75, 77–79, 84].

Interval Analysis, as particular case of Set-Valued Analysis, was introduced by Moore [58], motivated by its applications in computational mathematics (i.e. numerical analysis). The interval-valued multifunctions and multimeasures are involved in various applied sciences, including statistics, biology, theory of games, economics, social sciences and software development. They are Also used, for example, in signal and image processing, since the discretization of a continuous signal causes different sources of uncertainty and ambiguity, as we can see in [23, 24, 46, 49, 52, 83, 86].

In this chapter, we study the Riemann-Lebesgue integral with respect to an arbitrary set function, not necessarily countably additive. We present some classical properties of the integral together with some relationships between this Riemann-Lebesgue integral and some other integrals known in the literature as the Birkhoff simple and the Gould integrals. We describe also some convergence theorems (e.g. Lebesgue type convergence theorems, Fatou type theorem) for sequences of Riemann-Lebesgue integrable functions. Then these results are extended to the case of Riemann-Lebesgue integrable interval-valued multifunctions.

The chapter is organized as follows. In Section 2 we present some preliminaries which are necessary in what follows. In Section 3 we introduce the Riemann Lebesgue integral and we present different properties for the RL integral of a real function with respect to a non-additive set function. We also establish some comparative results among this Riemann-Lebesgue integral, the Birkhoff simple and the Gould integrals. Then some convergence theorems for sequences of Riemann-Lebesgue integrable are pointed out. Section 3 ends with a Hölder, Minkowski type inequalities for Riemann-Lebesgue integrable functions. The results of Section 3 are then applied in Section 4 to the case of interval-valued functions and set functions. We highlight that we present some generalizations of the Riemann-Lebesgue integral in two directions: the non-additive setting and the interval-valued setting. Some difficulties arise in these approaches: the techniques of the classical measure theory can no longer be used; the set spaces are more difficult to use. The motivation of it is mainly due the fact that, though finite or countable additivity is a fundamental concept in measure theory, it can be useless in some modeling problems of decision making, data mining, economy, computer science, game theory, subjective evaluation, and fuzzy logic, as shown in [21, 22, 46, 49–52, 61, 63, 81, 82].

2. Definitions and basic facts

Suppose S is a nonempty set and $\mathcal{C}$ a σ-algebra of subsets of S, while $\mathcal{P}(S)$ denotes the family of all subsets of S. For every $E \subset S$, as usual, let $E^c = S \setminus E$ and let χ_E be the characteristic function of E.

Following [12, Definition 1] a finite (countable, resp.) partition of S is a finite (countable, resp.) family of nonempty sets $P = \{E_i\}_{i=1}^n$ ($\{E_n\}_{n \in \mathbb{N}}$, resp.) $\subset \mathcal{C}$ such that $E_i \cap E_j = \emptyset, i \neq j$ and $\bigcup_{i=1}^n E_i = S$ ($\bigcup_{n \in \mathbb{N}} E_n = S$, resp.).

- If P and P' are two partitions of S, then P' is said to be *finer than* P, if every set of P' is included in some set of P

$$P \leq P' \text{ (or } P' \geq P). \tag{1}$$

- The *common refinement* of two finite or countable partitions $P = \{E_i\}$ and $P' = \{G_j\}$ is the partition $P \wedge P' := \{E_i \cap G_j\}$.
- A countable tagged partition of S is a family $\{(E_n, s_n), n \in \mathbb{N}\}$ such that $(E_n)_n$ is a countable partition of S and $s_n \in E_n$ for every $n \in \mathbb{N}$.

All over this chapter, without any additional assumptions, $\nu : \mathcal{C} \to [0, \infty)$ will be a set function, such that $\nu(\emptyset) = 0$. As in [12, Definitions 2 and 3], $\nu : \mathcal{C} \to [0, \infty)$ is said to be:

- *monotone* if $\nu(A) \leq \nu(B)$, for every $A, B \in \mathcal{C}$, with $A \subseteq B$ (such non additive measures are called also capacities or fuzzy measures);
- *subadditive* if $\nu(A \cup B) \leq \nu(A) + \nu(B)$, for every $A, B \in \mathcal{C}$, with $A \cap B = \emptyset$;

- *a submeasure* (in the sense of Drewnowski [35]) if ν is monotone and subadditive;
- *σ-subadditive* if $\nu(A) \leq \sum_{n=0}^{+\infty} \nu(A_n)$, for every sequence of (pairwise disjoint) sets $(A_n)_{n\in\mathbb{N}} \subset \mathcal{C}$, with $A = \bigcup_{n=0}^{+\infty} A_n$;
- *finitely additive* if $\nu(A \cup B) = \nu(A) + \nu(B)$, for every disjoint sets $A, B \in \mathcal{C}$;
- *σ-additive* if $\nu(\bigcup_{n=0}^{+\infty} A_n) = \sum_{n=0}^{+\infty} \nu(A_n)$, for every sequence of pairwise disjoint sets $(A_n)_{n\in\mathbb{N}} \subset \mathcal{C}$;
- *order-continuous* (or, *o-continuous*) if $\lim_{n\to+\infty} \nu(A_n) = 0$, for every decreasing sequence of sets $(A_n)_{n\in\mathbb{N}} \subset \mathcal{C}$, with $\bigcap_{n=0}^{+\infty} A_n = \emptyset$ (denoted by $A_n \searrow \emptyset$);
- *exhaustive* if $\lim_{n\to+\infty} \nu(A_n) = 0$, for every sequence of pairwise disjoint sets $(A_n)_{n\in\mathbb{N}} \subset \mathcal{C}$;
- *null-additive* if for every $A, B \in \mathcal{C}$, $\nu(A \cup B) = \nu(A)$ when $\nu(B) = 0$.

Moreover a set function $\nu : \mathcal{C} \to [0, \infty)$ satisfies:

($\boldsymbol{\sigma}$) the property $\boldsymbol{\sigma}$ if for every $\{E_n\}_n \subset \mathcal{C}$ with $\nu(E_n) = 0$, for every $n \in \mathbb{N}$ we have $\nu(\cup_{n=0}^{\infty} E_n) = 0$;

(E) the condition **(E)** if for every double sequence $(B_n^m)_{n,m\in\mathbb{N}^*} \subset \mathcal{C}$, such that for every $m \in \mathbb{N}^*$, $B_n^m \searrow B^m$ $(n \to \infty)$ and $\nu(\cup_{m=1}^{\infty} B^m) = 0$, there exist two increasing sequences $(n_p)_p, (m_p)_p \subset \mathbb{N}$ such that $\lim_{k\to\infty} \nu(\bigcup_{p=k}^{\infty} B_{n_p}^{m_p}) = 0$.

The property $\boldsymbol{\sigma}$ is a consequence of the countable subadditivity and it will be needed in some of our results. Observe that the condition **(E)** was given, for example, in [51], in order to give sufficient and necessary conditions to obtain Egoroff's Theorem for suitable non additive measures. See also [60] for null additive set functions and related questions. An example of a set function that satisfies the condition **(E)** can be found in [51, Example 3.3].

A property (P) holds ν-almost everywhere (denoted by ν-a.e.) if there exists $E \in \mathcal{C}$, with $\nu(E) = 0$, so that the property (P) is valid on $S \setminus E$. A set $A \in \mathcal{C}$ is said to be an atom of a set function $\nu : \mathcal{C} \to [0, \infty)$ if $\nu(A) > 0$ and for every $B \in \mathcal{C}$, with $B \subseteq A$, we have $\nu(B) = 0$ or $\nu(A \setminus B) = 0$.

A Survey on the Riemann-Lebesgue Integrability in Non-additive Setting 63

We associate to $\nu : S \to [0, \infty)$ the following set functions. (see [12, Definition 4]).

- The variation $\overline{\nu}$ of ν is the set function $\overline{\nu} : \mathcal{P}(S) \to [0, +\infty]$ defined by

$$\overline{\nu}(E) = \sup\{\sum_{i=1}^{n} \nu(A_i)\},$$

for every $E \in \mathcal{P}(S)$, where the supremum is extended over all finite families of pairwise disjoint sets $\{A_i\}_{i=1}^{n} \subset \mathcal{C}$, with $A_i \subseteq E$, for every $i \in \{1, \ldots, n\}$. The set function ν is said to be *of finite variation* (on $\mathcal{C}$) if $\overline{\nu}(S) < +\infty$.
- The semivariation $\widetilde{\nu}$ of ν is the set function : $\mathcal{P}(S) \to [0, +\infty]$ defined for every $A \subseteq S$, by

$$\widetilde{\nu}(A) = \inf\{\overline{\nu}(B); \ A \subseteq B, \ B \in \mathcal{C}\}.$$

Remark 1. ([27, Remark 1]) Let $\nu : \mathcal{C} \to [0, +\infty)$ be a non additive measure. Then $\overline{\nu}$ is monotone and super-additive on $\mathcal{P}(S)$, that is

$$\overline{\nu}(\bigcup_{i \in I} A_i) \geq \sum_{i \in I} \overline{\nu}(A_i),$$

for every finite or countable partition $\{A_i\}_{i \in I}$ of S. If ν is finitely additive, then $\overline{\nu}(A) = \nu(A)$, for every $A \in \mathcal{C}$. If ν is subadditive (σ-subadditive, resp.), then $\overline{\nu}$ is finitely additive (σ-additive, resp.). Moreover, for every $\nu, \nu_1, \nu_2 : \mathcal{C} \to \mathbb{R}$ and every $\alpha \in \mathbb{R}$,

- $\overline{\nu_1 \pm \nu_2} \leq \overline{\nu_1} + \overline{\nu_2}; \qquad \overline{\alpha \nu} = |\alpha| \, \overline{\nu}.$

For all unexplained definitions, see e.g. [9, 11].

Let $\mathcal{M}(S)$ be the set of all non negative submeasures on $(S, \mathcal{C})$. Let $(\nu_n)_n \subset \mathcal{M}(S)$, we will use the symbol $\nu_n \uparrow$ to indicate that $\nu_n \leq \nu_{n+1}$ for every $n \in \mathbb{N}$.

Definition 1. A sequence $(\nu_n)_n \subset \mathcal{M}(S)$ setwise converges to $\nu \in \mathcal{M}(S)$ if for every $A \in \mathcal{C}$

$$\lim_{n \to \infty} \overline{\nu_n - \nu}(A) = 0. \tag{2}$$

In the σ-additive case, the setwise convergence is given by $\lim_{n \to \infty} \nu_n(A) = \nu(A)$ for every $A \in \mathcal{C}$, (see e.g. [31]). Since $|\nu_n(A) - \nu(A)| \leq \overline{\nu_n - \nu}(A)$ for every $A \in \mathcal{C}$, the convergence given in Definition 1 implies the one of [31];

the converse does not hold in general. Neverthless, the two definitions co-incide, from [12, Remark 1], if ν, ν_n, for all $n \in \mathbb{N}$, are finitely additive and non negative.

Finally, if S is a locally compact Hausdorff topological space, we denote by $\mathcal{K}$ the lattice of all compact subsets of S, $\mathcal{B}$ the Borel σ-algebra (i.e., the smallest σ-algebra containing $\mathcal{K}$) and $\mathcal{O}$ the class of all open sets.

Definition 2. A set function $\nu : \mathcal{B} \to [0, \infty)$ is called regular if for every set $A \in \mathcal{B}$, and every $\varepsilon > 0$ there exist $K \in \mathcal{K}$ and $D \in \mathcal{O}$ such that $K \subseteq A \subseteq D$ and $\nu(D \setminus K) < \varepsilon$.

3. The Riemann-Lebesgue integrability

Let X be a Banach space over $\mathbb{R}$. As in Kadets and Tseytlin [48, Definition 4.5] (for scalar functions) and Potyrala [66, Definition 7], and Kadets and Tseytlin [47] (for vector functions), we introduce the following definition:

Definition 3. ([12, Definition 5]) A vector function $f : S \to X$ is called *absolutely (unconditionally* resp.) *Riemann-Lebesgue* ($|RL|$) (*RL* resp.) ν-*integrable* (on S) if there exists $a \in X$ such that for every $\varepsilon > 0$, there exists a countable partition P_ε of S, so that for every countable partition $P = \{A_n\}_{n \in \mathbb{N}}$ of S with $P \geq P_\varepsilon$, the following hold:

- f is bounded on every A_n, with $\nu(A_n) > 0$;
- for every $s_n \in A_n$, $n \in \mathbb{N}$, the series $\sum_{n=0}^{+\infty} f(s_n)\nu(A_n)$ is absolutely (unconditionally resp.) convergent and

$$\left\| \sum_{n=0}^{+\infty} f(s_n)\nu(A_n) - a \right\| < \varepsilon.$$

The vector a is called *the absolute (unconditional) Riemann-Lebesgue ν-integral of f on S* and it is denoted by $(|RL|) \int_S f \, d\nu \left((RL) \int_S f \, d\nu \text{ resp. } \right)$.

We denote by the symbol $|RL|_\nu^1(X)$ the class of all X-valued function that are $|RL|$ integrable with respect to ν and in an analogous way we denote the class of all functions that are RL ν-integrable.

Remark 2. (see [12, Remark 2]) Obviously if a exists, then it is unique. Moreover, if h is $|RL|$ ν-integrable, then h is RL ν-integrable and if

X is finite dimensional, then $|RL|$ ν-integrability is equivalent to RL ν-integrability. In this case, it is denoted by RL.

We remember also the following in the countably additive case:

2.a) Kadets and Tseytlin in [48] introduced the $|RL|$ ν-integral and the RL ν-integral for functions with values in a Banach space relative to a measure. They proved that if $(S, \mathcal{C}, \nu)$ is a finite measure space, then the following implications hold:
$$L^1_\nu(X) \subset |RL|^1_\nu(X) \subset RL^1_\nu(X) \subset P_\nu(X),$$
where $L^1_\nu(X)$, and $P_\nu(X)$ denotes respectively the Bochner and the Pettis integrability.

2.b) If X is a separable Banach space, then
$$L^1_\nu(X) = |RL|^1_\nu(X) \subset RL^1_\nu(X) = P_\nu(X).$$

2.c) If $(S, \mathcal{C}, \nu)$ is a σ-finite measure space, then the Birkhoff integrability coincides with RL ν-integrability [66].

2.d) If $h : [a, b] \to \mathbb{R}$ is Riemann integrable, then h is RL-integrable ([66, Corollary 17]). The converse is not valid: for example the function $h : [0, 1] \to \mathbb{R}$, $h = \chi_{[0,1] \cap \mathbb{Q}}$ is RL-integrable, but it is not Riemann integrable ([66, Example 19]).

3.1. *Some properties of RL ν-integrability*

In this section we present some results contained in [9, 12] regarding the Riemann-Lebesgue integrability of vector functions with respect to an arbitrary non-negative set function, pointing out its remarkable properties. We begin with a characterization of $|RL|$-integrability.

Theorem 1. *Let* $g, h \in |RL|^1_\nu(X)$ *and* $\alpha, \beta \in \mathbb{R}$. *Then:*

(1.a) *If h is $|RL|$ ν-integrable on S, then h is $|RL|$ ν-integrable on every $E \in \mathcal{C}$ (see [12, Theorem 1.a]);*

(1.b) *h is $|RL|$ ν-integrable on every $E \in \mathcal{C}$ if and only if $h\chi_E$ is $|RL|$ ν-integrable on S. In this case, by [12, Theorem 1.b],*
$$(|RL|) \int_E h \, d\nu = (|RL|) \int_S h\chi_E \, d\nu.$$

(The same holds for RL-integrability). Moreover, by [12, Theorem 3],

(1.c) *$\alpha g + \beta h \in |RL|^1_\nu(X)$ and*
$$(|RL|) \int_S (\alpha g + \beta h) \, d\nu = \alpha \cdot (|RL|) \int_S g \, d\nu + \beta \cdot (|RL|) \int_S h \, d\nu.$$

(1.d) $h \in |RL|^1_{\alpha\nu}(X)$ *for* $\alpha \in [0, +\infty)$ *and*

$$(|RL|) \int_S h \, \mathrm{d}(\alpha\nu) = \alpha(|RL|) \int_S h \, \mathrm{d}\nu.$$

(1.e) *Suppose* $h \in |RL|^1_{\nu_i}(X)$ *for* $i = 1, 2$. *By* ([12, Theorem 4]) $h \in |RL|^1_{\nu_1+\nu_2}(X)$ *and*

$$(|RL|) \int_S h \, \mathrm{d}(\nu_1 + \nu_2) = (|RL|) \int_S h \, \mathrm{d}\nu_1 + (|RL|) \int_S h \, \mathrm{d}\nu_2.$$

Similar results also hold for the RL ν-integrability.

Proof. We report here only the proofs of (1.a) and (1.b).

Fix any $A \in \mathcal{C}$ and denote by J the integral of h on S; then, fixed any $\varepsilon > 0$, there exists a partition P_ε of S, such that, for every finer partition $P' := \{A_n\}_{n \in \mathbb{N}}$ it is

$$\left\| \sum_{n=0}^{+\infty} h(t_n)\nu(A_n) - J \right\| \leq \varepsilon.$$

Now, denote by P_0 any partition finer than P' and also finer than $\{A, S \setminus A\}$, and by P_A the partition of A consisting of all the elements of P_0 that are contained in A.

Next, let Π_A and Π'_A denote two partitions of A finer than P_A, and extend them with a common partition of $S \setminus A$ (also with the same *tags*) in such a way that the two resulting partitions, denoted by Π and Π', are both finer than P'. So, if we denote by

$$\sigma(h, \Pi) := \sum_{n=0}^{\infty} h(t_n)\nu(A_n), \quad A_n \in \Pi, \tag{3}$$

then

$$\|\sigma(h, \Pi) - \sigma(h, \Pi')\| \leq \|\sigma(h, \Pi) - J\| + \|J - \sigma(h, \Pi')\| \leq 2\varepsilon.$$

Now, setting:

$$\alpha_1 := \sum_{I \in \Pi_A} h(t_I)\nu(I), \quad \alpha_2 := \sum_{I \in \Pi'_A} h(t'_I)\nu(I), \quad \beta := \sum_{I \in \Pi, I \subset A^c} h(\tau_I)\nu(I),$$

(with obvious meaning of the symbols), we now have

$$2\varepsilon \geq \|\alpha_1 + \beta - (\alpha_2 + \beta)\| = \|\alpha_1 - \alpha_2\|.$$

By the arbitrariness of Π_A and Π'_A, this means that the sums $\sigma(h, \Pi_A)$ satisfy a Cauchy principle in X, and so the first claim follows by completeness. Now, let us suppose that f is $|RL|$ ν-integrable on $A \in \mathcal{C}$.

A Survey on the Riemann-Lebesgue Integrability in Non-additive Setting   67

Then for every $\varepsilon > 0$ there exists a partition $P_A^\varepsilon \in \mathcal{P}_A$ so that for every partition $P_A = \{B_n\}_{n\in\mathbb{N}}$ of A with $P_A \geq P_A^\varepsilon$ and for every $s_n \in B_n, n \in \mathbb{N}$, we have

$$\left\| \sum_{n=0}^{\infty} h(s_n)\nu(B_n) - (|RL|) \int_A h \, d\nu \right\| < \varepsilon. \tag{4}$$

Let us consider $P_\varepsilon = P_A^\varepsilon \cup \{S \setminus A\}$, which is a partition of S. If $P = \{A_n\}_{n\in\mathbb{N}}$ is a partition of S with $P \geq P_\varepsilon$, then without any loss of generality we may write $P = \{C_n, D_n\}_{n\in\mathbb{N}}$ with pairwise disjoint C_n, D_n such that $A = \cup_{n=0}^{\infty} C_n$ and $\cup_{n=0}^{\infty} D_n = S \setminus A$. Now, for every $u_n \in A_n$, $n \in \mathbb{N}$, by (6), we get:

$$\left\| \sum_{n=0}^{\infty} h\chi_A(u_n)\nu(A_n) - (|RL|) \int_A h \, d\nu \right\|$$
$$= \left\| \sum_{n=0}^{\infty} h\chi_A(t_n)\nu(C_n) + \sum_{n=0}^{\infty} h\chi_A(s_n)\nu(D_n) - (|RL|) \int_A h \, d\nu \right\|$$
$$= \left\| \sum_{n=0}^{\infty} h(t_n)\nu(C_n) - (|RL|) \int_A h \, d\nu \right\| < \varepsilon,$$

where $t_n \in C_n, s_n \in D_n$, for every $n \in \mathbb{N}$, which says that $f\chi_A$ is $|RL|$ m-integrable on S and $(|RL|) \int_S h\chi_A \, d\nu := (|RL|) \int_A h \, d\mu$.

Finally, suppose that $f\chi_A$ is $|RL|$ ν-integrable on S. Then for every $\varepsilon > 0$ there exists $P_\varepsilon = \{B_n\}_{n\in\mathbb{N}} \in \mathcal{P}$ so that for every $P = \{C_n\}_{n\in\mathbb{N}}$ partition of S with $P \geq P_\varepsilon$ and every $t_n \in C_n, n \in \mathbb{N}$, we have

$$\left\| \sum_{n=0}^{\infty} h\chi_A(t_n)\nu(C_n) - (|RL|) \int_S h\chi_A \, d\nu \right\| < \varepsilon. \tag{5}$$

Let us consider $P_A^\varepsilon = \{B_n \cap A\}_{n\in\mathbb{N}}$, which is a partition of A. Let $P_A = \{D_n\}_{n\in\mathbb{N}}$ be an arbitrary partition of A with $P_A \geq P_A^\varepsilon$ and $P = P_A \cup \{S \setminus A\}$. Then P is a countable partition finer than P_ε. Let us take $t_n \in D_n, n \in \mathbb{N}$ and $s \in S \setminus A$. By (7) we obtain

$$\left\| \sum_{n=0}^{\infty} h(t_n)\nu(D_n) - (|RL|) \int_S h\chi_A \, d\nu \right\|$$
$$= \left\| \sum_{n=0}^{\infty} h\chi_A(t_n)\nu(D_n) + h\chi_A(s)\nu(S \setminus A) - (|RL|) \int_S h\chi_A \, d\nu \right\| < \varepsilon,$$

which assures that f is $|RL|$ ν-integrable on A and

$$(|RL|) \int_A h \, d\nu := (|RL|) \int_S h\chi_A \, d\nu.$$

$\square$

In particular the $|RL|$ ν-integrability with respect to a set function of finite variation allows to obtain the following properties.

Theorem 2. ([12, Proposition 1, Theorems 2 and 5, Corollary 2]) *Let* $\nu :$ $S \to [0, \infty)$ *be of finite variation. If we suppose that* $h : S \to X$ *is bounded, then:*

(2.a) $h \in |RL|_\nu^1(X)$ *and*

$$\left\| (|RL|) \int_S h \, d\nu \right\| \leq \sup_{s \in S} \|h(s)\| \cdot \overline{\nu}(S).$$

(2.b) *If* $h = 0$ ν-*a.e., then* $h \in |RL|_\nu^1(X)$ *and* $(|RL|) \int_S h \, d\nu = 0$.

Moreover let $g, h : S \to X$ *be vector functions.*

(2.c) *If* $\sup_{s \in S} \|g(s) - h(s)\| < +\infty$, $g \in |RL|_\nu^1(X)$ *and* $g = h$ ν-*a.e., then* $h \in |RL|_\nu^1(X)$ *and*

$$(|RL|) \int_S g \, d\nu = (|RL|) \int_S h \, d\nu.$$

(2.d) *If* $g, h \in |RL|_\nu^1(X)$ *then*

$$\left\| (|RL|) \int_S g \, d\nu - (|RL|) \int_S h \, d\nu \right\| \leq \sup_{s \in S} \|g(s) - h(s)\| \cdot \overline{\nu}(S).$$

Proof. We prove (2.b) here. From the boundedness of h let $M \in [0, \infty)$ so that $\|h(s)\| \leq M$, for every $s \in S$. If $M = 0$, then the conclusion is obvious. Suppose $M > 0$. Let us denote $A = \{s \in S : h(s) \neq 0\}$. Since $h = 0$ ν-ae, we have $\widetilde{\nu}(A) = 0$. Then, for every $\varepsilon > 0$, there exists $B_\varepsilon \in \mathcal{C}$ so that $A \subseteq B_\varepsilon$ and $\overline{\nu}(B_\varepsilon) < \varepsilon/M$. Let us take the partition $P_\varepsilon = \{C_n\}_{n \in \mathbb{N}}$ of B_ε, and let $C_0 = S \setminus B_\varepsilon$ and add C_0 to P_ε.

Let $P = \{A_n\}_{n \in \mathbb{N}}$ be an arbitrary partition of S so that $P \geq P_\varepsilon$. Without any loss of generality, we suppose that $P = \{D_n, E_n\}_{n \in \mathbb{N}} \subset \mathcal{C}$, with pairwise disjoint sets D_n, E_n such that

$$\bigcup_{n \in \mathbb{N}} D_n = C_0, \qquad \bigcup_{n \in \mathbb{N}} E_n = B_\varepsilon.$$

Let $t_n \in D_n, s_n \in E_n$, for every $n \in \mathbb{N}$, Then we can write

$$\left\| \sum_{n=0}^{\infty} h(t_n)\nu(D_n) + \sum_{n=0}^{\infty} h(s_n)\nu(E_n) \right\| = \left\| \sum_{n=0}^{\infty} h(s_n)\nu(E_n) \right\|$$

$$\leq \sum_{n=0}^{\infty} \|h(s_n)\|\nu(E_n) \leq M \cdot \overline{\nu}(B_\varepsilon) < \varepsilon,$$

which ensures that h is $|RL|$ ν-integrable and $(|RL|) \int_T h \, d\nu = 0$. $\qquad \square$

The next theorem shows that the integral of a real function is monotone with respect to the integrands and to the set functions (see [12, Theorems 6 and 7]) in the following way.

Theorem 3. *Let* $g, h \in |RL|^1_\nu(\mathbb{R})$ *such that* $g(s) \leq h(s)$, *for every* $s \in S$, *then*

(3.a) $\qquad (|RL|) \int_S g \, d\nu \leq (|RL|) \int_S h \, d\nu.$

Let $\nu_1, \nu_2 : \mathcal{C} \to [0, +\infty)$ *be set functions such that* $\nu_1(A) \leq \nu_2(A)$, *for every* $A \in \mathcal{C}$ *and* $h \in |RL|^1_{\nu_i}(\mathbb{R}^+_0)$ *for* $i = 1, 2$ *Then*

(3.b) $\qquad (|RL|) \int_S h \, d\nu_1 \leq (|RL|) \int_S h \, d\nu_2.$

Proof. We prove here (3.a). Let $\varepsilon > 0$ be arbitrary. Since $g, h \in |RL|^1_\nu(\mathbb{R})$, there exists a countable partition P_0 so that for every $P = \{C_n\}_{n \in \mathbb{N}}$, $P \geq P_0$ and every $t_n \in C_n, n \in \mathbb{N}$, the series $\sum_{n=0}^{\infty} g(t_n)\nu(C_n)$, $\sum_{n=0}^{\infty} h(t_n)\nu(C_n)$ are absolutely convergent and

$$\max \left\{ \left| (|RL|) \int_S g \, d\nu - \sum_{n=0}^{\infty} g(t_n)\nu(C_n) \right|, \left| (|RL|) \int_S h \, d\nu - \sum_{n=0}^{\infty} h(t_n)\nu(C_n) \right| \right\}$$

$$< \frac{\varepsilon}{3}.$$

Therefore

$$(|RL|) \int_S g \, d\nu - (|RL|) \int_S h \, d\nu =$$

$$(|RL|) \int_S g \, d\nu - \sum_{n=0}^{\infty} g(t_n)\nu(C_n) + \sum_{n=0}^{\infty} g(t_n)\nu(C_n) +$$

$$- \sum_{n=0}^{\infty} h(t_n)\nu(C_n) + \sum_{n=0}^{\infty} h(t_n)\nu(C_n) - (|RL|) \int_S h \, d\nu$$

$$< \frac{2\varepsilon}{3} + \left[\sum_{n=0}^{\infty} g(t_n)\nu(C_n) - \sum_{n=0}^{\infty} h(t_n)\nu(C_n) \right] \le \varepsilon$$

since, by the hypothesis, $\sum_{n=0}^{\infty} g(t_n)\nu(C_n) \le \sum_{n=0}^{\infty} h(t_n)\nu(C_n)$.
Consequently,

$$(|RL|) \int_S g \, d\nu - (|RL|) \int_S h \, d\nu \le 0.$$

$\square$

For every $h : S \to X$ that is $|RL|$ (RL resp.) ν-integrable on every set $E \in \mathcal{C}$, we consider the $|RL|$ integral operator $T_h : \mathcal{C} \to X$, defined for every $E \in \mathcal{C}$ by,

$$T_h(E) = (|RL|) \int_E h \, d\nu \quad \left(T_h(E) = (RL) \int_E h \, d\nu \quad \text{resp.} \right). \tag{6}$$

We point out that, even without the additive condition for the set function ν, the indefinite integral is additive thanks to Theorem 1.

In the next theorem we present some properties of the set function T_h.

Theorem 4. ([12, Theorem 8]) *Let $h \in |RL|_\nu^1(X)$. If h is bounded, and ν is of finite variation then:*

(4.a)

- T_h *is of finite variation too;*
- $\overline{T_h} \ll \overline{\nu}$ *in the $\varepsilon - \delta$ sense;*
- *If $\overline{\nu}$ is o-continuous (exhaustive resp.) also, then T_h is also o-continuous (exhaustive resp.).*

(4.b) *If $h : S \to [0, \infty)$ is nonnegative and ν is scalar-valued and monotone, then the same holds for T_h.*

Proof. (4.a). In order to prove that T_h is of finite variation let $\{A_i\}_{i=1,\dots,n}$ be a pairwise partition of S and $M = \sup_{s \in S} \|h(s)\|$. By Theorem 2.a and Remark 1, we have

$$\sum_{i=1}^{n} \|T_h(A_i)\| \leq M \cdot \sum_{i=1}^{n} \overline{\nu}(A_i) \leq M \cdot \overline{\nu}(S).$$

So $\overline{T_h}(S) \leq M \, \overline{\nu}(S)$, that yields $\overline{T_h}(S) < +\infty$. Now the absolute continuity in the $\varepsilon - \delta$ sense follows from Theorem 2.a).

Let M as before. If $M = 0$, then $h = 0$, hence $T_h = 0$.

If $M > 0$, by Theorem 2.a) we have $\|T_h(A)\| \leq M \cdot \overline{\nu}(A)$, for every $A \in \mathcal{C}$. So the o-continuity of T_h follows from that of $\overline{\nu}$. The proof of the exhaustivity is similar.

For (4.b) let $A, B \in \mathcal{C}$ with $A \subseteq B$ and $\varepsilon > 0$. Since h is ν-integrable on A, there exists a countable partition $P_1 = \{C_n\}_{n \in \mathbb{N}}$ of A so that for every other finer countable partition $P = \{A_n\}_{n \in \mathbb{N}}$, of A and every $t_n \in A_n, n \in \mathbb{N}$, the series $\sum_{n=0}^{\infty} h(t_n)\nu(A_n)$ is absolutely convergent and

$$\left| T_h(A) - \sum_{n=0}^{\infty} h(t_n)\nu(A_n) \right| < \frac{\varepsilon}{2}. \tag{7}$$

Since f is ν-integrable on B, let $P_2 = \{D_n\}_{n \in \mathbb{N}}$ a countable partition of B with the same meaning as before for the set A.

Let $\widetilde{P}_1 = \{C_n, B \setminus A\}_{n \in \mathbb{N}}$ and $\widetilde{P}_1 \wedge P_2$ (both countable partitions of B).

Let $P = \{E_n\}_{n \in \mathbb{N}}$ be an arbitrary countable partition of B, with $P \geq \widetilde{P}_1 \wedge P_2$.

We observe that $P'' = \{E_n \cap A\}_{n \in \mathbb{N}}$ is also a partition of A and $P'' \geq P_1$.

If $t_n \in E_n \cap A, n \in \mathbb{N}$ we have

$$\max \left\{ |T_h(B) - \sum_{n=0}^{\infty} h(t_n)\nu(E_n)|, |T_h(A) - \sum_{n=0}^{\infty} h(t_n)\nu(E_n \cap A)| \right\} < \frac{\varepsilon}{2}.$$

Therefore

$$T_h(A) - T_h(B) \le \left| T_h(A) - \sum_{n=0}^{\infty} h(t_n)\nu(E_n \cap A) \right|$$

$$+ \left[\sum_{n=0}^{\infty} h(t_n)\nu(E_n \cap A) - \sum_{n=0}^{\infty} h(t_n)\nu(E_n) \right]$$

$$+ \left| \sum_{n=0}^{\infty} h(t_n)\nu(E_n) - T_h(B) \right|$$

$$< \varepsilon + \left[\sum_{n=0}^{\infty} h(t_n)\nu(E_n \cap A) - \sum_{n=0}^{\infty} h(t_n)\nu(E_n) \right].$$

Since, by the hypotheses, $\sum_{n=0}^{\infty} h(t_n)\nu(E_n \cap A) \le \sum_{n=0}^{\infty} h(t_n)\nu(E_n)$, then $T_h(A) \le T_h(B)$. $\qquad\square$

3.2. *Comparison with other types of integrability*

In the non additive case there are other types of integral that can be considered. We present here some comparative results with the Gould and Birkhoff simple ones. We recall that:

Definition 4. ([9, Definition 3.2]) A vector function $h : S \to X$ is called *Birkhoff simple ν-integrable (on S)* if there exists $b \in X$ such that for every $\varepsilon > 0$, there exists a countable partition P_ε of S so that for every other countable partition $P = \{A_n\}_{n\in\mathbb{N}}$ of S, with $P \ge P_\varepsilon$ and every $s_n \in A_n, n \in \mathbb{N}$, it holds

$$\limsup_{n\to+\infty} \left\| \sum_{k=0}^{n} h(s_k)\nu(A_k) - b \right\| < \varepsilon.$$

The vector b is denoted by $(Bs)\displaystyle\int_S h \, d\nu$ and it is called *the Birkhoff simple integral* of h (on S) with respect to ν.

Let $(\mathcal{P}, \le)$ be the family of all finite partitions of S ordered by the relation $\le$ given in (1). Given a vector function $h : S \to X$, we denote by $\sigma(P)$ the finite sum: $\sigma(P) := \sum_{i=1}^{n} h(s_i)\nu(E_i)$, for every finite partition of S, $P = \{E_i\}_i^n \in \mathcal{P}$ and every $s_i \in E_i, i \in \{1, \dots, n\}$.
Following [43] we have:

Definition 5. A function $h : S \to X$ is called *Gould ν-integrable*(on S) if there exists $a \in X$ such that for every $\varepsilon > 0$, there exists a finite partition P_ε of S, so that for every other finite partition $P = \{E_i\}_{i=1}^n$ of S, with $P \geq P_\varepsilon$ and every $s_i \in E_i, i \in \{1, \ldots, n\}$, we have $\|\sigma(P) - a\| < \varepsilon$. The vector a is called *the Gould integral of h with respect to ν*, denoted by $(G) \int_S h \, d\nu$.

Observe that $h : S \to X$ is Gould ν-integrable (on S) if and only if the net $(\sigma(P))_{P \in (\mathcal{P}, \leq)}$ is convergent in X, The limit of $(\sigma(P))_P$ is exactly the integral $(G) \int_S h \, d\nu$.

Let $Bs_\nu^1(S)$ and $G_\nu^1(S)$ be respectively the families of Birkhoff simple, Gould integrable functions. In general $RL_\nu^1(S) \subset Bs_\nu^1(S)$ and the two integrals coincide, this is proved in [12, Theorem 9], while for what concernes the comparison between RL and Gould integrability for bounded functions we have the following relations:

- $RL_\nu^1(X) = G_\nu^1(X)$ when ν is of finite variation and defined on a complete σ-additive measure by [12, Proposition 2];
- $RL_\nu^1(\mathbb{R}) = G_\nu^1(\mathbb{R})$ when ν is of finite variation, monotone and σ-subadditive [12, Theorem 10];
- $RL_\nu^1(\mathbb{R}) \subset G_\nu^1(\mathbb{R})$ on each atom $A \in \mathcal{C}$ when ν is monotone, null additive and satisfies property (σ), (see [12, Theorem 11]).

In all the cases, the two integrals coincide.

Without the σ-additivity of ν, the second equivalence $RL_\nu^1(\mathbb{R}) = G_\nu^1(\mathbb{R})$ for bounded functions does not hold. Suppose $S = \mathbb{N}$, with $\mathcal{C} = \mathcal{P}(\mathbb{N})$ and

$$\nu(A) = \begin{cases} 0, \ \mathrm{card}(A) < +\infty, \\ 1, \ \mathrm{card}(A) = +\infty, \end{cases} \quad \text{for every } A \in \mathcal{C}.$$

Then, the constant function $h = 1$ is RL integrable and then Birkhoff simple integrable and $(Bs) \int_S h \, d\nu = 0$. However, h is not Gould-integrable. In fact, if P_ε is any finite partition of $\mathbb{N}$, some of its sets are infinite, so the quantity $\sigma(P_\varepsilon)$ is exactly the number of the infinite sets belonging to P_ε. So the quantity $\sigma(P)$ is unbounded when P runs over the family of all finer partitions of P_ε.

3.3. Convergence results

In this subsection, we want to quote some sufficient conditions in order to obtain, under suitable hypotheses, a convergence result of this type

$$\lim_{n\to\infty} (|RL|) \int_S h_n \, d\nu = (|RL|) \int_S \lim_{n\to\infty} h_n \, d\nu,$$

for sequences of Riemann-Lebesgue integrable functions. We assume ν is of finite variation unless otherwise specified. Let $p \in [1, \infty)$ be fixed. For every real valued function $h : S \to \mathbb{R}$, with $|h|^p \in RL_\nu^1(\mathbb{R})$, we associate the following number:

$$\|h\|_p = \left((|RL|) \int_S |h|^p \, d\nu \right)^{\frac{1}{p}}. \tag{8}$$

Theorem 5. *Let* $h, h_n : S \to X$, $\nu : \mathcal{C} \to [0, +\infty)$ *and* $p \in [1, +\infty)$.

([27, Theorem 5]) *If* $h, h_n \in |RL|_\nu^1(X)$ *for every* $n \in \mathbb{N}$ *and* h_n *converges uniformly to* h;

([27, Theorem 6]) *If* $X = \mathbb{R}$, $\sup_{s \in S, n \in \mathbb{N}} \{h(s), h_n(s)\} < +\infty$ *and* $h_n \xrightarrow{\widetilde{\nu}} h$, *or*;

([27, Theorem 8]) *If* $X = \mathbb{R}$, ν *is monotone and* $\widetilde{\nu}$ *satisfies condition* **(E)**, $\sup_{s \in S, n \in \mathbb{N}} \{h(s), h_n(s)\} < +\infty$ *and* $h_n \xrightarrow{\nu-ae} h$

then

$$\lim_{n\to\infty} (|RL|) \int_S h_n \, d\nu = (|RL|) \int_S h \, d\nu. \tag{9}$$

Finally

([27, Theorem 7]) *If* $X = \mathbb{R}$, ν *is countable subadditive (not necessarily of finite variation)*, $\chi_E \cdot |h_n - h|^p \in |RL|_\nu^1(\mathbb{R})$, *for every* $E \in \mathcal{C}$ *and* $\|h_n - h\|_p \to 0$. *Then* $h_n \xrightarrow{\widetilde{\nu}} h$.

Proof. We give only the proof of [27, Theorem 6]. According to [12, Proposition 1] $g, g_n, g_n - g \in |RL|_\nu^1(\mathbb{R})$, for every $n \in \mathbb{N}$. Let $\alpha \in (0, +\infty)$ such that:

$$\sup_{s \in S, n \in \mathbb{N}} \{|g(s)|, |g_n(s) - g(s)|\} < \alpha.$$

Let $\varepsilon > 0$ be fixed. By hypothesis, there is $n_0(\varepsilon) \in \mathbb{N}$ such that for every $n \geq n_0(\varepsilon)$

$$\widetilde{\nu}(\{s \in S; \ |g_n(s) - g(s)| \geq \varepsilon/4\overline{\nu}(S)\}) < \varepsilon/4\alpha.$$

Then, there exists $A_n \in \mathcal{C}$ such that $\{s \in S; |g_n(s) - g(s)| \geq \varepsilon/4\overline{\nu}(S)\} \subset A_n$ and $\widetilde{\nu}(A_n) = \overline{\nu}(A_n) < \varepsilon/4\alpha$. Using [12, Theorem 3 and Corollary 1], for every $n \geq n_0(\varepsilon)$, it holds that:

$$\left| (|RL|) \int_S g_n d\nu - (|RL|) \int_S g \, d\nu \right| \leq \left| (|RL|) \int_{A_n} (g_n - g) \, d\nu \right|$$

$$+ \left| (|RL|) \int_{A_n^c} (g_n - g) \, d\nu \right|$$

$$\leq \overline{\nu}(A_n) \cdot \sup_{s \in A_n} |g_n(s) - g(s)| + \overline{\nu}(A_n^c) \cdot \sup_{s \in A_n^c} |g_n(s) - g(s)| < \varepsilon,$$

which yields the (9). $\qquad\qquad\qquad\qquad\qquad\qquad\qquad\qquad\qquad\qquad\square$

The following theorem establishes a Fatou type result for sequences of Riemann-Lebesgue integrable functions.

Theorem 6. ([27, Theorem 9]) *Suppose $\nu : \mathcal{C} \to [0, +\infty)$ is a monotone set function of finite variation such that $\widetilde{\nu}$ satisfies* **(E)**. *For every $n \in \mathbb{N}$, let $h_n : S \to \mathbb{R}$ be such that (h_n) is uniformly bounded. Then*

$$(|RL|) \int_S (\liminf_n h_n) \, d\nu \leq \liminf_n \left((|RL|) \int_S h_n \, d\nu \right). \tag{10}$$

Its consequence is:

Corollary 1. ([27, Theorem 10]) *Suppose $p \in (1, +\infty)$ and $\nu : \mathcal{C} \to [0, +\infty)$ is a monotone set function of finite variation such that $\widetilde{\nu}$ satisfies* (E). *Let $h, h_n : S \to \mathbb{R}$ be such that h is bounded and $(h_n)_n$ is pointwise convergent to h. Let*

$$g_n = 2^{p-1}(|h_n|^p + |h|^p) - |h_n - h|^p,$$

such that (g_n) is uniformly bounded, $|h|^p, |h_n|^p, |h_n - g|^p, g_n, \inf_{k \geq n} g_k \in |RL|_\nu^1(\mathbb{R})$, for every $n \in \mathbb{N}$ and $\|h_n\|_p \longrightarrow \|h\|_p$. Then

$$\|h_n - h\|_p \longrightarrow 0.$$

3.4. *Hölder and Minkowski type inequalities*

At the end of this section, we expose a result on the reverse inequalities of Hölder and Minkowski type in Riemann-Lebesgue integrability. First of all we need that ν satisfies the following property:

Definition 6. The set function $\nu : \mathcal{C} \to [0, \infty)$ is called RL-integrable if for all $E \in \mathcal{C}, \chi_E \in RL_\nu^1(\mathbb{R})$ and $\int_S \chi_E \, d\nu = \nu(E)$.

Theorem 7. ([27, Theorem 4] and [28, Theorem 3.4]) *Let $\nu : \mathcal{C} \to [0, \infty)$ be a countable subadditive RL-integrable set function and let $g, h : S \to \mathbb{R}$ be measurable functions.*
Let $p, q \in (1, \infty)$, with $p^{-1} + q^{-1} = 1$.

(7.a) *If $g \cdot h \in RL_\nu^1(\mathbb{R})$, then*

$$\|g \cdot h\|_1 \le \|g\|_p \cdot \|h\|_q \quad \text{(Hölder Inequality)}.$$

(7.b) *Let $p \in [1, \infty)$. If $|g+h|^p, |g+h|^{q(p-1)}, |g|^p$ and $|h|^p$ are in $\in RL_\nu^1(\mathbb{R})$, then*

$$\|g + h\|_p \le \|g\|_p + \|h\|_p \quad \text{(Minkowski Inequality)}.$$

Let p, q be such that $0 < p < 1$ and $p^{-1} + q^{-1} = 1$.

(7.c) *If $g \cdot h, |g|^p, |h|^q \in RL_\nu^1(\mathbb{R})$ and $0 < (RL) \int_S |h|^q \, d\nu$, then*

$$\|g \cdot h\|_1 \ge \|g\|_p \cdot \|h\|_q \quad \text{(Reverse Hölder Inequality)}.$$

(7.d) *If $|g + h|^p, |g + h|)^{(p-1)q}, |g|^p$ and $|h|^p$ are RL-integrable, then*

$$\big\| \, |g| + |h| \, \big\|_p \ge \|g\|_p + \|h\|_p \quad \text{(Reverse Minkowski Inequality)}.$$

According to [27, Remark 4], for $p \in [1, \infty)$, the function $\|\cdot\|_p$ defined in (8) is a seminorm on the linear space of measurable RL-integrable functions.

Proof. We give here only the proof of the reverse part.

(7.c) If $(RL) \int_S g^p \, d\nu = 0$, then according to [27, Theorem 3] it follows $g \cdot h = 0 \; \nu - a.e.$ In this case, the inequality of integrals is satisfied. Consider $(RL) \int_S g^p \, d\nu > 0$.
We replace $a = \dfrac{|g|}{((RL) \int_S |g|^p \, d\nu)^{\frac{1}{p}}}$ and $b = \dfrac{|h|}{((RL) \int_S |h|^q \, d\nu)^{\frac{1}{q}}}$ in the reverse
Young inequality $ab \ge \dfrac{a^p}{p} + \dfrac{b^q}{q}$, for every $a, b > 0$ and for every $0 < p < 1$ with $p^{-1} + q^{-1} = 1$ (see for example [1, 19]). Then

$$\frac{|gh|}{((RL) \int_S |g|^p \, d\nu)^{\frac{1}{p}} ((RL) \int_S |h|^q \, d\nu)^{\frac{1}{q}}} \ge$$

$$\frac{|g|^p}{p((RL) \int_S |g|^p \, d\nu)} + \frac{|h|^q}{q((RL) \int_S |h|^q \, d\nu)}.$$

Applying [12, Theorems 3 and 6], the following holds:

$$\frac{(RL)\int_S |gh|\,d\nu}{\left((RL)\int_S |g|^p\,d\nu\right)^{\frac{1}{p}}\left((RL)\int_S |h|^q\,d\nu\right)^{\frac{1}{q}}} \geq$$

$$\frac{(RL)\int_S |g|^p\,d\nu}{p\left(\int_S |g|^p\,d\nu\right)} + \frac{(RL)\int_S |h|^q\,d\nu}{q\left((RL)\int_S |h|^q\,d\nu\right)} = 1$$

and the conclusion yields.

(7.d) By (7.c), it results:

$$(RL)\int_S (|g|+|h|)^p\,d\nu = (RL)\int_S (|g|+|h|)^{p-1}(|g|+|h|)\,d\nu$$

$$\geq ((RL)\int_S (|g|+|h|)^{q(p-1)}\,d\nu)^{\frac{1}{q}}((RL)\int_S |g|^p\,d\nu)^{\frac{1}{p}}$$

$$+ ((RL)\int_S (|g|+|h|)^{q(p-1)}\,d\nu)^{\frac{1}{q}}(\int_S |h|^p\,d\nu)^{\frac{1}{p}}$$

$$= ((RL)\int_S (|g|+|h|)^{q(p-1)}\,d\nu)^{\frac{1}{q}}(\|g\|_p + \|h\|_p).$$

Dividing the above inequality by $((RL)\int_S(|g|+|h|)^{q(p-1)}\,d\nu)^{\frac{1}{q}}$, we obtain the Reverse Minkowski inequality.

$\square$

4. Interval-valued Riemann-Lebesgue integral

In the past few years, much attention was given to the study of interval-valued multifunctions and multimeasures because of their applications in statistics, biology, theory of games, economics, social sciences and software development. Interval-valued multifunctions have been applied in new ways, including signal and image processing.

Motivated by the large number of fields in which the interval-valued multifunctions can be involved, we present some classic properties for the Riemann-Lebesgue integral of an interval-valued multifunction with respect to an interval-valued set multifunction.

We begin by recalling some preliminaries. The symbol $ck(\mathbb{R})$ denotes the family of all non-empty, convex, compact subsets of $\mathbb{R}$, by convention, $\{0\} = [0,0]$.

We consider on $ck(\mathbb{R})$ (see [45]), the Minkowski addition

$$A \oplus B := \{a + b \mid a \in A,\ b \in B\}, \quad \text{for every } A, B \in ck(\mathbb{R})$$

and the multiplication by scalars

$$\lambda A = \{\lambda a \mid a \in A\}, \quad \text{for every } \lambda \in \mathbb{R}, A \in ck(\mathbb{R}).$$

d_H denotes the Hausdorff distance in $ck(\mathbb{R})$ and it is defined for every $A, B \in ck(\mathbb{R})$, in the following way:

$$d_H(A, B) = \max\{e(A, B), e(B, A)\},$$

where $e(A, B) = \sup\{d(x, B), x \in A\}$. We use the symbol $\|A\|_H$ to denote $d_H(A, \{0\})$. In particular, for closed intervals we have

$$d_H([r, s], [x, y]) = \max\{|x - r|, |y - s|\}, \quad \text{for every } r, s, x, y \in \mathbb{R};$$
$$d_H([0, s], [0, y]) = |y - s|, \quad \text{for every } s, y \in \mathbb{R}_0^+;$$
$$\|[r, s]\|_{\mathcal{H}} = s, \quad \text{for every } r, s \in \mathbb{R}_0^+.$$

In the subfamily $L(\mathbb{R})$ of intervals in $ck(\mathbb{R})$, the following operations are also considered, for every $r, s, x, y \in \mathbb{R}$ (see [58]):

i) $[r, s] \cdot [x, y] = [rx, sy]$;
ii) $[r, s] \subseteq [x, y]$ if and only if $x \le r \le s \le y$;
iii) $[r, s] \preceq [x, y]$ if and only if $r \le x$ and $s \le y$; (weak interval order, (see [44]))
iv) $[r, s] \wedge [x, y] = [\min\{r, x\}, \min\{s, y\}]$;
v) $[r, s] \vee [x, y] = [\max\{r, x\}, \max\{s, y\}]$.

In general, there is no relation between the weak interval order and the inclusion; they only coincide on the subfamily of intervals $[0, s], s \ge 0$.
For every pair of sequences of real numbers $(u_n)_n, (v_n)_n$ such that $0 \le u_n \le v_n$, for every $n \in \mathbb{N}$, we define:

vi) $\inf_n [u_n, v_n] = [\inf_n u_n, \inf_n v_n]$;
vii) $\sup_n [u_n, v_n] = [\sup_n u_n, \sup_n v_n]$;
viii) $\liminf_n [u_n, v_n] = [\liminf_n u_n, \liminf_n v_n]$.

We consider $(ck(\mathbb{R}_0^+), d_H, \preceq)$, namely the space $ck(\mathbb{R}_0^+)$ is endowed with the Hausdorff distance and the weak interval order.

For two set functions $\nu_1, \nu_2 : \mathcal{C} \to \mathbb{R}_0^+$ with $\nu_1(\emptyset) = \nu_2(\emptyset) = 0$ and $\nu_1(A) \le \nu_2(A)$ for every $A \in \mathcal{C}$ the set multifunction $\Gamma : \mathcal{C} \to L(\mathbb{R}_0^+)$ defined by

$$\Gamma(A) = [\nu_1(A), \nu_2(A)], \quad \text{for every } A \in \mathcal{C}, \tag{11}$$

is called an interval-valued set function. In this case, Γ is of finite variation if and only if ν_2 is of finite variation.

Let $\Gamma : \mathcal{C} \to L(\mathbb{R}_0^+)$. We say that Γ is an interval-valued multisubmeasure if:

- $\Gamma(\emptyset) = \{0\}$;
- $\Gamma(A) \preceq \Gamma(B)$ for every $A, B \in \mathcal{C}$ with $A \subseteq B$ (monotonicity);
- $\Gamma(A \cup B) \preceq \Gamma(A) \oplus \Gamma(B)$ for every disjoint sets $A, B \in \mathcal{C}$. (subadditivity).

By [62, Remark 3.6], Γ is a multisubmeasure with respect to $\preceq$ if and only if ν_i, $i = 1, 2$ are submeasures.

Definition 7. It is said that Γ is a d_H-multimeasure if for every sequence of pairwice disjoint sets $(A_n)_n \subset \mathcal{C}$ such that $\cup_n^\infty A_n = A$,

$$\lim_{n \to \infty} d_H \Big(\sum_{k=1}^{n} \Gamma(A_k), \Gamma(A) \Big) = 0.$$

4.1. The interval-valued RL-integral and its properties

In what follows, all the interval-valued set functions we consider are multisubmeasures.

Given $h_1, h_2 : S \to \mathbb{R}_0^+$ with $h_1(s) \le h_2(s)$ for all $s \in S$, let $H : S \to L(\mathbb{R}_0^+)$ be the interval-valued multifunction defined by

$$H(s) := \big[h_1(s), h_2(s) \big], \qquad \text{for every } s \in S. \tag{12}$$

H is bounded if and only if h_2 is bounded. If $H, G : S \to L(\mathbb{R}_0^+)$ are as in (12) so that $G \preceq H$ or $G \subset H$, and H is bounded, then G is bounded too. For every countable tagged partition $P = \{(B_n, s_n), n \in \mathbb{N}\}$ of S, we denote by

$$\sigma_{H,\Gamma}(P) = \sum_{n=1}^{\infty} H(s_n) \cdot \Gamma(B_n) = \sum_{n=1}^{\infty} \big[h_1(s_n) \nu_1(B_n), h_2(s_n) \nu_2(B_n) \big]$$

$$= \Big\{ \sum_{n=1}^{\infty} y_n, y_n \in \big[h_1(s_n) \nu_1(B_n), h_2(s_n) \nu_2(B_n) \big], n \in \mathbb{N} \Big\}.$$

The set $\sigma_{H,\Gamma}(P)$ is closed and convex in $\mathbb{R}_0^+$, so it is an interval $\big[h_{1,H,\Gamma}^P, h_{2,H,\Gamma}^P \big]$.

Definition 8. A multifunction $H : S \to L(\mathbb{R}_0^+)$ is called Riemann-Lebesgue (RL in short) integrable with respect to Γ (on S) if there exists $[c, d] \in$

$L(\mathbb{R}_0^+)$ such that for every $\varepsilon > 0$, there exists a countable partition P_ε of S, so that for every tagged partition $P = \{(B_n, s_n)\}_{n \in \mathbb{N}}$ of S with $P \geq P_\varepsilon$:

- the series $\sigma_{H,\Gamma}(P)$ is convergent with respect to the Hausdorff distance d_H;
- $d_H(\sigma_{H,\Gamma}(P), [c, d]) < \varepsilon$.

The interval $[c, d]$ is called the Riemann-Lebesgue integral of H with respect to Γ and it is denoted

$$[c, d] = (RL) \int_S H \, d\Gamma.$$

The symbol $RL_\Gamma^1(L(\mathbb{R}_0^+))$ denotes the class of all interval-valued functions that are Riemann-Lebesgue integrable with respect to Γ on S.

Example 1. ([27, Example 5]) Suppose $S = \{s_n \mid n \in \mathbb{N}\}$ is countable, $\{s_n\} \in \mathcal{C}$, for every $n \in \mathbb{N}$, and let $H : S \to L(\mathbb{R}_0^+)$ be such that the series $\sum_{n=0}^{\infty} h_i(s_n)\nu_i(\{s_n\})$, $i \in \{1, 2\}$, are convergent. Then H is RL integrable with respect to Γ and

$$(RL) \int_S H \, d\Gamma = \left[\sum_{n=0}^{\infty} h_1(s_n)\nu_1(\{s_n\}), \sum_{n=0}^{\infty} h_2(s_n)\nu_2(\{s_n\}) \right].$$

Observe moreover that, in this case, the RL-integrability of H with respect to Γ implies that the product $H \cdot H$ is integrable in the same sense, where $(H \cdot H)(s) = [h_1^2(s), h_2^2(s)]$, for every $s \in S$. In particular, if H is a discrete or countable interval-valued signal, then the integral $(RL) \int_S H \cdot H \, d\Gamma$ represents the energy of the signal, (see e.g. [22, Example 2]).

If Γ is of finite variation and $H : S \to L(\mathbb{R}_0^+)$ is bounded and such that $H = \{0\}$ Γ-a.e., then H is Γ-integrable and $(RL) \int_S H \, d\Gamma = \{0\}$.

Continuing on, some properties of interval-valued RL integrable multifunctions (see [22]) are presented for Γ as in (11) and the multifunctions as in (12). The following theorem shows a characterization of the RL integrability.

Theorem 8. ([22, Proposition 2]) *An interval-valued multifunction $H = [h_1, h_2]$ is RL integrable with respect to Γ on S if and only if h_1 and h_2 are*

RL integrable with respect to ν_1 and ν_2 respectively and

$$(RL)\int_S H \, \mathrm{d}\Gamma = \left[(RL)\int_S h_1 \, \mathrm{d}\nu_1, (RL)\int_S h_2 \, \mathrm{d}\nu_2\right].$$

Proof. By the *RL* integrability of H there exists $[a,b]$ such that for every $\varepsilon > 0$, there exists a countable partition P_ε of S, so that for every tagged partition $P = \{(A_n, t_n)\}_{n\in\mathbb{N}}$ of S with $P \geq P_\varepsilon$, the series $\sigma_{H,\Gamma}(P)$ is convergent and

$$\max\{|\sum_{n=1}^{\infty} h_1(t_n)\nu_1(A_n) - a|, |\sum_{n=1}^{\infty} h_2(t_n)\nu_2(A_n) - b|\}$$

$$= d_H\left(\sum_{n=1}^{\infty} \left[h_1(s_n)\nu_1(B_n), h_2(s_n)\nu_2(B_n)\right]\right) < \varepsilon.$$

Then h_i are *RL* integrable with respect to ν_i, $i = 1, 2$. So the first implication follows from the convexity of the *RL* integral.

For the converse, for every $\varepsilon > 0$, let $P_{\varepsilon,h_i}, i = 1, 2$ be two countable partitions verifying the *RL* integrability definition for $h_i, i = 1, 2$ respectively. Let $P_\varepsilon \geq P_{\varepsilon,h_1} \wedge P_{\varepsilon,h_2}$ be a countable partition of S, then, for every finer partition $P := (B_n)_n$ and for every $t_n \in B_n$ it is

$$\left|\sum_{n=0}^{+\infty} h_i(t_n)\nu_i(B_n) - (RL)\int_S h_i d\nu_i\right| < \varepsilon, \quad i = 1, 2.$$

Since $h_i, i = 1, 2$ are selections of H then

$$d_H\left(\sigma_{H,\Gamma}(P), \left[(RL)\int_S h_1 d\nu_1, (RL)\int_S h_2 d\nu_2\right]\right) \leq \varepsilon$$

and then the "only if" assertion follows. $\qquad\square$

According to Theorem 2 and the previous theorem, if ν_2 is of finite variation and h_2 is bounded, then H is RL integrable. Another consequence of the previous theorem, together with Theorem 1.b) is the inheritance of the *RL* integral on the subsets $E \in \mathcal{C}$. In fact, the following holds:

Corollary 2. *Let $H \in RL^1_\Gamma(L(\mathbb{R}^+_0))$, then H is RL integrable with respect to Γ on every $E \in \mathcal{C}$. Moreover, H is RL integrable with respect to Γ on $E \in \mathcal{C}$ if and only if $H\chi_E$ is RL integrable with respect to Γ on S. In this case, for every $E \in \mathcal{C}$,*

$$(RL)\int_E H \, \mathrm{d}\Gamma = (RL)\int_S H\chi_E \, \mathrm{d}\Gamma.$$

Moreover, the RL integral is homogeneous with respect to both interval-valued multifunctions H and Γ.

Theorem 9. ([22, Remark 7, Theorem 1 and Proposition 8]) *If* H, H_1, $H_2 \in RL^1_\Gamma(L(\mathbb{R}^+_0))$ *then for every* $\alpha \in [0, \infty)$:

(9.a) $\alpha H \in RL^1_\Gamma(L(\mathbb{R}^+_0))$ *and*

$$(RL) \int_S \alpha H \, d\Gamma = \alpha (RL) \int_S H \, d\Gamma.$$

(9.b) $H \in RL^1_{\alpha\Gamma}(L(\mathbb{R}^+_0))$ *and*

$$(RL) \int_S H \, d(\alpha\Gamma) = \alpha \int_S H \, d\Gamma.$$

(9.c) $H_1 \oplus H_2 \in RL^1_\Gamma(L(\mathbb{R}^+_0))$ *and*

$$(RL) \int_S (H_1 \oplus H_2) \, d\Gamma = (RL) \int_S H_1 \, d\Gamma \oplus (RL) \int_S H_2 \, d\Gamma.$$

Proof. We give here the proof of (9.c). Namely, we prove that for every pair of interval-valued multifunctions H_1, H_2, which are RL integrable with respect to Γ we have that

$$(RL) \int_S (H_1 \oplus H_2) \, d\Gamma = (RL) \int_S H_1 \, d\Gamma \oplus (RL) \int_S H_2 \, d\Gamma. \tag{13}$$

Let $\varepsilon > 0$ be fixed. Since H_1, H_2 are RL integrable with respect to Γ, there exists a countable partition $P_\varepsilon \in \mathcal{P}$ such that for every $P = \{A_n\}_{n \in \mathbb{N}} \geq P_\varepsilon$ and every $t_n \in A_n$, $n \in \mathbb{N}$, the series $\sigma_{H_i, \Gamma}(P)$, $i = 1, 2$ are convergent and

$$d_H \left(\sigma_{H_i, \Gamma}(P), (RL) \int_S H_i \, d\Gamma \right) < \frac{\varepsilon}{2}, \qquad i = 1, 2.$$

Then $\sigma_{H_1 \oplus H_2, \Gamma}(P)$ is convergent and, by [45, Proposition 1.17],

$$d_H \left(\sigma_{H_1 \oplus H_2, \Gamma}(P), (RL) \int_S H_1 \, d\Gamma \oplus (RL) \int_S H_2 \, d\Gamma \right) < \varepsilon.$$

So $H_1 \oplus H_2$ is RL integrable with respect to Γ and formula (13) is satisfied. $\square$

If $H \in RL^1_\Gamma(L(\mathbb{R}^+_0))$, then we may consider $T_H : \mathcal{C} \to L(\mathbb{R}^+_0)$ defined by

$$T_H(E) = (RL) \int_E H \, d\Gamma, \qquad \text{for every } E \in \mathcal{C}. \tag{14}$$

In the following theorem we present some properties of the interval-valued integral set operator T_H.

A Survey on the Riemann-Lebesgue Integrability in Non-additive Setting 83

Theorem 10. *Let* $\Gamma : \mathcal{C} \to L(\mathbb{R}_0^+)$ *be so that* ν_2 *is of finite variation and* $H : S \to L(\mathbb{R}_0^+)$ *is bounded. Then the following properties hold:*

(10.a) T_H *is a finitely additive multimeasure, i.e. for every* $A, B \in \mathcal{C}$, *with* $A \cap B = \emptyset$ *it is* $T_H(A \cup B) = T_H(A) \oplus T_H(B)$ *see ([22, Theorem 1]).*

(10.b) *Let* $G, H \in RL^1_\Gamma(L(\mathbb{R}_0^+))$. *Then, for every* $E \in \mathcal{C}$, *by* [22, Propositions 4 and 5], *if* $G \preceq H$, *then* $T_G(E) \preceq T_H(E)$; *if* $G \subseteq H$, *then* $T_G(E) \subseteq T_H(E)$.

Moreover, by [22, Corollary 1], *for every* $E \in \mathcal{C}$:

$$T_{G \wedge H}(E) \preceq T_G(E) \wedge T_H(E); \quad T_G(E) \vee T_H(E) \preceq T_{G \vee H}(E).$$

Finally from [22, Propositions 6 and 7, Theorem 2], *we have:*

(10.c)

- $\|T_H(S)\|_{\mathcal{H}} = (RL) \int_S h_2 \, d\nu_2 = (RL) \int_S \|H\|_{\mathcal{H}} \, d\|\Gamma\|_{\mathcal{H}}.$
- $\overline{T}_H(S) = (RL) \int_S h_2 \, d\nu_2.$
- $T_H \ll \overline{\Gamma}$ *(in the* ε - δ *sense) and* T_H *is of finite variation.*
- *If moreover* Γ *is o-continuous (exhaustive resp.), then* T_H *is also o-continuous (exhaustive resp.).*
- *If* Γ *is monotone, then* T_H *is monotone too.*
- *If* Γ *is a* d_H-*multimeasure, then* T_H *is countably additive.*

Proof. We point out that the additivity of T_H is indipendent of the additivity of H. In fact, by Corollary 2 we have that $T_H(A) \in L(\mathbb{R}_0^+)$ for every $A \in \mathcal{C}$. Moreover for every $A, B \in \mathcal{C}$ with $A \cap B = \emptyset$, by Theorem 9.c)

$$T_H(A \cup B) = (RL) \int_S H \chi_{A \cup B} \, d\Gamma = (RL) \int_S (H \chi_A \oplus H \chi_B) \, d\Gamma$$

$$= (RL) \int_S H \chi_A d\Gamma \oplus (RL) \int_S H \chi_B \, d\Gamma = T_H(A) \oplus T_H(B).$$

$\square$

The RL integral is additive and monotone with respect to the weak interval order and the inclusion one relative to Γ, as we can see in the following theorem.

Theorem 11. *([22, Theorems 3 and 4]) Let* $\Gamma_1, \Gamma_2 : \mathcal{A} \to L(\mathbb{R}_0^+)$ *be multisubmeasures of finite variation, with* $\Gamma_1(\emptyset) = \Gamma_2(\emptyset) = \{0\}$ *and suppose* $H, G : S \to L(\mathbb{R}_0^+)$ *are bounded multifunctions. Then the following properties hold for every* $E \in \mathcal{C}$:

(11.a) *If* $\Gamma := \Gamma_1 \oplus \Gamma_2$, *then*

$$(RL) \int_E H \, d\Gamma = (RL) \int_E H \, d\Gamma_1 \oplus (RL) \int_E H \, d\Gamma_2.$$

(11.b) *If* $\Gamma_1 \preceq \Gamma_2$, *then*

$$(RL) \int_E H \, d\Gamma_1 \preceq (RL) \int_E H \, d\Gamma_2.$$

(11.c) *If* $\Gamma_1 \subseteq \Gamma_2$, *then*

$$(RL) \int_E H \, d\Gamma_1 \subseteq (RL) \int_E H \, d\Gamma_2.$$

(11.d)

$$d_H \left((RL) \int_S G \, d\Gamma, (RL) \int_S H \, d\Gamma \right) \leq \sup_{s \in S} d_H(G(s), H(s)) \cdot \overline{\Gamma}(S).$$

4.2. Convergence results

In the following we present some results of [22, 26, 27] regarding convergent sequences of Riemann-Lebesgue integrable interval-valued multifunctions. Firstly, we recall the definitions of convergence almost everywhere and convergence in measure for interval-valued multimeasures.

Definition 9. Let $\nu : \mathcal{C} \to [0, \infty)$ be a set function with $\nu(\emptyset) = 0$, $H : S \to L(\mathbb{R}_0^+)$ a multifunction and a sequence of interval-valued multifunctions $H_n : S \to L(\mathbb{R}_0^+)$, for every $n \in \mathbb{N}$. We recall that:

(9.i) $(H_n)_n$ converges ν-almost everywhere to H on S ($H_n \xrightarrow{\nu-a.e.} H$) if there exists $B \in \mathcal{C}$ with $\nu(B) = 0$ and $\lim_{n \to \infty} d_H(H_n(s), H(s)) = 0$, for every $s \in S \setminus B$.

(9.ii) (H_n) ν-converges to H on S ($H_n \xrightarrow{\nu} H$) if for every $\delta > 0$, $B_n(\delta) = \{s \in S; d_H(H_n(s), H(s)) \geq \delta\} \in \mathcal{C}$ and $\lim_{n \to \infty} \nu(B_n(\delta)) = 0$.

Theorem 12. ([27, Theorem 11]) *Let* $\Gamma : \mathcal{C} \to L(\mathbb{R}_0^+)$, $\Gamma = [\nu_1, \nu_2]$, *so that* ν_2 *is of finite variation. Let* $H = [h_1, h_2]$, $H_n = [h_1^{(n)}, h_2^{(n)}] : S \to L(\mathbb{R}_0^+)$ *be multifunctions such that* $\sup\{h_2(s), h_2^{(n)}(s), s \in S, n \in \mathbb{N}\} < +\infty$ *and* $H_n \xrightarrow{\widetilde{\Gamma}} H$. *Then*

$$\lim_{n \to \infty} d_H \left((RL) \int_S H_n \, d\Gamma, (RL) \int_S H \, d\Gamma \right) = 0.$$

Theorem 13. ([27, Theorem 12]) *Suppose $\nu : \mathcal{C} \to [0, \infty)$ is monotone, of finite variation and $\tilde{\nu}$ satisfies* **(E)**. *Let $H = [h_1, h_2]$, $H_n = [h_1^{(n)}, h_2^{(n)}]$: $S \to L(\mathbb{R}_0^+)$ be multifunctions such that $\sup\{h_2(s), h_2^{(n)}(s), s \in S, n \in \mathbb{N}\} < +\infty$ and $H_n \xrightarrow{\nu - ae} H$, then*

$$\lim_{n \to \infty} d_H \left((RL) \int_S H_n \, \mathrm{d}\nu, (RL) \int_S H \, \mathrm{d}\nu \right) = 0.$$

Theorem 14. ([27, Theorem 13]) *Let $\Gamma := [\nu_1, \nu_2] : \mathcal{C} \to L(\mathbb{R}_0^+)$ with ν_1, ν_2 monotone set functions satisfying* **(E)** *and ν_2 of finite variation. Let $H = [h_1, h_2]$, $H_n = [h_1^{(n)}, h_2^{(n)}] : S \to L(\mathbb{R}_0^+)$ be multifunctions such that $\sup\{h_2(s), h_2^{(n)}(s), s \in S, n \in \mathbb{N}\} < +\infty$ and $H_n \xrightarrow{\tilde{\Gamma} - ae} H$, then*

$$\lim_{n \to \infty} d_H \left((RL) \int_S H_n \, \mathrm{d}\Gamma, (RL) \int_S H \, \mathrm{d}\Gamma \right) = 0.$$

A Fatou type theorem for sequences of RL integrable interval-valued multifunctions holds.

Theorem 15. ([27, Theorem 14]) *Suppose $\nu : \mathcal{C} \to [0, \infty)$ is monotone with $0 < \overline{\nu}(S) < \infty$ and $\tilde{\nu}$ satisfies* **(E)**. *For every $n \in \mathbb{N}$, let $H_n = [h_1^{(n)}, h_2^{(n)}]$ be such that $(h_2^{(n)})_n$ is uniformly bounded. Then*

$$(RL) \int_S (\liminf_n H_n) \, \mathrm{d}\nu \preceq (RL) \liminf_n \int_S H_n \, \mathrm{d}\nu.$$

Next, some Lebesgue type theorems are presented.

Theorem 16. (Monotone Convergence, [26, Proposition 1]) *Suppose $\Gamma = [\nu_1, \nu_2]$ with $\nu_i \in \mathcal{M}(S)$, $i \in \{1, 2\}$ of finite variation. For every $n \in \mathbb{N}$, let $H_n = [h_1^{(n)}, h_2^{(n)}]$ be a multifunction such that $(h_2^{(n)})$ is uniformly bounded and $H_n \preceq H_{n+1}$ for every $n \in \mathbb{N}$. Then*

$$(RL) \int_S \bigvee_n H_n \, \mathrm{d}\Gamma = \bigvee_n (RL) \int_S H_n \, \mathrm{d}\Gamma.$$

A convergence type theorem for varying multisubmeasures also holds.

Theorem 17. ([26, Theorem 4.2]) *Let $(H_n)_n := ([h_1^{(n)}, h_2^{(n)}])_n$ be a sequence of bounded multifunctions, and $(\Gamma_n)_n := ([\nu_1^{(n)}, \nu_2^{(n)}])_n$ a sequence of multisubmeasures. Suppose there exist an interval-valued multisubmeasure $\Gamma := [\nu_1, \nu_2]$, with ν_2 of finite variation, and a bounded multifunction $H := [h_1, h_2]$ such that:*

(17.a) $H_n \preceq H_{n+1}$ *for every* $n \in \mathbb{N}$ *and* $d_H(H_n, H) \to 0$ *uniformly on* S;

(17.b) $\Gamma_n \preceq \Gamma_{n+1} \preceq \Gamma$ *for every* $n \in \mathbb{N}$ *and* $(\Gamma_n)_n$ *setwise converges to* Γ *(namely* $\lim_n \Gamma_n(A) = \Gamma(A)$ *for every* $A \in \mathcal{C}$).

Then

$$\lim_{n \to \infty} d_H\left((RL)\int_S H_n \, d\Gamma_n, (RL)\int_S H \, d\Gamma\right) = 0.$$

Proof. By 17.b) we have $\nu_i^{(n)} \leq \nu_i^{(n+1)} \leq \nu_i$ for every $i = 1, 2$ and for every $n \in \mathbb{N}$. Moreover, $\lim_{n \to \infty} \overline{\nu}_i^{(n)}(A) = \overline{\nu}_i(A)$ for every $A \in \mathcal{C}$ and $i = 1, 2$. Since H_n, H are bounded and ν_2 is of finite variation then, by [12, Proposition 1] and [22, Proposition 2], $H_n, H \in RL_{\Gamma_k}^1(L(\mathbb{R}_0^+)) \cap RL_\Gamma^1(L(\mathbb{R}_0^+))$ for every $n, k \in \mathbb{N}$. Let $\varepsilon > 0$ be fixed and let $n(\varepsilon)$ be such that $d_H(H_n, H) \leq \varepsilon$ for every $n \geq n(\varepsilon)$. By [22, Theorem 3], for every $n \geq n(\varepsilon)$,

$$d_H\left((RL)\int_S H_n \, d\Gamma_n, (RL)\int_S H \, d\Gamma\right)$$

$$\leq d_H\left((RL)\int_S H_n \, d\Gamma_n, (RL)\int_S H \, d\Gamma_n\right) + d_H\left((RL)\int_S H \, d\Gamma_n, (RL)\int_S H \, d\Gamma\right)$$

$$\leq \varepsilon \overline{\Gamma}_n(S) + d_H\left((RL)\int_S H \, d\Gamma_n, (RL)\int_S H \, d\Gamma\right) \leq \varepsilon \overline{\nu}_2(S)$$

$$+ d_H\left(\left[(RL)\int_S h_1 d\nu_1^{(n)}, (RL)\int_S h_2 d\nu_2^{(n)}\right], \left[(RL)\int_S h_1 d\nu_1, (RL)\int_S h_2 d\nu_2\right]\right).$$

We have to evaluate

$$d_H\left(\left[(RL)\int_S h_1 d\nu_1^{(n)}, (RL)\int_S h_2 d\nu_2^{(n)}\right], \left[(RL)\int_S h_1 d\nu_1, (RL)\int_S h_2 d\nu_2\right]\right)$$

$$= \max_{i=1,2}\left\{(RL)\int_S h_i d\nu_i - (RL)\int_S h_i d\nu_i^{(n)}\right\}.$$

Using now [26, Lemma 4.1] the last term tends to 0 for $n \to \infty$ and so

$$\lim_{n \to \infty} d_H\left((RL)\int_S H_n \, d\Gamma_n, (RL)\int_S H \, d\Gamma\right) = 0.$$

$\square$

Analogously to [22, Remark 3], Theorem 17 can be extended to the bounded sequences $(H_n)_n$ converging $\overline{\Gamma}$-almost uniformly on S.

Corollary 3. ([26, Corollary 1]) *Let* $(H_n)_n := ([h_1^{(n)}, h_2^{(n)}])_n$ *be a sequence of bounded multifunctions and* $(\Gamma_n)_n := ([\nu_1^{(n)}, \nu_2^{(n)}])_n$, *be a sequence of multisubmeasures. Suppose there exist a multisubmeasure* $\Gamma = [\nu_1, \nu_2]$ *with*

ν_2 of finite variation and a bounded multifunction $H = [h_1, h_2]$ such that:

(3.a) $H_n \preceq H_{n+1}$ *for every $n \in \mathbb{N}$ and $d_H(H_n, H) \to 0$ Γ-almost uniformly on S;*

(3.b) $\Gamma_n \preceq \Gamma_{n+1} \preceq \Gamma$, *for every $n \in \mathbb{N}$ and (Γ_n) setwise converges to Γ.*

Then

$$\lim_{n \to \infty} d_H \left((RL) \int_S H_n \, d\Gamma_n, (RL) \int_S H \, d\Gamma \right) = 0.$$

Remark 3. We can observe that the results of Theorem 17 and Corollary 3 are still valid if we assume that $\Gamma_{n+1} \succeq \Gamma_n \succeq \Gamma$ for every $n \in \mathbb{N}$, with the additional hypothesis that $\sup_n \overline{\Gamma}_n(S) < +\infty$.

Moreover, in Corollary 3, if $\Gamma_n = \Gamma = \nu$ in the condition (3.a), then the monotonicity can be omitted.

In particular, in the finitely additive case, we obtain the following theorem.

Theorem 18. ([26, Theorem 4.4]) *Let $\nu : \mathcal{C} \to [0, \infty)$ be finitely additive and of finite variation. Let $H = [h_1, h_2]$, $H_n = [h_1^{(n)}, h_2^{(n)}] : S \to L(\mathbb{R}_0^+)$ be multifunctions such that $\sup\{h_2(s), h_2^{(n)}(s), s \in S, n \in \mathbb{N}\} < +\infty$ and $H_n \overset{\widetilde{\nu}}{\to} H$. Then*

$$\lim_{n \to \infty} d_H \left((RL) \int_S H_n \, d\nu, (RL) \int_S H \, d\nu \right) = 0.$$

4.3. Convergence results on atoms

Finally, the field of atoms in measure theory has many applications and has been studied by many authors (see e.g., [50, 60, 63]).

In order to obtain convergence results on atoms, we suppose S is a locally compact Hausdorff topological space. We denote by $\mathcal{K}$ the lattice of all compact subsets of S, $\mathcal{B}$ the Borel σ-algebra (i.e. the smallest σ-algebra containing $\mathcal{K}$) and $\mathcal{O}$ the class of all open sets.

Definition 10. The set multifunction $\Gamma : \mathcal{B} \to L(\mathbb{R}_0^+)$ is said to be regular if for every set $A \in \mathcal{B}$ and every $\varepsilon > 0$ there exist $K \in \mathcal{K}$ and $D \in \mathcal{O}$ such that $K \subseteq A \subseteq D$ and $\|\Gamma(D \setminus K)\|_{\mathcal{H}} < \varepsilon$.

We observe that the regularity of Γ is equivalent to the regularity of ν_2.

Definition 11. It is said that $B \in \mathcal{C}$ is an atom of an interval-valued multifunction $\Gamma : \mathcal{C} \to L(\mathbb{R}_0^+)$ if $\{0\} \preceq \Gamma(B), \{0\} \neq \Gamma(B)$ and for every $C \in \mathcal{C}$, with $C \subseteq B$, we have $\Gamma(C) = \{0\}$ or $\Gamma(B \setminus C) = \{0\}$.

Theorem 19. ([27, Theorem 15]) *Let* $\Gamma : \mathcal{B} \to L(\mathbb{R}_0^+)$ *be a regular multisubmeasure of finite variation and satisfying property* $(\boldsymbol{\sigma})$ *and let* $H : S \to L(\mathbb{R}_0^+)$ *be bounded. If* $B \in \mathcal{B}$ *is an atom of* Γ, *then* $(RL) \int_B H \, \mathrm{d}\Gamma = H(b) \cdot \Gamma(\{b\})$, *where* $b \in B$ *is the single point resulting from* [50, Corollary 4.7].

Proof. Firstly, we prove the uniqueness of $b \in B$. Because Γ is an interval-valued regular multisubmeasure, then the set functions ν_1 and ν_2 are null-additive an regular too. Suppose $B \in \mathcal{B}$ is an atom of Γ. Then, B is an atom of ν_1 and ν_2. According to [50, Corollary 4.7], for $\nu_i, i \in \{1,2\}$, there exists a unique point $b_i, i \in \{1,2\}$ such that $\nu_i(\{b_i\}) = \nu_i(B)$ and $\nu_i(B \setminus \{b_i\}) = 0$, for $i \in \{1,2\}$. We prove that $b_1 = b_2$. If it is not true then $\{b_1\} \subset B \setminus \{b_2\}$. By the monotonicity of ν_2 we have $\nu_2(\{b_1\}) \leq \nu_2(B \setminus \{b_2\}) = 0$. Since $\nu_1 \leq \nu_2$ then $\nu_1(\{b_1\}) = 0$, but $\nu_1(\{b_1\}) = \nu_1(B) > 0$, and we have a contradiction. Therefore, there is only one point $b \in B$ such that $\nu_i(\{b\}) = m_i(B)$ and $\nu_i(B \setminus \{b\}) = 0$, for $i \in \{1,2\}$.

By the RL_Γ-integrability of H, then h_1 is RL_{ν_1}-integrable and h_2 is RL_{ν_2}-integrable. According to [12, Theorem 11] and Subsection 3.2, h_1, h_2 are Gould integrable in the sense of [43], and moreover:

$$(RL_{\nu_1}) \int_B h_1 \mathrm{d}\nu_1 = (G) \int_B h_1 \mathrm{d}\nu_1, \quad (RL_{\nu_2}) \int_B h_2 \mathrm{d}\nu_2 = (G) \int_B h_2 \mathrm{d}\nu_2,$$

where $(G) \int_B h_1 \mathrm{d}\nu_1$, $(G) \int_B h_2 \mathrm{d}\nu_2$ are the Gould integrals of h_1, h_2 respectively. Applying now [10, Theorem 3] and [27, Remark 5], we have

$$(RL_\Gamma) \int_B H \mathrm{d}\Gamma = (G) \int_B H \mathrm{d}\Gamma = H(b)\Gamma(\{b\}).$$

$\square$

Theorem 20. ([27, Theorem 16]) *Let* $\Gamma : \mathcal{B} \to L(\mathbb{R}_0^+)$ *be a regular multisubmeasure of finite variation and satisfying property* $(\boldsymbol{\sigma})$. *Let* $H : S \to L(\mathbb{R}_0^+)$ *be bounded and, for every* $n \in \mathbb{N}$, *let* $H_n = [u_n, v_n]$ *be such that* $(v_n)_n$ *is uniformly bounded. If* $B \in \mathcal{B}$ *is an atom of* Γ *and* $H_n(b) \xrightarrow{d_H} H(b)$, *where* $b \in B$ *is the single point resulting by Theorem 19, then*

$$\lim_{n \to \infty} d_H \left((RL) \int_B H_n \, \mathrm{d}\Gamma, (RL) \int_B H \, \mathrm{d}\Gamma \right) = 0.$$

Proof. By Theorem 19, there exists a unique point $b \in B$ such that:

$$\Gamma(B \setminus \{b\}) = \{0\}, \quad (RL) \int_B H \mathrm{d}\Gamma = H(b) \cdot \Gamma(B).$$

Similarly, for every $n \in \mathbb{N}$, there is a unique $b_n \in B$ such that:

$$\Gamma(B \setminus \{b_n\}) = \{0\}, \quad {}_{(RL)}\int_B H_n \mathrm{d}\Gamma = H_n(b_n) \cdot \Gamma(B).$$

If there exists $n_0 \in \mathbb{N}$ such that $b_{n_0} \neq b$, this means that $\{b_{n_0}\} \subset B \setminus \{b\}$, and by the monotonicity of Γ, it follows that: $\Gamma(\{b_{n_0}\}) \preceq \Gamma(B \setminus \{b\}) = \{0\}$; however, this is not possible since $\Gamma(\{b_{n_0}\}) = \Gamma(B) \neq \{0\}$. Therefore, for every $n \in \mathbb{N}$, $b_n = b$. Then, for $n \to \infty$:

$$d_H \left({}_{(RL_\Gamma)}\int_B H_n \mathrm{d}\Gamma, {}_{(RL_\Gamma)}\int_B H \mathrm{d}\Gamma \right) \leq d_H(H_n(b), H(b)) \cdot \overline{\Gamma}(B) \longrightarrow 0.$$

$\square$

Funding. This research has been accomplished within the UMI Group TAA - "Approximation Theory and Applications", the G.N.AM.P.A. group of INDAM and the University of Perugia. This research was partly funded by: (1) Research project of MIUR (Italian Ministry of Education, University and Research) Prin 2022 "Nonlinear differential problems with applications to real phenomena" (Grant Number: 2022ZXZTN2, CUP J53D23003920006); (2) PRIN 2022 PNRR: "RETINA: REmote sensing daTa INversion with multivariate functional modelling for essential climAte variables characterization" funded by the European Union under the Italian National Recovery and Resilience Plan (NRRP) of NextGenerationEU, under the MUR (Project Code: P20229SH29, CUP: J53D23015950001); (3) Gnampa Project 2024 "Dynamical Methods: Inverse problems, Chaos and Evolution" and (4) Ricerca di Base, Universitá degli Studi di Perugia: Prof. A.R. Sambucini.

References

[1] R. A. Adams, J. J. F. Fournier, *Sobolev Spaces*, Vol. 140. Elsevier, (2003).

[2] G. Birkhoff, Integration of functions with values in a Banach space, *Trans. Amer. Math. Soc.* **38** (2), 357-378, (1935).

[3] M. Balcerzak, K. Musiał, A convergence theorem for the Birkhoff integral, *Funct. Approx. Comment. Math.* **50** (1), 161-168, (2014).

[4] A. Boccuto, D. Candeloro, A.R. Sambucini, Henstock multivalued integrability in Banach lattices with respect to pointwise non atomic measures, *Atti Accad. Naz. Lincei Rend. Lincei Mat. Appl.* **26** (4), 363-383, (2015). Doi: 10.4171/RLM/710.

[5] A. Boccuto, A.M. Minotti, A.R. Sambucini, Set-valued Kurzweil-Henstock integral in Riesz space setting, *PanAmerican Math. J.* **23** (1), 57-74, (2013).

[6] A. Boccuto, A. R. Sambucini, A note on comparison between Birkhoff and Mc Shane integrals for multifunctions, *Real Analysis Exchange* **37** (2), 3-15, (2012).

[7] A. Boccuto, A.R. Sambucini, Some applications of modular convergence in vector lattice setting, *Sampling Theory, Signal processing and Data Analysis* **20** (2), 12, (2022).

[8] A. Boccuto, A.R. Sambucini, Abstract integration with respect to measures and applications to modular convergence in vector lattice setting, *Results in Mathematics* **78** (1), 4, (2023).

[9] D. Candeloro, A. Croitoru, A. Gavriluţ, A.R. Sambucini, An extension of the Birkhoff integrability for multifunctions, *Mediterranean J. Math.* **13** (5), 2551-2575, (2016).

[10] D. Candeloro, A. Croitoru, A. Gavriluţ, A.R. Sambucini, Atomicity related to non-additive integrability, *Rend. Circ. Mat. Palermo, II. Ser.* **65**, 435-449, (2016). Doi: 10.1007/s12215-016-0244-z

[11] D. Candeloro, A. Croitoru, A. Gavriluţ, A.R. Sambucini, A multivalued version of the Radon-Nikodym theorem, via the single-valued Gould integral, *Australian J. of Mathematical Analysis and Applications*, **15** (2), art. 9, 1-16, (2018).

[12] D. Candeloro, A. Croitoru, A. Gavriluţ, A. Iosif, A.R. Sambucini, Properties of the Riemann-Lebesgue integrability in the non-additive case, *Rend. Circ. Mat. Palermo, Serie 2,* **69**, 577-589, (2020). Doi: 10.1007/s12215-019-00419-y.

[13] D. Candeloro, L. Di Piazza, K. Musiał, A.R. Sambucini, Gauge integrals and selections of weakly compact valued multifunctions, *J. Math. Anal. Appl.* **441** (1), 293-308, (2016). Doi: 10.1016/j.jmaa.2016.04.009.

[14] D. Candeloro, L. Di Piazza, K. Musiał, A.R. Sambucini, Relations among gauge and Pettis integrals for multifunctions with weakly compact convex values, *Annali di Matematica*, **197** (1), 171-183, (2018). Doi: 10.1007/s10231-017-0674-z

[15] D. Candeloro, A.R. Sambucini, Comparison between some norm and order gauge integrals in Banach lattices, *Pan American Math. J.* **25** (3), 1-16, (2015).

[16] D. Caponetti, V. Marraffa, K. Naralenkov, On the integration of Riemann-measurable vector-valued functions, *Monatsh. Math.* **182**, 513-536, (2017).

[17] B. Cascales, J. Rodríguez, The Birkhoff integral and the property of Bourgain, *Math. Ann.* **331**, 259-279, (2005).

[18] B. Cascales, J. Rodriguez, Birkhoff integral for multi-valued functions, *J. Math. Anal. Appl.* **297** (2), 540-560, (2004).

[19] G. Chen, Z. Chen, A functional generalization of the reverse Hölder integral inequality on time scales, *Mathematical and Computer Modelling* **54**, 2939-2942, (2011).

[20] K. Chichoń, M. Chichoń, B. Satco, Differential inclusions and multivalued integrals, *Discuss. Math. Differ. Incl. Control Optim.* **33**, 171-191, (2013).

[21] I. Chiţescu, M.G. Manea, T. Paraschiv, *Using the Choquet integral for the determination of the anxiety degree*, Doi: 10.21203/rs.3.rs-1027357/v1.

[22] D. Costarelli, A. Croitoru, A. Gavriluț, A. Iosif, A.R. Sambucini, The Riemann-Lebesgue integral of interval-valued multifunctions, *Mathematics* **8** (12), 2250, 1-17, (2020).

[23] D. Costarelli, M. Seracini, G. Vinti, A comparison between the sampling Kantorovich algorithm for digital image processing with some interpolation and quasi-interpolation methods, *Applied Mathematics and Computation* **374**, 125046, (2020).

[24] D. Costarelli, M. Seracini G. Vinti, A segmentation procedure of the pervious area of the aorta artery from CT images without contrast medium, *Mathematical Methods in the Applied Sciences* **43**, 114-133, (2020).

[25] A. Croitoru, A. Gavriluț, Comparison between Birkhoff integral and Gould integral, *Mediterr. J. Math.* **12**, 329-347, (2015), Doi: 10.1007/s00009-014-0410-5.

[26] A. Croitoru, A. Gavriluț, A. Iosif, A.R. Sambucini, A note on convergence results for varying interval valued multisubmeasures, *Math. Found. Comput.* **4** (4), 299-310, (2021).

[27] A. Croitoru, A. Gavriluț, A. Iosif, A.R. Sambucini, Convergence theorems in interval-valued Riemann-Lebesgue integrability, *Mathematics* **10** (3), 450, 1-15, (2022), Doi: 10.3390/math10030450.

[28] A. Croitoru, A. Gavriluț, A. Iosif, A.R. Sambucini, Inequalities in Riemann Lebesgue integrability, *Mathematics* **12** (1), 49, (2024), Doi: 10.3390/math12010049.

[29] L. Di Piazza, V. Marraffa, The Mc Shane, PU and Henstock integrals of Banach valued functions, *Czechoslovak Math. J.* **52** (127), 609-633, (2002).

[30] L. Di Piazza, V. Marraffa, K. Musiał, Variational Henstock integrability of Banach space valued functions, *Math. Bohem* **141** (2), 287-296, (2016).

[31] L. Di Piazza, K. Musiał, V. Marraffa, A.R. Sambucini, Convergence for varying measures, *J. Math. Anal. Appl.* **518**, (2023) Doi: 10.1016/j.jmaa.2022.126782.

[32] L. Di Piazza, K. Musiał, V. Marraffa, A.R. Sambucini, Convergence for varying measures in the topological case, *Annali di Matematica* **203**, 71-86, (2024). Doi: 10.1007/s10231-023-01353-8

[33] L. Di Piazza, V. Marraffa, B. Satco, Set valued integrability and measurability in non separable Frechet spaces and applications, *Math. Slovaca* **66** (5), 1119-1138, (2016).

[34] L. Di Piazza, D. Preiss, When do McShane and Pettis integrals coincide?, *Illinois Journal of Mathematics*, **47** (4), 1177-1187, (2003).

[35] L. Drewnowski, Topological rings of sets, continuous set functions, integration, I, II,III, *Bull. Acad. Polon. Sci. Ser. Math. Astron. Phys.* **20**, 277-286, (1972).

[36] A. Fernandez, F. Mayoral, F. Naranjo, J. Rodriguez, On Birkhoff integrability for scalar functions and vector measures, *Monatsh. Math.* **157**, 131-142, (2009).

[37] D.H. Fremlin, *The McShane and Birkhoff integrals of vector-valued functions*, University of Essex Mathematics Department Research Report 92-10,

version of 13.10.04, available at:
http://www.essex.ac.uk/maths/staff/fremlin/preprints.htm.

[38] S.G. Gal, On a Choquet-Stieltjes type integral on intervals, *Math. Slov.* **69** (4), 801-814, (2019).

[39] A. Gavriluţ, A. Petcu, A Gould type integral with respect to a submeasure, *An. Şt. Univ. Al. I. Cuza Iaşi* **53** (2), 351-368, (2007).

[40] A. Gavriluţ, A Gould type integral with respect to a multisubmeasure, *Math. Slovaca* **58**, 43-62, (2008).

[41] A. Gavriluţ, A generalized Gould type integral with respect to a multisubmeasure, *Math. Slovaca* **60**, 289-318, (2010).

[42] A. Gavriluţ, A. Iosif, A. Croitoru, The Gould integral in Banach lattices, *Positivity* **19** (1), 65-82, (2015), Doi: 10.1007/s11117-014-0283-7.

[43] G.G. Gould, On integration of vector-valued measures, *Proc. London Math. Soc.* **15**, 193-225, (1965).

[44] Guo, C., Zhang, D., On set-valued fuzzy measures, *Inform. Sci.* **160**, 13-25, (2004).

[45] S. Hu, N.S. Papageorgiou, *Handbook of Multivalued Analysis I and II*, Mathematics and Its Applications 419, Kluwer Academic Publisher, Dordrecht, (1997).

[46] A. Jurio, D. Paternain, C. Lopez-Molina, H. Bustince, R. Mesiar,G. Beliakov, *A Construction Method of Interval-Valued Fuzzy Sets for Image Processing*, 2011 IEEE Symposium on Advances in Type-2 Fuzzy Logic Systems, (2011), Doi: 10.1109/T2FUZZ.2011.5949554.

[47] V.M. Kadets, B. Shumyatskiy, R. Shvidkoy, L.M. Tseytlin, K. Zheltukhin, Some remarks on vector-valued integration, *Mat. Fiz. Anal. Geom.*, **9**, 48-65, (2002).

[48] V.M. Kadets, L.M. Tseytlin, On integration of non-integrable vector-valued functions, *Mat. Fiz. Anal. Geom.* **7**, 49-65, (2000).

[49] D. La Torre, R. Mendivil, F., E.R. Vrscay, Iterated function systems on multifunctions, *Math. Everywhere*, 125-138, (2007).

[50] J. Li, R. Mesiar, E. Pap, Atoms of weakly null-additive monotone measures and integrals, *Inf. Sci.* **257**, 183-192, (2014).

[51] J. Li, M. Yasuda, On Egoroff's theorems on finite monotone non-additive measure space, *Fuzzy Sets Syst.* **153**, 71-78, (2005).

[52] C. Lopez-Molina, B. De Baets, E. Barrenechea, H. Bustince, *Edge detection on interval-valued images*, In Eurofuse 2011. Advances in Intelligent and Soft Computing, Melo-Pinto P., Couto P., Serôdio C., Fodor J., De Baets B. (eds), **107** Springer, Berlin, Heidelberg, Doi: 10.1007/978-3-642-24001-0_30.

[53] M.A.S. Mahmoud Muhammad, S. Moraru, *Accident rates estimation modeling based on human factors using fuzzy c-means clustering algorithm*, World Academy of Science, Engineering and Technology (WASET), **64**, 1209-1219, (2012).

[54] V. Maraffa, A Birkhoff type integral and the Bourgain property in a locally convex space, *Real. Anal. Exchange* **32** (2), (2006-2007).

[55] V. Marraffa, A. R. Sambucini, Vitali Theorems for varying measures, *Symmetry*, **16** (8), 972, (2024) Doi: 10.3390/sym16080972.

A Survey on the Riemann-Lebesgue Integrability in Non-additive Setting   93

[56] S.B. Memetaj, Some convergence theorems for bk-integral in locally convex spaces, *Tatra Mt. Math. Publ.* **46** (1), 29-40, (2010).

[57] E. A. Milne, Note on Rosseland's integral for the stellar absorption coefficient, *Monthly Notices R Astron. Soc.* **85**, 979-984, (1925).

[58] R.E. Moore, *Interval Analysis*, Prentice Hall, Englewood Cliffs, NJ, (1966).

[59] E. Pap, An integral generated by decomposable measure, *Univ. u Novom Sadu Zh. Rad. Prirod. Mat. Fak. Ser. Mat.* **20**, 135-144, (1990).

[60] E. Pap, *Null-additive set functions*, Kluwer Academic Publishers, Dordrecht-Boston-London, (1995).

[61] E. Pap, *Pseudo-additive measures and their applications*, in: E. Pap (Ed.), Handbook of Measure Theory, II, Elscvicr, 1403-1465, (2002).

[62] E. Pap, A. Iosif, A. Gavriluţ, Integrability of an Interval-valued Multifunction with respect to an Interval-valued Set Multifunction, *Iranian Journal of Fuzzy Systems*, **15** (3), 47-63, (2018).

[63] E. Pap, I. Stajner, Generalized pseudo-convolution in the theory of probabilistic metric spaces, information, fuzzy numbers, optimization, system theory, *Fuzzy Sets and Systems* **102**, 393-415, (1999).

[64] M. Patriche, Minimax theorems for set-valued maps without continuity assumptions, *Optimization* **65** (5), 957-976, (2016). Doi: 10.1080/02331934.2015.1091822

[65] M.M. Popov, On integrability in F-spaces, *Studia Mathematica* **110** (3), 205-220, (1994).

[66] M. Potyrala, Some remarks about Birkhoff and Riemann-Lebesgue integrability of vector valued functions, *Tatra Mt. Math. Publ.* **35**, 97-106, (2007).

[67] M. Potyrala, The Birkhoff and variational McShane integral of vector valued functions, *Folia Mathematica, Acta Universitatis Lodziensis*, **13**, 31-40, (2006).

[68] A. Precupanu, A. Croitoru A Gould type integral with respect to a multimeasure I/II, *An. Şt. Univ. "Al. I. Cuza" Iaşi* **48**, 165-200, (2002) / **49**, 183-207, (2003).

[69] A. Precupanu, A. Gavriluţ, A. Croitoru A fuzzy Gould type integral, *Fuzzy Sets and Syst.* **161**, 661-680, (2010).

[70] A. Precupanu, B. Satco, The Aumann-Gould integral, *Mediterr. J. Math.* **5**, 429-441, (2008).

[71] J. Rodriguez, Convergence theorems for the Birkhoff integral, *Houston. J. Math.* **35**, 541-551, (2009).

[72] J. Rodriguez, Pointwise limits of Birkhoff integrable functions, *Proc. Amer. Math. Soc.* **137**, 235-245, (2009).

[73] J. Rodriguez, Some examples in vector integration, *Bull. of the Australian Math. Soc.*, **80** (3), 384-392, (2009).

[74] H. Román-Flores, Y. Chalco-Cano, Y., W.A. Lodwick, Some integral inequalities for interval-valued functions, *Comp. Appl. Math.* **37**, 1306-1318, (2018).

[75] A.R. Sambucini, A survey on multivalued integration, *Atti Sem. Mat. Fis. Univ. Modena* **50**, 53-63, (2002).

[76] G. Shafer, *A Mathematical Theory of Evidence*, Princeton University Press, Princeton, (1976).

[77] J. Sipos, Integral with respect to a pre-measure, *Math. Slovaca* **29**, 141-155, (1979).

[78] F.N. Sofian-Boca, Another Gould type integral with respect to a multisubmeasure, *An. Ştiinţ. Univ. "Al. I. Cuza" Iaşi* **57**, 13-30, (2011).

[79] N. Spaltenstein, A Definition of Integrals, *J. Math. Anal. Appl.* **195**, 835-871, (1995).

[80] C. Stamate, A. Croitoru, The general Pettis-Sugeno integral of vector multifunctions relative to a vector fuzzy multimeasure, *Fuzzy Sets Syst.* **327**, 123-136, (2017).

[81] V. Torra, Y. Narukawa, M. Sugeno, (eds) (2014) *Non-additive measures: Theory and applications*, Studies in Fuzziness and Soft Computing 310 (2014), Springer, Switzerland, Doi: 10.1007/978-3-319-03155-2.

[82] V. Torra, *Use and Applications of Non-Additive Measures and Integrals*, in *Non-Additive Measures, Theory and Applications*, (Torra, V., Narukawa, Y., Sugeno M. eds.), Studies in Fuzziness and Soft Computing, **310**, (2014), 1-33, Springer, Doi: 10.1007/978-3-319-03155-2_2

[83] E.R. Vrscay, A generalized class of fractal-wavelet transforms for image representation and compression, *Can. J. Elect. Comp. Eng.* **23**, 69-84, (1998).

[84] L. Yin, F. Qi, *Some integral inequalities on time scales*, ArXiv 1105.1566

[85] K. Weichselberger, The theory of interval-probability as a unifying concept for uncertainty, *Int. J. Approx. Reason.* **24**, 149-170, (2000).

[86] B. Wohlberg, G. De Jager, A review of the fractal image coding literature, *IEEE Transactions on Image Processing* **8** (12), 1716-1729, (1999).

© 2025 World Scientific Publishing Company
https://doi.org/10.1142/9789819812202_0005

Chapter 5

Some Nonlinear Integrals of Vector Multifunctions with Respect to a Submeasure

Cristina Stamate

Octav Mayer Institute of Mathematics,
Romanian Academy, Iaşi Branch,
Bd. Carol I, No. 8, Iaşi, Romania
Email: cstamate@ymail.com

Anca Croitoru

Alexandru Ioan Cuza University, Faculty of Mathematics,
Bd. Carol I, no. 11, Iaşi, Romania
Email: croitoru@uaic.ro

In this chapter, we present some generalizations for the Sugeno integral of vector multifunctions with respect to a submeasure. Also, properties for these integrals and relationships between them are established.

1. Introduction

Although (countable) additivity is one of the most important notion in measure theory, it can be inappropriate or useless in many problems modeling different real aspects in economics, computer science, psychology, game theory, fuzzy logic, data mining, decision making, subjective evaluation, etc. Such problems have led to the definition of nonadditive measures and nonadditive integrals. Various types of nonlinear integrals, their properties and applications have been introduced in the literature (e.g. [2], [3], [5]–[9], [14], [15], [16], [17], [21], [25], [27], [28], [32], [36]–[42], [47], [48]).

Comparisons between different nonlinear integrals can be found for instance in [22], [33], [39].

To model these real aspects, in 1974, Sugeno [43] has defined a nonadditive measure (named fuzzy measure) and a nonlinear integral for real functions relative to a fuzzy measure. Since many process (for example, deterministic problems, stochastic process, probability problems) happen

95

in vector case, it is important to define an integral for vector functions and multifunctions using the technique of Sugeno.

This integral was extended for vector valued functions by using the efficient points as in [36] and for real and vector multifunctions in different ways as we can see for example in [11], [12], [13], [24], [44], [48]. Another type of nonlinear integral which was extended for the case of vector multifunctions is the Choquet integral as we can see in [20], [29], [34], [42].

Taking into account that the scalarization is a way for studying the vector problems, the Pettis method ([30]) was studied for the case of nonlinear integrals. As we can see in Section 3, the classical Pettis method by using all the elements of the positive dual cone for the scalarization is not useful for defining vector nonlinear integrals and thus, we will replace X_+^* by an arbitrary set $\mathcal{G}$ of positive monotone real maps g with $g(0) = 0$. By this way, we may consider a general Pettis–Sugeno [41], Pettis–Choquet [35], Pettis–Aumann [40] and even a Pettis–Lebesgue [41] integral, as we can see in Section 4. Let us remark that for a real function f, if $\mathcal{G}$ contains only one bijective monotone real map with $g(0) = 0$, then the Pettis–Lebesgue integral of f with respect to a real Lebesgue measure is the pseudo-integral introduced and studied in [25], [26] and the Pettis–Sugeno and Pettis–Choquet are the aggregate operators from [4], [23].

The chapter is organized as follows. In Section 2 we present the preliminaries and the auxiliary results which will be useful in the following. In Section 3, we recall some definitions of nonlinear integrals of multifunctions introduced prior. In Section 4, we give a generalization of the Pettis–Sugeno type integral and present some classic properties. In Section 5, some comparative results concerning the general Pettis nonlinear integrals (Pettis–Lebesgue, Pettis–Choquet, Pettis–Aumann–Sugeno) are established. Finally, some concluding remarks are added.

2. Preliminaries

In what follows, we consider a locally convex space X ordered by a closed convex pointed cone X_+. We will write $a \leq b$ if $b - a \in X_+$, $a < b$ if $b - a \in X_+ \setminus \{0\}$ and $a \not< b$ if $b - a \notin X_+ \setminus \{0\}$. As usual, we denote the dual space as X^*, the dual cone as $X_+^* = \{x^* \in X^* \mid x^*(x) \geq 0, \ \forall x \in X_+\}$ and the strictly-dual cone as $X_+^{\#} = \{x^* \in X_+^* \mid x^*(x) > 0, \ \forall x > 0\}$. We denote the smallest element as $-\infty$ and the biggest element as $+\infty$ and $\overline{X} = X \cup \{+\infty\} \cup \{-\infty\}$. The complementary set of $A \subset X$ will be denoted $A^c = X \setminus A$. We denote the positive cone of the euclidian space $\mathbb{R}^p$ as $\mathbb{R}^p_+$ and

$\mathbb{R}_+ = [0, +\infty)$, $\overline{\mathbb{R}}_+ = [0, +\infty]$, $\mathbb{N}^* = \mathbb{N} \setminus \{0\}$. The closure (the interior and the boundary respectively) of the set $A \subset X$ will be denoted as $\overline{A}$ (Int A, Fr A) and if Int $X_+ \neq \emptyset$, we denote by K the cone $K = \text{Int} X_+ \cup \{0\}$. The convex hull (respectively the linear hull) of a set A will be denoted as conv A (respectively lin A. The cone generated by A will be denoted as cone A. As usual, a fundamental system of neighborhoods for a point x will be denoted by $\mathcal{V}(x)$. If the interior of the cone X_+ is nonempty, we can consider a fundamental system of neighborhoods for $-\infty$, denoted $\mathcal{V}(-\infty)$ (for $+\infty$ respectively, denoted $\mathcal{V}(+\infty)$), given by the sets $V = [-\infty, a) = \{x \in X \mid x < a\} \cup \{-\infty\}$ ($V = (a, +\infty] = \{x \in X \mid x > a\} \cup \{+\infty\}$, respectively). The principal efficient points by respect to a cone $Y_+ \subseteq X_+$ will be presented in the following definition.

Definition 1. (*Efficient points*)
1. $MIN\ (A \mid Y_+) = \{a \in A \mid a - b \notin Y_+ \setminus \{0\},\ \forall b \in A\}$ and
$(MAX\ (A \mid Y_+) = \{a \in A \mid a - b \notin Y_+ \setminus \{0\},\ \forall b \in A\})$.
2. [31] $INF_1\ (A \mid Y_+) = \{y \in \overline{X} \mid y - a \notin Y_+ \setminus \{0\},\ \forall a \in A$, and if $y' - y \in Y_+ \setminus \{0\} \Rightarrow \exists a \in A,\ y' - a \in Y_+ \setminus \{0\}\}$ and
$(SUP_1\ (A \mid Y_+) = \{y \in \overline{X} \mid y - a \notin -Y_+ \setminus \{0\},\ \forall a \in A$, and if $y' - y \in -Y_+ \setminus \{0\} \Rightarrow \exists a \in A,\ y' - a \in -Y_+ \setminus \{0\}\})$.
3. [31] $INF\ (A \mid Y_+) = \{y \in \overline{X} \mid y \not> a\ \forall a \in A\} = (A + Y_+ \setminus \{0\})^c \cup \{-\infty\}$ and
$(SUP\ (A \mid Y_+) = \{y \in \overline{X} \mid y \not< a\ \forall a \in A\} = (A - Y_+ \setminus \{0\})^c \cup \{+\infty\})$.

For $Y_+ = X_+$, if no confusion holds, we denote simply $INF_1 A$, $INF\ A$, $SUP_1 A$ and $SUP\ A$. We recall from [38] Theorem 3.1 which will be useful in what follows.

Theorem 1. *Let $A \subset \overline{X}$ be a nonempty set of a locally convex space X ordered by a convex pointed closed cone with nonempty interior.*
Then, $\emptyset \neq INF_1(A \mid K) \subset X$ if and only if $INF\ (A \mid K) \neq \{-\infty\}$. In this case, the following "domination" properties (DP) holds:
$A \subseteq (INF_1(A \mid K) + K) \cup \{+\infty\}\ INF\ (A \mid K) = (INF_1(A \mid K) - K) \cup \{-\infty\}.$

For a nonvoid set $Y \subseteq X$, we denote by $\mathcal{P}(Y)$ the family of all subsets of Y and by $\mathcal{P}_0(Y)$ the family of all nonvoid subsets of Y. Also, $\mathcal{P}_f(\mathbb{N})$ will denote the finite parts of $\mathbb{N}$. If T is a nonempty set, $\mathcal{C}$ will denote a σ-algebra of subsets of T and $\mathcal{F}(T, \mathbb{R})$ will be the set of functions $f : T \to \mathbb{R}$. A multifunction (also called a set-valued function) from T to X (denoted by $F : T \rightrightarrows X$) is a function $F : T \to \mathcal{P}_0(X)$.

Definition 2. [9] Let X be a topological space. A multifunction $F :$ $T \Longrightarrow X$ is called:
1. *Measurable* if $F^{-1}(U) = \{t \in T \mid F(t) \cap U \neq \emptyset\} \in \mathcal{C}$, for every closed set $U \subseteq X$.
2. *Weak-measurable, in short w-measurable,* if $x^* \circ F$ is measurable for each $x^* \in X_+^*$.

Remark 1. Generally, a measurable multifunction may be defined as a multifunction $F : T \Longrightarrow X$ with $F_\alpha \in \mathcal{C}$, for each $\alpha > 0$ where $F_\alpha = \{t \in T \mid F(t) \cap (\alpha + X_+) \neq \emptyset\}$. For a real function, this is the concept of measurability considered in [43].

Definition 3. [18] A set function $\mu : \mathcal{C} \to [0, +\infty]$ is called:
1. *Monotone* if $\mu(A) \leq \mu(B)$, $\forall A, B \in \mathcal{C}$, $A \subseteq B$.
2. Fuzzy measure (or *submeasure*) if μ is monotone and $\mu(\emptyset) = 0$.

Definition 4. A multifunction $\mu : \mathcal{C} \Longrightarrow X_+ \cup \{+\infty\}$ (a function $\lambda : \mathcal{C} \to X_+ \cup \{+\infty\}$ respectively) with $\mu(\emptyset) = \{0\}$ ($\lambda(\emptyset) = 0$, respectively) is called a *vector multisubmeasure* (a *vector submeasure,* respectively) if $A \subset B$ implies $\mu(A) \subset \mu(B) - X_+$ and $\mu(B) \subset \mu(A) + X_+$, $\forall A, B \in \mathcal{C}$ ($\lambda(A) \leq \lambda(B)$ respectively).

Remark 2. In [19] a multisubmeasure is defined to be a set-valued set function $\mu : \mathcal{C} \to \mathcal{P}_0(X)$ with $\mu(\emptyset) = \{0\}$ satisfying the following condition

$$A \subseteq B \ \text{ implies } \ \mu(A) \subseteq \mu(B) \ \forall A, \ B \in \mathcal{C}. \tag{1}$$

If μ satisfies (1), then $A \subset B$ implies $\mu(A) \subset \mu(B) - X_+$.

Suppose $\mu(A) = \{\lambda(A)\}$ for every $A \in \mathcal{C}$ where $\lambda : \mathcal{C} \to [0, +\infty)$ is a non-negative set function.
(i) If $A \subset B$ implies $\mu(A) \subset \mu(B) - X_+$ for every $A, \ B \in \mathcal{C}$, then λ is monotone.
(ii) If $A \subset B$ implies $\mu(B) \subset \mu(A) + X_+$ for every $A, \ B \in \mathcal{C}$, then λ is monotone.

3. Nonlinear integrals for multifunctions

For real functions, Choquet [10] and Sugeno [43] introduced some nonlinear integrals as in the following definition.

Definition 5. Let $\mu : \mathcal{C} \to \mathbb{R}_+ \cup \{+\infty\}$ be a fuzzy measure and let $f : T \to \mathbb{R}_+$ be a measurable function.

- The *Sugeno integral* of f with respect to μ on $A \in \mathcal{C}$ is

$$(S) \int_A \mathrm{f d}\mu = \sup_{\alpha \geq 0} \min\{\alpha, \mu(A \cap F_\alpha)\}.$$

- The *Choquet integral* of f with respect to μ on $A \in \mathcal{C}$ is

$$(C) \int_A \mathrm{f d}\mu = \int_0^{+\infty} \mu(F_\alpha \cap A)\mathrm{d}\alpha.$$

These definitions were extended for the case of real and vector multifunctions. We present in this section some of these definitions introduced and studied until now.

Definition 6. [11] Let $F : T \rightrightarrows X$ be a measurable multifunction. *The weak fuzzy integral* of F on an arbitrary set $A \in \mathcal{C}$ with respect to a fuzzy measure μ, denoted by $(w) \int_A F d\mu$, is defined by

$$(w) \int_A F d\mu = \sup_{\alpha \in [0,+\infty]} \min\{\alpha, \mu(A \cap |F|^{-1}([\alpha, +\infty)))\}$$

where $|\cdot|$ is a *set-norm* on $\mathcal{P}_0(X)$.

Definition 7. [12] Let $F : T \rightrightarrows \mathbb{R}_+$ be a measurable multifunction and $\mu : \mathcal{C} \to [0, +\infty]$ a fuzzy measure. The *strong fuzzy integral* of F on an arbitrary set $A \in \mathcal{C}$ with respect to μ, denoted by $(s) \int_A F d\mu$, is defined by

$$(s) \int_A F d\mu = \sup_{\alpha \in [0,+\infty]} \inf\{\alpha, \mu(A \cap F^{-1}([\alpha, +\infty)))\}.$$

Different concepts of a set-valued integral were obtained by Aumann selections. For real multifunctions, Zhang and Guo introduced in [45] the G-fuzzy integral for a multifunction F by using the Sugeno integral for selections maps of F with respect to a fuzzy measure μ.

Definition 8. [45] Suppose $F : T \rightrightarrows \mathbb{R}$ is a multifunction and $\mu : \mathcal{C} \to \mathbb{R}_+$ is a fuzzy measure. The *G-fuzzy integral* of F on $A \in \mathcal{C}$ with respect to μ is defined as

$$(G) \int_A F d\mu = \{\int_A f d\mu \mid f \in S(F)\},$$

where $S(F)$ is the set of all measurable selections of F and the integral from the right side is the Sugeno integral for the real map f.

By using the efficient points from Definition 1, which replaces the real supremum and infimum, we have the following integral.

Definition 9. [38] Let $\mu : \mathcal{C} \rightrightarrows X_+ \cup \{+\infty\}$ be a multisubmeasure and $F : T \rightrightarrows X_+$ be a measurable multifunction. The *vector fuzzy integral* of F with respect to μ on $A \in \mathcal{C}$ is defined by

$$\int_A F d\mu = SUP_1 \bigcup_{E \in \mathcal{C}} INF_1(F(E) \cup \mu(A \cap E)).$$

We also present in [36] a vector integral defined by $T_{F,A} = SUP_1 \bigcup_{E \in \mathcal{C}} \alpha_E$ where $\alpha_E \in INF \ (F(E) \bigcup \mu(A \bigcap E))$.

Let remark that $\int_A F d\mu = \int_A INF_1 F d\mu$ and $\int_A F d\mu = \int_A IMIN \ F d\mu$, if $IMIN \ F$ exists. Using this result, we observe that for a real multifunction, the integrals presented in Definitions 7, 9 reduce to the real Sugeno integral for the function inf F which is the strong integral defined in [12]. The principal properties for these integrals may be found in [11], [12], [36], [37].

For the case of the Choquet integral, we find in the literature an integral introduced by Sambucini for vector functions and an integral for vector functions as a particular case for the Choquet integral introduced by Park for vector multifunctions by scalarization.

Definition 10. [34] Let $f : T \to \mathbb{R}^n_+$, $f = (f_1, f_2, ..., f_n)$, f_i measurable functions. The Choquet integral of f with respect to μ on $A \in \mathcal{C}$ is defined by

$$\int_A f d\mu = (\int_A f_1 d\mu, \int_A f_2 d\mu, ..., \int_A f_n d\mu)$$

Definition 11. [29] Let X be a Banach space with dual X^*. A vector function $f : T \to X$ is called Pettis–Choquet integrable if for each $x^* \in X^*$, the function $x^* \circ f$ is Choquet integrable and for every $A \in \mathcal{C}$ there exists $x_A \in X$ such that $x^*(x_A) = \int_A x^* \circ f d\mu$ for each $x^* \in X^*$. In this case x_A is called the *Pettis–Choquet integral* of f on A and it is denoted $(CP) \int_A f d\mu$.

The Pettis–Choquet integral of Park is only positive homogeneous, but not homogeneous. Agahi and Mesiar [1] gave an example which shows that the existence of Pettis–Choquet integral forces, in some sense, the homogeneity, which is a contradiction. The linearity of the Pettis–Choquet integral holds for σ-additive measures but in this case, the Choquet integral reduces to the Lebesgue integral.

The scalarization is thus, the other method which allow us to study a vector problem. Thus we may construct some integrals for vector functions and multifunctions by using the Pettis method as follows.

Definition 12. Let $\mu : \mathcal{C} \Longrightarrow X_+ \cup \{+\infty\}$ be a multisubmeasure and let $F : T \Longrightarrow X_+$ be a measurable multifunction. The *Pettis–Sugeno integral* of F with respect to μ on $A \in \mathcal{C}$ will be denoted $I_{F,A}$ and it will be the subset of X such that:
1. $x^*(I_{F,A}) = \int_A x^* \circ F d\, x^* \circ \mu$, for each $x^* \in X_+^* \setminus \{0\}$;
2. $\sup x^*(I_{F,A}) = \int_A \sup x^* \circ F d\, x^* \circ \mu$, for each $x^* \in X_+^* \setminus \{0\}$;
3. $\inf x^*(I_{F,A}) = \int_A x^* \circ F d\, x^* \circ \mu$ for each $x^* \in X_+^* \setminus \{0\}$;
4. $\sup x^*(I_{F,A}) = \int_A x^* \circ F d\, x^* \circ \mu$ for each $x^* \in X_+^* \setminus \{0\}$, if this set exists and the empty set else.

The integral from the right side is one of the integrals defined for real multifunctions as we have seen in Definitions 7, 8, 9.

For example, let remark that if in the right side of 1) we have the integral from Definition 8 we get the Aumann–Pettis–Sugeno integral from [40]. Also, if we consider in 1) the integral from Definition 7 or 9 the right side will be the Sugeno integral for the real function $f = \inf x^* \circ F$ and $I_{F,A}$ will contains only an element in this case.

The following examples will offer some information concerning the opportunity of the Definition 12.

Example. Let $X = \mathbb{R}^2$, $f = (f_1, f_2)$, $f : T \to \mathbb{R}^2$, $F = f + \mathbb{R}_+^2$ and let μ be a real fuzzy measure. Following Definition 12, 1., if F is Pettis–Sugeno integrable by respect to μ, we find $I = (I_1, I_2) \in \mathbb{R}^2$ such that for each $x^* = (\alpha, \beta) \in \mathbb{R}_+^2$ we have $\alpha I_1 + \beta I_2 = \int_A (\alpha f_1 + \beta f_2) d\mu$. If we consider $\alpha = 0$ (respectively $\beta = 0$) we obtain $I = (\int_A f_1 d\mu, \int_A f_2 d\mu)$. Following Definition 12, 3. (respectively 4.) if F is Pettis–Sugeno integrable by respect to μ, for each $I = (I_1, I_2) \in I_{F,A}$ we get $I_1 \geq \int_A f_1 d\mu$ and $I_2 \geq \int_A f_2 d\mu$ (respectively $I_{F,A} = (\int_A f_1 d\mu, \int_A f_2 d\mu) - \mathbb{R}^2$). Similarly, for $F = f - \mathbb{R}_+^2$, Definition 12, 2. gives $I_{F,A} = (\int_A f_1 d\mu, \int_A f_2 d\mu) - \mathbb{R}^2$ and for each case

$$\int_A (\alpha f_1 + \beta f_2) d\mu = \alpha \int_A f_1 d\mu + \beta \int_A f_2 d\mu.$$

But, in general, this is not the case for the real nonlinear integrals which are nonadditive integrals. For example, in Proposition 1, the authors obtained some necessary and sufficient conditions for the additivity of the Sugeno integral. For the Choquet integral, the additivity is still an open problem; it holds for comonotonic maps or for a Lebesgue measure.

Proposition 1. *[21] Let $u, v : T \to \mathbb{R}_+$, $\mu : \mathcal{C} \to \mathbb{R}_+$, $a, b \in \mathbb{R}_+$, $\mu(T) < +\infty$. We have $\int_T (au + bv) d\mu = a \int_T u d\mu + b \int_T v d\mu$ for all $u, v : T \to X$ and $a, b \in \mathbb{R}_+$ if and only if μ is additive and $\mu(A) \in \{0, \mu(T)\}$, for all $A \subset T$.*

For this reason, we conclude that the scalarization with all the elements of X_+^* is not appropriate for defining a nonlinear vector integral, so in the following we will replace X_+^* by a family of real functions $\mathcal{G}$ with some special properties and we will define the $\mathcal{G}$-integral.

4. A general Pettis type integral for vector multifunctions relative to a submeasure

In this section we introduce a generalization for the Pettis type integral of vector multifunctions relative to a vector submeasure and present some classical properties and calculus rules.

Let $\mu : \mathcal{C} \to X_+ \cup \{+\infty\}$ be a submeasure, $F : T \rightrightarrows X_+$ be a vector multifunction and $\mathcal{G} \subset \mathcal{F}(X, \mathbb{R})$.

Definition 13. The $\mathcal{G}$-*integral* of F with respect to μ on $A \in \mathcal{C}$ is denoted by $\mathcal{G} \int_A F d\mu$ and is the maximal subset of X such that

$$g(\mathcal{G} \int_A F d\mu) = \int_A g \circ F dg \circ \mu$$

for all $g \in \mathcal{G}$ if this set exists and the empty set else, where the integral from the right side is a nonlinear integral for a real multifunction as in Section 3.

Similarly to Definition 12, we may consider the following definitions.

Definition 14. The $\mathcal{G}$-integral of F with respect to μ on $A \in \mathcal{C}$ is denoted by $\mathcal{G} \int_A F d\mu$ and is the maximal subset of X such that:

1. $\sup g(\mathcal{G} \int_A F d\mu) = \int_A \sup g \circ F dg \circ \mu$;
2. $\sup g(\mathcal{G} \int_A F d\mu) = \int_A g \circ F dg \circ \mu$;
3. $\inf g(\mathcal{G} \int_A F d\mu) = \int_A g \circ F dg \circ \mu$;

for all $g \in \mathcal{G}$ if this set exists and the empty set else, where the integral from the right side is a nonlinear integral for a real multifunction as in Section 3.

Some Nonlinear Integrals of Vector Multifunctions with Respect to a Submeasure 103

In the following, $\mathcal{G}$ satisfies some additional hypothesis.

(H_0^A) **Hypothesis.**

Let $\mathcal{G} \subset \mathcal{F}(X, \mathbb{R})$, $Y_+ \subset X_+$ a cone and suppose $\mathcal{G}$ has the following properties:

$(H_0^A)_1$ $x, y \in A$, $g(x) \leq g(y)$, $\forall g \in \mathcal{G} \Rightarrow x - y \in Y_+$;

$(H_0^A)_2$ $x, y \in A$, $g(x) < g(y)$, $\forall g \in \mathcal{G} \Rightarrow x - y \in Y_+ \setminus \{0\}$;

$(H_0^A)_3$ $g(0) = 0, \forall g \in \mathcal{G}$.

Remark 3. If X_+ is pointed, then (H_0^A) is equivalent with $(H_0^{A'})$:

$(H_0^{A'})_1$ $x, y \in A$, $g(x) \leq g(y)$, $\forall g \in \mathcal{G} \Longleftrightarrow x - y \in Y_+ \setminus \{0\}$;

$(H_0^{A'})_2$ $x, y \in A$, $g(x) < g(y)$, $\forall g \in \mathcal{G} \Longleftrightarrow x - y \in Y_+$;

$(H_0^{A'})_3$ $g(0) = 0, \forall g \in \mathcal{G}$.

(H_0^A) will also be equivalent with $(H_0^{A''})$:

$(H_0^{A''})_1$ $x, y \in A$, $g(x) \leq g(y)$, $\forall g \in \mathcal{G} \Longleftrightarrow x - y \in Y_+$;

$(H_0^{A''})_2$ $x, y \in A$, $g(x) < g(y)$, $\forall g \in \mathcal{G} \Longleftrightarrow x - y \in Y_+ \setminus \{0\}$;

$(H_0^{A''})_3$ $g(x) = g(y)$, $\forall g \in \mathcal{G} \Longleftrightarrow x = y$.

If $A = X$, we denote simply (H_0), (H_0'), (H_0'').

Examples.

1. If $X_+^\# \neq \emptyset$, then $\mathcal{G} = X_+^\#$ satisfies (H_0) for $Y_+ = X_+$.

2. If $\overset{\circ}{X}_+ \neq \emptyset$ and $Y_+ = \overset{\circ}{X}_+ \cup \{0\}$, then $\mathcal{G} = X_+^*$ satisfies (H_0).

3. Let $X = \mathbb{R}^2$, $Y_+ = \mathbb{R}^2$, $\mathcal{G} = \{g_1, g_2\}$, $x = (x_1, x_2) \in \mathbb{R}^2$, $C = \{x \mid x_1 \geq x_2\}$.

3_1. For $g_1(x) = x_1$, $g_2(x) = x_2$, $\mathcal{G}$ satisfies (H_0).

3_2. For $g_1(x) = \min(x_1, x_2)$, $g_2 = \max(x_1, x_2) \Rightarrow \mathcal{G}$, satisfies (H_0^C).

3_3. For $g_1(x) = x_1 1_E + x_2 1_F$, $g_2(x) = x_2 1_E + x_1 1_F$, where $\emptyset \neq E, F \subset C$, $E \cap F = \emptyset$, $\mathcal{G}$ satisfies (H_0^C).

3_4. If $\mathcal{G} = \{g_1, g_2\}$ satisfies (H_0^C) and $(g_1(x) - g_2(x))(g_1(y) - g_2(y)) \geq 0$, $\forall x, y \in E \cup F$, then $\widetilde{\mathcal{G}} = \{G_1, G_2\}$, $G_1 = g_1 1_E + g_2 1_F$, $G_2 = g_2 1_E + g_1 1_F$ satisfies (H_0^C).

Using these families we obtain some examples of $\mathcal{G}$-integrals.

Examples.

1. Let $X = \mathbb{R}^n$, $\mathcal{G} = \{e_1, e_2, ..., e_n\}$, $f = (f_1, f_2, ..., f_n)$, μ a fuzzy measure. The integral obtained from Definition 13 will be

$$(\mathcal{G}) \int_A f \, d\mu = \left(\int_A f_1 d\mu, ..., \int_A f_n d\mu \right), \, \forall A \in \mathcal{C}.$$

2. If X is locally convex space ordered by a cone X_+ which has a base $\mathcal{B} = \{x_\alpha, \ \alpha \in \mathbb{N}\} \subset X_+$ (lin $\mathcal{B} = X$), then $\mathcal{G} = \{x_\alpha^* \mid \alpha \in \mathbb{N}\}$ given by $x_\alpha^*(x_\beta) = \delta_{\alpha\beta} = \begin{cases} 1, \alpha = \beta, \\ 0, \alpha \neq \beta, \end{cases}$, $x_\alpha^* : X \to \mathbb{R}$, $x_\alpha^*(x) = x_\alpha^* \left(\sum_{\beta \in I \subset P_f(\mathbb{N})} \mu_\beta x_\beta \right) = \sum \mu_\beta \delta_{\alpha\beta}$ is a base of X^* and $\mathcal{G} \subset X_+^*$. More, if convcone $\mathcal{B} = X_+$, then convcone $\mathcal{G} = X_+^*$.

Definition 13 gives $(\mathcal{G}) \int_A f \, d\mu = \left(\int_A f_\alpha d\mu \right)_{\alpha \in \mathbb{N}}$, $\forall A \in \mathcal{C}$. In this case, $X_+ - X_+ = X$ and $X^* = X_+^* - X_+^*$.

For example, $X = \ell_\infty([0, \infty))$ satisfies these properties.

3. Let $X = \mathbb{R}$, $\mathcal{G} = \{g\}$ where g is a strictly increasing surjective map, $g(0) = 0$. Then $\mathcal{G} \int_T f d\mu = g^{-1}(\int_T g \circ f d\mu)$, the aggregate operator from [4].

4. Let $X = \mathbb{R}^2$ and $\mathcal{G} = \{g_1, g_2\}$, $x = (x_1, x_2)$, $C = \{x \mid x_1 \geq x_2\}$.

4_1. For $g_1(x) = \min(x_1, x_2)$, $g_2 = \max(x_1, x_2) \Rightarrow \mathcal{G}$ satisfies (H_0^C) and

$$\mathcal{G} \int_T f d\mu = (\int_T \min(f_1, f_2) d\mu, \int_T \max(f_1, f_2) d\mu).$$

4_2. For $g_1(x) = x_1 1_E + x_2 1_F$, $g_2(x) = x_2 1_E + x_1 1_F$, where $\emptyset \neq E, F \subset C$, $E \cap F = \emptyset$, $\mathcal{G}$ satisfies (H_0^C) and

$$\mathcal{G} \int_T f d\mu = (\min(\int_T s_1 d\mu, \int_T s_2 d\mu), \max(\int_T s_1 d\mu, \int_T s_2 d\mu)),$$

where $s_1 = f_1 1_A + f_2 1_B$, $s_2 = f_2 1_A + f_1 1_B$, $A = f^{-1}(E)$, $B = f^{-1}(F)$.

5. If X is a reflexive space and $\mathcal{G} \subset X_+^*$ has (H_0) then $\overline{\text{convcone } \mathcal{G}} = X_+^*$. If X is a Hilbert space, then $\overline{\text{lin } \mathcal{G}} = X^*$.

Definition 15.

1. A nonempty subset of X is called a $s\mathcal{G}$-*set* if $\sup g(A) = +\infty$ for each $g \in \mathcal{G}$ implies $SUP_1 A = \{+\infty\}$.

2. A set $A \subset X$ is called $\mathcal{G}$-*separable* if, for each $x_0 \notin A$, there exists $g \in \mathcal{G}$ such that $g(x_0) < \inf g(A)$.

Some Nonlinear Integrals of Vector Multifunctions with Respect to a Submeasure 105

Examples.

I. Every convex set $A \subset X$ is sX_+^*. Indeed, if $SUP \ A \neq \{+\infty\}$, there exists $u \in X$ such that $A \cap (u + X_+) = \emptyset$. Thus, we find $x^* \in X_+^*$ such that $x^*(a) \leq x^*(u)$ for each $a \in A$ and $\sup x^*(A) \neq +\infty$, false. We deduce that for each $u \in X$ there exists $a \in A$ such that $a \geq u$ and since $SUP_1 A \subset SUP \ A$ we get $SUP_1 A = +\{\infty\}$.

II. Every closed convex set $A \subset X$ is $X^* \setminus \{0\}$ separable.

Remark 4. In Definition 13, if in the right side we have only one element, if $\mathcal{G} \int_A F d\mu$ exists, it contains only an element. Indeed, if it contains two different elements u, v, we have $g(u) = g(v)$ for all $g \in \mathcal{G}$ which implies $u = v$, false.

In what follows we present some properties of the general Pettis–Sugeno integral. We begin with the case of the constant multifunctions.

Theorem 2. *Let consider* $M \subset X$, $F : T \rightrightarrows X$ *such that* $F(t) = M$ *for all* $t \in T$ *and* $\mu : \mathcal{C} \to X_+ \cup \{+\infty\}$ *be a submeasure.*

1. *If* $M \cup \mu(A)$ *is a* $\mathcal{G}$-*separable set, then the integral* $\mathcal{G} \int_A F d\mu$ *exists if and only if there exists* $IMIN(M \cup \mu(A))$ *and* $\mathcal{G} \int_A F d\mu = IMIN(M \cup \mu(A))$.

2. *If* $\mathcal{G} \subset X_+^*$, *then* $\mathcal{G} \int_A F d\mu$ *exists if and only if* $IMIN(\overline{\text{conv} \ M \cup \mu(A)})$ *exists and* $\mathcal{G} \int_A F d\mu = IMIN(\overline{\text{conv} \ M \cup \mu(A)})$.

Proof. Suppose that $\mathcal{G} \int_A F d\mu$ exists. For all $g \in \mathcal{G}$ we have $g(\mathcal{G} \int_A F d\mu) = \int_A g \circ F dg \circ \mu$. Following Proposition 3.4 [43] we have $\int_A g \circ F dg \circ \mu = \int_A \inf g(M) dg \circ \mu = \inf(g(M) \cup g \circ \mu(A)) = \inf g(M \cup \mu(A))$. Thus

$$g(\mathcal{G} \int_A F d\mu) = \inf g(M \cup \mu(A)),$$

there exists $IMIN(\overline{\text{conv} \ M \cup \mu(A)})$ and

$$\mathcal{G} \int_A F d\mu = IMIN(\overline{\text{conv} \ M \cup \mu(A)}).$$

For the converse, we have

$$g(IMIN(\overline{\text{conv} \ M \cup \mu(A)})) = \inf g(M \cup \mu(A)) = \int_A g \circ F dg \circ \mu, \ \forall g \in \mathcal{G},$$

which means that $\mathcal{G} \int_A F d\mu$ exists. $\diamond$

In the next theorem we study the case of simple multifunctions.

Theorem 3.

1. *Let* $h : T \to X$, $T = \bigcup_{i=\overline{1,n}} E_i$, $E_i \cap E_j = \emptyset$, $h(t) = a_i$, $\forall t \in E_i$, $a_i \in X_+$, $a_1 \leq a_2 \leq ... \leq a_n$ *and* $\mu : T \to X_+$ *be a submeasure. If* $\int_A hd\mu$ *is* $\mathcal{G}$-*separable then there exists* $IMAX \int_A hd\mu$ *and* $(\mathcal{G}) \int_A hd\mu = IMAX \int_A hd\mu$.

2. *If* $\mathcal{G} \subset X_+^*$, *then there exists* $IMAX\overline{\text{conv}} \int_A hd\mu$,

$$(\mathcal{G}) \int_A hd\mu = IMAX\overline{\text{conv}} \int_A hd\mu.$$

For $F_i = \bigcup_{j \geq i} E_j$, $(\mathcal{G}) \int_A hd\mu = \text{IMAX} \bigcup_{i=\overline{1,n}} INF_1(a_i \cup \mu(A \cap F_i))$.

Proof. Using the same argument as in Theorem 3.10 [47], we have $\int_A g \circ hdg \circ \mu = \sup_{i=\overline{1,n}} \inf\{g(a_i), g \circ \mu(A \cap F_i)\}$ where $F_i = \bigcup_{j \geq i} E_j$. We have $g((\mathcal{G}) \int_A hd\mu - K \setminus \{0\}) = g((\mathcal{G}) \int_A hd\mu) - \mathbb{R}_+^* = g(\bigcup_{i=\overline{1,n}} INF_1(a_i \cup \mu(A \cap F_i)) - K \setminus \{0\}) = g(SUP_1(\bigcup_{i=\overline{1,n}} INF_1(a_i \cup \mu(A \cap F_i)) - K \setminus \{0\}) = g(\int_A hd\mu - K \setminus \{0\})$ for all $g \in \mathcal{G}$ and the conclusion follows. $\diamond$

The calculus of the integral for the characteristic multifunction is given in the next two results.

Theorem 4. *Let* $F \subset T$, $\alpha \in \mathbb{R}$, *a fuzzy measure* $\mu : \mathcal{C} \to \mathbb{R}_+$ *and*

$$\chi_F(t) = \begin{cases} \alpha, & t \in F, \\ 0, & t \notin F. \end{cases}$$

The Sugeno integral will be

$$\int_A \chi_F d\mu = \begin{cases} \mu(F), & \mu(F) \leq \alpha, \\ \alpha, & \mu(F) > \alpha. \end{cases}$$

Proof. $\int_A \chi_F d\mu = \sup_{E \subset A} \inf(\chi_F(E) \cup \mu(E))$.

We have

$$\chi_F(E) = \begin{cases} \{0\}, & E \cap F = \emptyset, \\ \{0, \alpha\}, & E \cap F \neq \emptyset, E \not\subset F, \\ \{\alpha\}, & E \subseteq F. \end{cases}$$

Thus,

$$\inf(\chi_F(E) \cup \mu(E)) = \begin{cases} 0, & E \cap F = \emptyset, \\ \inf\{0, \alpha, \mu(E)\}, & E \cap F \neq \emptyset, E \not\subset F, \\ \inf\{\alpha, \mu(E)\}, & E \subset F. \end{cases}$$

Some Nonlinear Integrals of Vector Multifunctions with Respect to a Submeasure 107

If there exists $E \subset F$ such that $\mu(E) > \alpha$, then $\mu(F) > \alpha$ and

$$\sup_E \inf(\chi_F(E) \cup \mu(E))$$

$$= \sup \left\{ \sup_{E \cap F = \emptyset} \{0\}, \quad \sup_{E \cap F \neq \emptyset E \not\subset F} \{0\}, \quad \sup_{E \subset F \mu(E) < \alpha} \mu(E), \quad \sup_{E \subset F \mu(E) \geq \alpha} \alpha \right\} = \alpha.$$

If $\mu(E) \leq \alpha$ for all $E \subset F$, we have $\mu(F) \leq \alpha$ and

$$\sup_E \inf(\chi_F(E) \cup \mu(E)) = \sup_{E \subset F} \{0, \ 0, \ \mu(E)\} = \mu(F).$$

Thus

$$\int_A \chi_F d\mu = \begin{cases} \mu(F), & \mu(F) \leq \alpha, \\ \alpha, & \mu(F) > \alpha. \end{cases}$$

$\diamond$

Theorem 5. *Let consider* $\operatorname{Int} X_+ \neq \emptyset$, $\emptyset \neq B \subset T$, $\emptyset \neq D \subset X$, *a submeasure* $\mu : \mathcal{C} \to X_+ \cup \{+\infty\}$ *and*

$$H_B(t) = \begin{cases} D, & t \in B, \\ \{0\}, & t \notin B. \end{cases}$$

If $\mu(B) \leq d$ for all $d \in D$, then $(\mathcal{G}) \int_A H_B d\mu = \mu(B)$.

If D is $\mathcal{G}$-separable and there exists $d \in D$ such that $d < \mu(B)$ then, $(\mathcal{G}) \int_A H_B d\mu$ exists if and only if $IMIN \ D$ exists and $(\mathcal{G}) \int_A H_B d\mu = IMIN \ D$.

If $\mathcal{G} \subset X_+^*$ and there exists $d \in D$ such that $d < \mu(B)$ then, $(\mathcal{G}) \int_A H_B d\mu$ exists if and only if $IMIN\overline{\operatorname{conv}D}$ exists and $(\mathcal{G}) \int_A H_B d\mu = IMIN\overline{\operatorname{conv}D}$.

Proof. For $g \in \mathcal{G}$, we have $\int_A g \circ H_B dg \circ \mu = \int_A \inf g \circ H_B dg \circ \mu$.

$$\int_A \inf g \circ H_B dg \circ \mu = \begin{cases} g \circ \mu(B), & g \circ \mu(B) \leq \inf g(D), \\ \inf g(D), & g \circ \mu(B) > \inf g(D). \end{cases}$$

If $\mu(B) \leq d$ for all $d \in D$, then $g \circ \mu(B) \leq \inf g(D)$ and $g((\mathcal{G}) \int_A H_B d\mu) = g(\mu(B))$ for all $g \in \mathcal{G}$ which implies $(\mathcal{G}) \int_A H_B d\mu = \mu(B)$.

If there exists $d < \mu(B)$, then $g(d) < g(\mu(B))$ and $\inf g(D) < g(\mu(B))$. Thus $\int_A g \circ H_B dg \circ \mu = \inf g(D)$. Since the integral $(\mathcal{G}) \int_A H_B d\mu$ exists if and only if $IMIN\overline{D}$ exists then $(\mathcal{G}) \int_A H_B d\mu = IMIN\overline{D}$.

$\diamond$

In the next 3 theorems, we present some properties concerning the monotonicity about the set and the multifunction.

108   *C. Stamate, A. Croitoru*

Theorem 6. *Let $F : T \rightrightarrows X$ be a multifunction and $\mu : \mathcal{C} \rightrightarrows X_+ \cup \{+\infty\}$ be a multisubmeasure, $A, B \in \mathcal{C}$.*
1. *If $A \subset B$ then $(\mathcal{G}) \int_A F\mu \leq (\mathcal{G}) \int_B F d\mu$.*
2. $(\mathcal{G}) \int_A F d\mu \cup (\mathcal{G}) \int_B F d\mu \subset (\mathcal{G}) \int_{A \cup B} F d\mu - X_+$.
3. $(\mathcal{G}) \int_A F d\mu \cup (\mathcal{G}) \int_B F d\mu \subset (\mathcal{G}) \int_{A \cap B} F d\mu + X_+$.

Proof. 1. Following Definition 13 and Theorem 3.4 [43], for all $g \in \mathcal{G}$ we have $g((\mathcal{G}) \int_A F d\mu) = \int_A g \circ F dg \circ \mu = \int_A \inf g \circ F dg \circ \mu \leq \int_B \inf g \circ F dg \circ \mu = \int_B g \circ F dg \circ \mu = g \circ (\mathcal{G}) \int_B F d\mu$. Thus $g((\mathcal{G}) \int_A F d\mu) \leq g((\mathcal{G}) \int_B F d\mu$ for each $g \in \mathcal{G}$ which implies $(\mathcal{G}) \int_A F d\mu \leq (\mathcal{G}) \int_B F d\mu$.

2. Using 1., since $A \subset A \cup B$ and $B \subset A \cup B$ we have $(\mathcal{G}) \int_A F d\mu \leq (\mathcal{G}) \int_{A \cup B} F d\mu$ and $(\mathcal{G}) \int_B F d\mu \leq (\mathcal{G}) \int_{A \cup B} F d\mu$. Thus, the conclusion follows.

3. Using 1., since $A \cap B \subset A$ and $A \cap B \subset B$ we have $(\mathcal{G}) \int_A F d\mu \geq (\mathcal{G}) \int_{A \cap B} F d\mu$, $(\mathcal{G}) \int_B F d\mu \geq (\mathcal{G}) \int_{A \cap B} F d\mu$ and the conclusion follows obviously. $\diamond$

Theorem 7. *Let consider two multifunctions $H_1, H_2 : T \rightrightarrows X$, $A \in \mathcal{C}$ and a submeasure $\mu : \mathcal{C} \rightrightarrows X_+ \cup \{+\infty\}$. If $H_1(t) \subset H_2(t) + X_+$ for each $t \in T$, then $(\mathcal{G}) \int_A H_1 d\mu \leq (\mathcal{G}) \int_A H_2 d\mu$.*

Proof. Using Definition 13 and Theorem 3.2 [43], $g((\mathcal{G}) \int_A H_1 d\mu) = \int_A g \circ H_1 dg \circ \mu = \int_A \inf g \circ H_1 dg \circ \mu \leq \int_A \inf g \circ H_2 dg \circ \mu = g((\mathcal{G}) \int_A H_2 d\mu)$. Thus $g((\mathcal{G}) \int_A H_1 d\mu) \leq g((\mathcal{G}) \int_A H_2 d\mu)$ for $g \in (\mathcal{G})$ which gives

$$(\mathcal{G}) \int_A H_1 d\mu \leq (\mathcal{G}) \int_A H_2 d\mu.$$

$\diamond$

Theorem 8. *Let consider* $\text{Int} X_+ \neq \emptyset$, $H_1, H_2 : T \rightrightarrows X$ *be two multifunctions and* $\mu : \mathcal{C} \rightrightarrows X_+ \cup \{+\infty\}$ *be a multisubmeasure, $A \in \mathcal{C}$. Then,* $INF_1((\mathcal{G}) \int_A H_1 d\mu \cup (\mathcal{G}) \int_A H_2 d\mu) \subseteq (\mathcal{G}) \int_A INF_1(H_1 \cup H_2) d\mu) + X_+$.

Proof. We have

$$g((\mathcal{G}) \int_A INF_1(H_1 \cup H_2) d\mu) = \int_A g(INF_1(H_1 \cup H_2)) d\mu$$
$$= \int_A \inf(g(INF_1(H_1 \cup H_2))) d\mu = \int_A \inf g(H_1 \cup H_2) d\mu$$
$$= \int_A \inf(\inf g \circ H_1, \inf g \circ H_2) d\mu$$
$$\leq \inf(\int_A \inf g \circ H_1 d\mu, \int_A \inf g \circ H_2 d\mu)$$
$$= \inf(g((\mathcal{G}) \int_A H_1 d\mu), g((\mathcal{G}) \int_A H_2 d\mu)).$$

Thus,

$$(\mathcal{G}) \int_A H_1 d\mu \cup (\mathcal{G}) \int_A H_2 d\mu \subset (\mathcal{G}) \int_A INF_1(H_1 \cup H_2) d\mu + X_+$$

which implies $INF_1((\mathcal{G}) \int_A H_1 d\mu \cup (\mathcal{G}) \int_A H_2 d\mu) + \mathrm{Int}\, X_+ \subset (\mathcal{G}) \int_A H_1 d\mu \cup (\mathcal{G}) \int_A H_2 d\mu + \mathrm{Int}\, X_+ \subset (\mathcal{G}) \int_A INF_1(H_1 \cup H_2) d\mu + \mathrm{Int}\, X_+$. By taking the closure we obtain the conclusion. $\diamond$

The last proposition of this section generalizes the classical positivity property of the integral.

Proposition 2. *Let $f : T \to X_+$ be a positive map and let $\mu : \mathcal{C} \to X_+$ be a vector submeasure such that $(\mathcal{G}) \int_A f d\mu = 0$. Then, $f = 0$, μ-a.p.t.*

Proof. We have $0 = \int_A g \circ f dg \circ \mu$ and since $g \circ f \geq 0$ we obtain $g \circ \mu(\{x \mid g \circ f(x) > 0\}) = 0$. In this case, $\{x \mid f(x) > 0\} = \{x \mid g \circ f(x) > 0, \forall g \in \mathcal{G}\}$ which implies $g \circ \mu(\{x \mid f(x) > 0\}) = 0$ for all $g \in \mathcal{G}$ and finally, $\mu(\{x \mid f(x) > 0\}) = 0$. $\diamond$

5. Comparative results

In this section some comparative results on the integral introduced in Definition 13 are established. We firstly present the definition of the Pettis–Lebesgue integral.

Definition 16. [46] Let $A \subset X$, $h : T \to X$ be a map and $\mu : \mathcal{C} \to X_+ \cup \{+\infty\}$ be a vector measure. The *Pettis–Lebesgue integral* is defined to be a subset of X denoted $(PL) \int_A f d\mu$ such that $x^* \circ (PL) \int_A f d\mu = (L) \int_A x^* \circ f dx^* \circ \mu$ for all $x^* \in \mathcal{G} \subset X_+^*$ if this set exists and $\emptyset$ else.

Remark 5. Let $h : T \to X$, $T = \bigcup_{i=\overline{1,n}} E_i$, $E_i \cap E_j = \emptyset$, $h(t) = a_i$, $\forall t \in E_i$, $a_i \in X_+$, $a_1 \leq a_2 \leq \dots \leq a_n$ and $\mu : T \to \mathbb{R}_+$ be a real measure. It is obvious that $(PL) \int_A h d\mu = \sum_{i=\overline{1,n}} a_i \mu(E_i \cap A)$.

Remark 6. If $\mathcal{G} \subset X_+^* \setminus \{0\}$ has property (H_0), then $\tilde{\mathcal{G}} = \{x^*/\|x^*\|, x^* \in \mathcal{G}\}$ has property (H_0). The integral defined with $\tilde{\mathcal{G}}$ will be denoted by $(\tilde{P}S)$, $(\tilde{P}L)$.

In the following we will present some comparison results between the Pettis–Lebesgue and the Pettis–Sugeno integrals.

Theorem 9. *Let* $(X, \|\cdot\|)$ *be a reflexive space,* $\mu : T \to X_+$ *be a measure such that* $\mu(T) < +\infty$, $h : T \to X$ *such that* $\|h\| = \sup_{t \in T} \|h(t)\| \leq 1$. *Then,*

$$\|(\tilde{P}L) \int hd\mu - (\tilde{P}S) \int hd\mu\| \leq \frac{1}{4}.$$

Proof. Following Theorem 3.11 [47], we have $-(\frac{1-\inf x^* \circ h}{2})^2 \leq (S) \int_T x^* \circ hd\mu - (L) \int_T x^* \circ hd\mu \leq (\frac{\sup x^* \circ h}{2})^2 \leq \frac{1}{4}$ for each $x^* \in \tilde{\mathcal{G}}$. This implies that $-\frac{1}{4} \leq x^*((\tilde{P}L) \int_T hd\mu - (\tilde{P}S) \int_T hd\mu) \leq \frac{\|x^*\|\|h\|^2}{2} \leq \frac{1}{4}$ for each $x^* \in \tilde{\mathcal{G}}$. Since convcone$\tilde{\mathcal{G}} = X_+^*$, the inequality does hold for each $x^* \in X_+^*$ with $\|x^*\| = 1$ and the conclusion follows. $\diamond$

Now, following [41] we present necessary and sufficient conditions for the equality between the Pettis–Lebesgue and the Pettis–Sugeno vector integrals.

Theorem 10. *Let* $h : T \to X_+$, $\mu : \mathcal{C} \to \mathbb{R}_+$, $a \in \mathbb{R}_+$, $\mu(T) < +\infty$. *We have* $(\tilde{P}S) \int_T ahd\mu = a(\tilde{P}S) \int_T hd\mu$ *if and only if* $\mu(A) \in \{0, \mu(T)\}$, *for all* $A \subset T$.

Proof. Let suppose that there exists $A \subset T$ such that $\mu(A) = \alpha_0 \notin \{0, \mu(T)\}$. Let $x^* \in \tilde{\mathcal{G}}$ and $e \in X$ such that $x^*(e) = 1$, $h = \frac{1}{2}e\alpha_0\chi_A$ and $a = 1 + \frac{\mu(T)}{\alpha_0}$.

We have $x^* \circ h = f = \frac{1}{2}\alpha_0\chi_A$ and $\mu(f \geq \alpha) = \begin{cases} \mu(T), & \alpha = 0, \\ \alpha_0, & 0 < \alpha \leq \frac{1}{2}\alpha_0, \\ 0, & \alpha > \frac{1}{2}\alpha_0. \end{cases}$

Thus, $\int_T x^* \circ hd\mu = \frac{1}{2}\alpha_0$.

Now, $\mu(af \geq \alpha) = \begin{cases} \mu(T), & \alpha = 0, \\ \alpha_0, & 0 < \alpha \leq \frac{1}{2}(\mu(T) + \alpha_0), \text{ and } \int_T afd\mu = \alpha_0 \\ 0, & \alpha > \frac{1}{2}(\mu(T) + \alpha_0), \end{cases}$

which implies $\int_T ax^* \circ fd\mu \neq a \int_T x^* \circ fd\mu$ and obviously $(PS) \int_T ahd\mu \neq a(PS) \int_T hd\mu$. For the converse, let $a > 0$ and $\beta_0 = \sup\{\alpha \mid \mu(f \geq \alpha) = \mu(T)\}$. We have $\int_T fd\mu = \beta_0$ and $\{af \geq \alpha\} = \begin{cases} \{f \geq \frac{\alpha}{a}\}, & \alpha \leq a, \\ \emptyset, & \alpha > 0, \end{cases}$ and $\sup\{\alpha \mid \mu(af \geq \alpha) = \mu(T)\} = a\beta_0$. Thus, $\int_T afd\mu = a \int_T fd\mu$ which implies $x^* \int_T ahd\mu = x^*(a \int_T hd\mu)$ for all $x^* \in \tilde{\mathcal{G}}$ which implies $\int_T ahd\mu = a \int_T hd\mu$. $\diamond$

Some Nonlinear Integrals of Vector Multifunctions with Respect to a Submeasure 111

Theorem 11. *Let $h : T \to X_+$, $\mu : \mathcal{C} \to \mathbb{R}_+$ a submeasure, $c \in \mathbb{R}_+$, $\mu(T) < +\infty$. We have $(\tilde{PS}) \int_T h + cd\mu = (\tilde{PS}) \int_T hd\mu + c\min\{1, \mu(T)\}$ for all $h : T \to X$ and $c \in X_+$ if and only if $\mu(A) \in \{0, \mu(T)\}$, for all $A \subset T$.*

Proof. If $(\tilde{PS}) \int_T h + cd\mu = (\tilde{PS}) \int_T hd\mu + c\min\{1, \mu(T)\}$ for all $h : T \to X$ and $c \in X_+$, let $h = 0$ and $x^* \in \tilde{\mathcal{G}}$.
We get $x^*((\tilde{PS}) \int_T cd\mu) = x^*(c)\min\{1, \mu(T)\}$ and thus

$$\int_T x^*(c)d\mu = x^*(c)\min\{1, \mu(T)\}.$$

Now, from $(PS) \int_T h + cd\mu = (PS) \int_T hd\mu + c\min\{1, \mu(T)\}$ we deduce $\int_T x^* \circ h + x^*(c)d\mu = \int_T x^* \circ hd\mu + x^*(c)\min\{1, \mu(T)\} = \int_T x^* \circ hd\mu + \int_T x^*(c)d\mu$ and following the corresponding result from the real case we obtain $\mu(A) \in \{0, \mu(T)\}$, for all $A \subset T$.
For the converse, from the real result and the precedent proposition we get for all $x^* \in \tilde{\mathcal{G}}$, $h : T \to X$ and $c \in X_+$, $\int_T x^* \circ h + x^*(c)d\mu = \int_T x^* \circ hd\mu + \int_T x^*(c)d\mu = \int_T x^* \circ hd\mu + x^*(c) \int_T 1d\mu = \int_T x^* \circ hd\mu + x^*(c)\min\{1, \mu(T)\}$ which implies $(\tilde{PS}) \int_T h + cd\mu = (\tilde{PS}) \int_T hd\mu + c\min\{1, \mu(T)\}$ for all $h : T \to X$ and $c \in X_+$. $\diamond$

Theorem 12. *Suppose the submeasure $\mu : \mathcal{C} \to \mathbb{R}_+$ satisfies $\mu(T) < +\infty$. We have $(\tilde{PS}) \int_T au + bvd\mu = a(\tilde{PS}) \int_T ud\mu + b(PS) \int_T vd\mu$ for all $u, v : T \to X_+$ and $a, b \in \mathbb{R}_+$ if and only if μ is finitely additive and $\mu(A) \in \{0, \mu(T)\}$, for all $A \subset T$.*

Proof. The proof follows easily by using the similar result from the real case. $\diamond$

Theorem 13. *Let $h : T \to X_+$, $\mu : \mathcal{C} \to \mathbb{R}_+$ a submeasure with $\mu(T) < +\infty$. We have $(\tilde{PL}) \int_T hd\mu = (\tilde{PS}) \int_T hd\mu$ if and only if μ is finitely additive and $\mu(A) \in \{0, \mu(T)\}$.*

Proof. If $(\tilde{PL}) \int_T hd\mu = (\tilde{PS}) \int_T hd\mu$ then for all $u, v : T \to X_+$ and $a, b \in \mathbb{R}_+$ we have $(\tilde{PS}) \int_T au + bvd\mu = a(\tilde{PS}) \int_T ud\mu + b(PS) \int_T vd\mu$ since the Pettis–Lebesgue integral is a linear integral. Thus, following the precedent theorem we obtain that μ is finitely additive and $\mu(A) \in \{0, \mu(T)\}$. For the converse, if μ is finitely additive and $\mu(A) \in \{0, \mu(T)\}$, from Proposition 3 and Corollary 1 [21] we obtain $(L) \int_T x^* \circ hd\mu = (S) \int_T x^* \circ hd\mu$ for each $x^* \in X_+^*$. Thus, $(\tilde{PL}) \int_T hd\mu = (\tilde{PS}) \int_T hd\mu$. $\diamond$

Using this theorem, since the Pettis–Lebesgue is a linear integral, Theorem 12 may be reformulated for real constants a, b.

Corollary 1. *Let $\mu : \mathcal{C} \to \mathbb{R}_+$ be a submeasure with $\mu(T) < +\infty$. We have $(\tilde{PS}) \int_T au + bvd\mu = a(\tilde{PS}) \int_T ud\mu + b(\tilde{PS}) \int_T vd\mu$ for all $u, v : T \to X_+$ and $a, b \in \mathbb{R}$ if and only if μ is finitely additive and $\mu(A) \in \{0, \mu(T)\}$, for all $A \subset T$.*

6. Aumann–Pettis–Sugeno integral

As we have seen in Section 3, an integral for a real multifunction was defined by Zhang and Guo, the G-fuzzy integral for a multifunction F by using the Sugeno integral for selections maps of F with respect to a fuzzy measure μ.

For this integral, Zhang and Wang proved some properties including the transfer of the convexity and closure property from the multifunction's values to the G-fuzzy integral.

Theorem 14. *Let $F : T \Longrightarrow \mathbb{R}$ be a measurable set-valued function and $\mu : \mathcal{C} \to \mathbb{R}_+$ be a fuzzy measure. Then the following holds:*

- *If F has convex values, then $(G) \int_A F d\mu$ is a convex set.*
- *conv $(G) \int_A F d\mu = (G) \int_A$ conv $F d\mu$.*
- *If F has closed values, then $(G) \int_A F d\mu$ is a closed set.*
- *Let $F(x) = [f(x), g(x)]$ with $f, g : T \to \mathbb{R}$ being measurable functions such that $f \le g$. Then $\int_A F d\mu = [\int_A f d\mu, \int_A g d\mu]$.*

Using the $\mathcal{G}$-integral for $\mathcal{G}$-integrable selections we define the Aumann–Pettis integrals for vector multifunctions relative to a vector submeasure. We denote by $\mathcal{S}_{F,B}$ the set of $\mathcal{G}$-integrable selections of F on B relative to μ, respectively by $\mathcal{S}_{F,B}^{\mathcal{G}}$ the set of integrable selections for $g \circ F$ (with $g \in \mathcal{G}$) on B relative to μ. If $\mathcal{G} = X_+^*$, we denote $\mathcal{S}_{F,B}^{\mathcal{G}}$ with $\mathcal{S}_{F,B}^*$.

Definition 17. Let $\mu : \mathcal{C} \to X_+ \cup \{+\infty\}$ be a vector submeasure and let $F : T \Longrightarrow X_+$ be a multifunction. The *Aumann–Pettis–Sugeno integrals* of F with respect to μ on $B \in \mathcal{C}$ are defined as follows.

1. If there exists $f \in \mathcal{S}_{F,B}^{\mathcal{G}}$ such that $I_{f,B} \ne \emptyset$, the integral will be denoted by $I_{F,B}$ and it is given by

$$I_{F,B} = \{(\mathcal{G}) \int_B f d\mu \mid f \in \mathcal{S}_{F,B}^{\mathcal{G}}\}.$$

Else, $I_{F,B} = \emptyset$.

2. The integral denoted $I'_{F,B}$ is the maximal set which satisfy for each $g \in \mathcal{G}$,

$$g(I'_{F,B}) = (G) \int_B g \circ F dg \circ \mu.$$

Some Nonlinear Integrals of Vector Multifunctions with Respect to a Submeasure 113

If this set does not exist, the integral will be the empty set.

Remark 7. For each $g \in \mathcal{G}$ we have $g(I_{F,B}) \subset g(I'_{F,B})$. If $\mathcal{G} \subset X_+^*$ then $\overline{\text{conv } I_{F,B}} \subseteq \overline{\text{conv } I'_{F,B}}$.

Some properties for these integrals will be presented in the following theorems.

Theorem 15. *Let $M, N \in \mathcal{C}$, $\mu : \mathcal{C} \to X_+ \cup \{+\infty\}$ be a vector submeasure, $\mathcal{G} \subset X_+^*$ and let $F : T \Longrightarrow X_+$ be a multifunction. Then, the following properties holds:*

1. *If $M \subset N$, then $I'_{F,M} \subseteq \overline{\text{conv } I'_{F,N}} - X_+$.*
In addition, if $S_{F,M}^* \cap S_{F,N}^* \neq \emptyset$ then,

$$I_{F,M} \bigcap (I_{F,N} - X_+) \neq \emptyset.$$

2. $\overline{\text{conv } I'_{F,M}} \bigcup \overline{\text{conv } I'_{F,N}} \subseteq \overline{\text{conv } I'_{F,M\cup N}} - X_+.$
If $S_{F,M}^* \cap S_{F,M\cup N}^* \neq \emptyset$ and $S_{F,N}^* \cap S_{F,M\cup N}^* \neq \emptyset$ then,

$$(I_{M,F} \bigcup I_{F,N}) \bigcap (I_{F,M\cup N} - X_+) \neq \emptyset.$$

3. $\overline{\text{conv } I'_{F,M\cap N}} \subseteq \overline{\text{conv } I'_{F,M}} - X_+ \bigcap \overline{\text{conv } I'_{F,N}} - X_+.$
If $S_{F,N}^* \cap S_{F,M}^* \cap S_{F,M\cup N}^* \neq \emptyset$, then

$$I_{F,M\cap N} \subseteq (I_{F,M} - X_+) \bigcap (I_{F,N} - X_+).$$

Proof. The assertions of the theorem follow by using the monotonicity properties for the Pettis–Sugeno integral of the selections and Theorem 12.

$\diamond$

Theorem 16. *Let $\mu : \mathcal{C} \to X_+ \cup \{+\infty\}$ be a vector submeasure, $\mathcal{G} \subset X_+^*$ and $M \in \mathcal{C}$ be an arbitrary set. Then the following properties hold:*
1. *Let $H_1, H_2 : T \Longrightarrow X$ be such that $H_1 \subset H_2$. Then*

$$I_{H_1,M} \subset I_{H_2,M},$$

$$\overline{\text{conv } I'_{H_1,M}} \subset \overline{\text{conv } I'_{H_2,M}}.$$

2. *Let $H : T \Longrightarrow X$ and $F(x) = H(x) + X_+$, for all $x \in X$. Then*

$$I_{H,M} \subset I_{F,M},$$

$$\overline{\text{conv } I'_{H,M}} \subset \overline{\text{conv } I'_{F,M}} \subset \overline{\text{conv } I'_{H,M}} + X_+.$$

3. Let $H : T \Longrightarrow X$ and we suppose that there exists $IMINH(x) = h(x)$ for each $x \in X$. Then $I_{h,M} \in I_{H,M}$ and $I'_{h,M} = I_{h,M} \in \overline{\text{conv } I'_{H,M}}$.

4. Let $H : T \Longrightarrow X$ be a vector multifunction. If $\mu(M) = \{0\}$, then $I_{H,M} = I'_{H,M} = \{0\}$.

5. Let $H : T \Longrightarrow X$ be a vector multifunction. If $0 \in I_{H,M}$ or $0 \in I'_{H,M}$, then $\mu(\{t \in T \mid 0 \notin H(t)\}) = \{0\}$.

Example. Let $H : T \Longrightarrow X$, $H(t) = [a,b]$, where $a, b \in X$, $a < b$ and let $\mu : \mathcal{C} \to X_+ \cup \{+\infty\}$ be a vector submeasure. Then,

$$INF_1(\{a\} \cup \mu(M)) \subset I_{H,M} - X_+ \subset (\{b\} \cup \mu(M)) - X_+,$$

$$\{b\} \cup \mu(M) \subset I_{H,M} + X_+ \subset INF_1(\{a\} \cup \mu(M)) + X_+,$$

$$\overline{\text{conv } (I'_{H,M} - \text{Int } X_+)} = \overline{\text{conv } INF(\{b\} \cup \mu(M))},$$

$$\overline{\text{conv } (I'_{H,M} + \text{Int } X_+)} = \overline{\text{conv } (INF_1(\{a\} \cup \mu(M)) + \text{Int } X_+)}.$$

Indeed, let $f \in \mathcal{S}^*_{H,M}$; we have

$$\int_M x^*(a)dx^* \circ \mu \le \int_M x^* \circ f dx^* \circ \mu \le \int_M x^*(b)dx^* \circ \mu.$$

For each $x^* \in \mathcal{G} \subset X^*_+$ we obtain

$$\sup x^*(I'_{H,M}) = \inf x^*(\{b\} \cup \mu(M)) = \sup x^*(INF(\{b\} \cup \mu(M)))$$

and $\inf x^*(I'_{H,M}) = \inf x^*(\{a\} \cup \mu(M))$ which implies the inclusions for $I'_{H,M}$ from the example. Also, for $f \in \mathcal{S}^*_{H,M}$ we have $\int_M x^*(a)dx^* \circ \mu \le x^*(I_{f,M}) \le \int_M x^*(b)dx^* \circ \mu$ which gives that $\alpha \le I_{f,M} \le u$ for each $\alpha \in INF(\{a\} \cup \mu(M))$ and $u \in \{b\} \cup \mu(M)$. Thus, the inclusion for $I_{H,M}$ from the example follows.

Theorem 17. Let $F : T \Longrightarrow X$ be a vector multifunction and $\mu : \mathcal{C} \to X_+$ be a vector submeasure such that F is $\mathcal{G}$-integrable. Then for each $M \in \mathcal{C}$ we have

$$I_{F,M} \subset (\mathcal{G}) \int_M F d\mu + X_+.$$

In the following, we extend the results concerning the convexity and the closure of the real Aumann–Sugeno integral to the vector case of the Aumann–Pettis–Sugeno integral.

Theorem 18. Let $\mu : \mathcal{C} \to X_+ \cup \{+\infty\}$ be a vector submeasure, $\mathcal{G} \subset X^*_+$, $F : T \Longrightarrow X$ be a multifunction and $M \in \mathcal{C}$.

1. If F has convex values, then $I'_{F,M}$ is a convex set.

Some Nonlinear Integrals of Vector Multifunctions with Respect to a Submeasure 115

2. $\overline{\text{conv } I'_{F,M}} = I'_{\text{conv } F,M}$.

3. If F has weak closed values, then $I'_{F,M}$ is a weak closed set.

Proof. 1. The proposition's hypothesis ensure that $(G) \int_M g \circ F dg \circ \mu$ is a convex set and thus $g(I'_{F,M})$ is a convex set, too. If $u, v \in I'_{F,M}$, then $g(\lambda u + (1-\lambda)v) = \lambda g(u) + (1-\lambda)g(v) \in (G) \int_M g \circ F dg \circ \mu = g(I'_{F,M})$. Since $I'_{F,M}$ is a maximal set, we obtain that $\lambda u + (1-\lambda)v \in I'_{F,M}$ for each $\lambda \in [0,1]$, which implies that $I'_{F,M}$ is a convex set.

2. Using the corresponding property for the real case, we get $x^*(\text{conv } I'_{F,M}) = x^*(I'_{\text{conv } F,M})$ and by Theorem 12 we obtain the conclusion. $\diamond$

Theorem 19. Let $\mu : \mathcal{C} \to X_+ \cup \{+\infty\}$ be a vector submeasure and $F(t) = \text{conv } \{f(t), h(t)\}$, for all $t \in T$, where $f, h : T \to X$ are $\mathcal{G}$-integrable maps, $\mathcal{G} \subset X^*_+$. Then $I'_{F,M} = \text{conv } \{(\mathcal{G}) \int_M f d\mu, (\mathcal{G}) \int_M f d\mu\}$ for each $M \in \mathcal{C}$.

Proof. Since $F(t) = \text{conv } \{f(t), g(t)\}$, then $g \circ F(t) = \text{conv } \{g \circ f(t), g \circ h(t)\}$ which implies that $g(I'_{F,M}) = g(\text{conv } \{I_{f,M}, I_{g,M}\})$ for each $g \in \mathcal{G} \subset X^*_+$. Using Theorem 12 and the precedent theorem we obtain the conclusion. $\diamond$

Example. Let $F(t) = [a, b]$ for each $t \in T$ and let $\mu : \mathcal{C} \to X_+ \cup \{+\infty\}$ be a vector submeasure. It results for every $M \in \mathcal{C}$ that

$$I'_{F,M} = (\text{conv } \{a, b, \mu(M)\} + X_+) \bigcap (\text{conv } \{a, b\} - X_+) \bigcap (\mu(M) - X_+).$$

7. Estimation results

In this section, some estimations between the Aumann-Lebesgue and Aumann-Sugeno integrals, respectively the Aumann–Pettis–Sugeno and the Aumann–Pettis–Lebesgue integrals are established. We begin with some definitions.

Definition 18.
1. Let $F : T \rightrightarrows \mathbb{R}$ be a real multifunction and $\mu : \mathcal{C} \to \mathbb{R}_+$ be a measure. The *Aumann-Lebesgue integral* of F on $M \in \mathcal{C}$, denoted $(AL) \int_M F d\mu$, is the set defined by

$$(AL) \int_M F d\mu = \{(L) \int_M f d\mu \mid f \in (L)\mathcal{S}_{F,M}\},$$

where $(L) \int_M f d\mu$ means the Lebesgue integral of f by respect to μ and $(L)\mathcal{S}_{F,M}$ is the set of Lebesgue integrable selections of F on M.

2. Let $f : T \to X$ be a vector function and $\mu : \mathcal{C} \to X_+$ be a $\star$-measure (i.e. $x^* \circ \mu$ is a Lebesgue measure for each $x^* \in X^*$). The *Pettis–Lebesgue integral* of f on $M \in \mathcal{C}$ is the set denoted $(PL) \int_M f d\mu$, which satisfies $x^*((PL) \int_M f d\mu) = (L) \int_M x^* \circ f dx^* \circ \mu$, for each $x^* \in X_+^*$.

3. Let $F : T \rightrightarrows X$ be a vector multifunction and $\mu : \mathcal{C} \to X_+$ be a $\star$-measure. The *Aumann–Pettis–Lebesgue integral* of F on $M \in \mathcal{C}$ denoted $(APL) \int_M F d\mu$, is the set given by

$$(APL) \int_M F d\mu = \{(PL) \int_M f d\mu \mid f \in (L)\mathcal{S}_{F,M}^*\},$$

where $(L)\mathcal{S}_{F,M}^*$ is the set of Pettis–Lebesgue integrable selections of F.

For the next theorem, h is the *Hausdorff–Pompeiu metric*, i.e. for every $A \subset X$, $B \subset X$, $h(A,B) = \max\{e(A,B), e(A,B)\}$ where $e(A,B) = \sup_{x \in A} d(x,B)$ is the excess of A over B and $d(x,B) = \inf_{y \in B} \|x - y\|$ is the distance from x to B.

Theorem 20.

1. *Let $F : T \rightrightarrows \mathbb{R}$ be a real multifunction and $\mu : \mathcal{C} \to \mathbb{R}_+$ be a measure such that $\sup \mu(T) < +\infty$ and $\sup_{t \in T} |f(t)| < 1$ for each $f \in (L)\mathcal{S}_{F,M} \cap \mathcal{S}_{F,M} \neq \emptyset$. Then*

$$h((AL) \int_M F d\mu, (AS) \int_M F d\mu) \leq \frac{1}{4}.$$

2. *Suppose that $(X, \|\cdot\|)$ is a normed space, $F : T \rightrightarrows X$ is a vector multifunction, $\mu : \mathcal{C} \to \mathbb{R}_+$ is a measure such that $\sup \mu(T) < +\infty$ and suppose that $\sup_{t \in T} \|f(t)\| < 1$ for each $f \in (L)\mathcal{S}_{F,M}^* \cap \mathcal{S}_{F,M}^* \neq \emptyset$. Then,*

$$h((APL) \int_M F d\mu, (APS) \int_M F d\mu) \leq \frac{1}{4}.$$

Proof. The proof follows from the corresponding results of the real maps, Theorem 3.1 [36] and Proposition 5.3 [35] for the vector case. $\diamond$

8. Convergence Results

In this section we recall some convergence results for the strong and weak integrals which will be useful for obtaining a convergence result for the $\mathcal{G}$-integral.

Theorem 21. [12] (Monotone Convergence). *Suppose $\mu : \mathcal{C} \to [0, +\infty)$ is a continuous fuzzy measure. Let $F_n : T \rightrightarrows \mathbb{R}_+$ be a measurable multifunction with compact values, for every $n \in \mathbb{N}^*$, such that $F_n(t) \supseteq F_{n+1}(t)$, for*

Some Nonlinear Integrals of Vector Multifunctions with Respect to a Submeasure 117

all $t \in T, n \in \mathbb{N}^*$, and let $F(t) = \lim\limits_{n \to \infty} F_n(t) = \bigcap\limits_{n=1}^{\infty} F_n(t), \forall t \in T$. Then

$$\lim\limits_{n \to \infty} (s) \int_A F_n d\mu = (s) \int_A F d\mu, \forall A \in \mathcal{C}.$$

Theorem 22. [11] (Monotone Convergence) *Let "$\leq$" be an order relation on $\mathcal{P}_0(X)$ so that the set-norm $|\cdot|$ is monotone on $(\mathcal{P}_0(X), \leq)$.*

I. *Suppose μ is a continuous fuzzy measure and let $F_n : T \Longrightarrow X$ be measurable multifunctions such that $F_n(t) \geq F_{n+1}(t)$, for all $t \in T$, $n \in \mathbb{N}$, $F_n \xrightarrow{h} F$ (i.e. $\lim\limits_{n \to \infty} h(F_n(t), F(t)) = 0$, for all $t \in T$), F is measurable and there exists $n_0 \in \mathbb{N}$ so that $\mu(\{t; |F_{n_0}(t)| > \int_T |F| d\mu\}) < +\infty$. Then*

$$\lim\limits_{n \to \infty} (w) \int_T F_n d\mu = (w) \int_T F d\mu.$$

II. *Suppose μ is continuous from below. Let $F_n : T \Longrightarrow X$ be measurable multifunctions such that $F_n(t) \leq F_{n+1}(t)$, for all $t \in T$, $n \in \mathbb{N}$, $F_n \xrightarrow{h} F$ and F is measurable. Then*

$$(w) \int_A F d\mu = \lim\limits_{n \to \infty} (w) \int_A F_n d\mu, \quad \text{for every } A \in \mathcal{C}.$$

Theorem 23. [11] (Uniform Convergence) *Let $F, F_n : T \Longrightarrow X$, $n \in \mathbb{N}$, be measurable multifunctions such that $F_n \xrightarrow{h} F$ uniformly on T. Then*

$$\lim\limits_{n \to \infty} (w) \int_T F_n d\mu = (w) \int_T F d\mu.$$

In the sequel, we recall some convergence results for the vector integral presented in [36]. Let recall that $\liminf A_n = \bigcap\limits_{n=1}^{\infty} \bigcup\limits_{k=n}^{\infty} A_k$ and $\limsup A_n = \bigcup\limits_{n=1}^{\infty} \bigcap\limits_{k=n}^{\infty} A_k$, for $\{A_n\}_{n \in \mathbb{N}^*} \subset \mathcal{C}$. Let $F_n, F : T \Longrightarrow X$ be multifunctions, $n \in \mathbb{N}^*$. We say that $\liminf(\text{SUP } F_n) = F$ if for each $E \subseteq T$, $\liminf(\text{SUP } F_n(E)) = F(E)$. A multifunction F has X_+-closed values if for each E, $F(E) + X_+$ is a closed set.

The following proposition will be useful for the next theorems.

Theorem 24. *Let $A_n^i \subseteq X$, for every $n \in \mathbb{N}, i \in I$. Then:*
1. $\liminf \bigcap\limits_{i \in I} A_n^i = \bigcap\limits_{i \in I} \liminf A_n^i.$
2. $\limsup \bigcap\limits_{i \in I} A_n^i \supset \bigcap\limits_{i \in I} \limsup A_n^i.$

Theorem 25. *Let F_n, F be multifunctions with X_+-closed values such that* $\liminf F_n = F$ *and μ is a multisubmeasure with X_+-closed values. Then* $\liminf \int_A F_n d\mu = \int_A F d\mu$, $\forall A \in \mathcal{C}$.

Similarly, we obtain the result concerning the $\limsup$ convergence.

Theorem 26. *Let F_n, F having X_+-closed values and $h(F_n, F) \to 0$. Then* $h(\int_A F_n d\mu, \int_A F d\mu) \to 0$ *for all $A \in \mathcal{C}$.*

The following theorem presents a result concerning the convergence of the scalar integral for vector functions introduced in Definition 6.

Theorem 27. *Let be the functions $f, f_n : T \to X$, $n \in \mathbb{N}^*$ such that $(f_n)_n$ is convergent in fuzzy measure to f. Then, $(\mathcal{G}I_{f_n, A} d\mu)_n$ is g-convergent to $\mathcal{G} \int_A f_n d\mu$ (i.e. $g(\mathcal{G} \int_A f_n d\mu)$ is convergent to $g(\mathcal{G} \int_A f_n d\mu)$ for every $g \in \mathcal{G}$). If X is a locally convex space, then $(\int_A f_n)_n$ is weakly convergent to $\int_A f_n$.*

9. Conclusions

We introduced and studied a general Pettis–Sugeno type integral of vector multifunctions relative to multisubmeasures taking values in a locally convex space X ordered by a closed convex pointed cone with nonempty interior. This definition extends the Pettis–Sugeno integral by using an arbitrary set of positive monotone real maps g with $g(0) = 0$ instead of usual X_+^*. Some classic properties of this integral and comparisons with other integrals (such as Pettis–Lebesgue, Aumann–Sugeno, Pettis–Choquet) are established. Our next research concerns the properties of the $\mathcal{G}$-Choquet integral and the relationships with other set-valued integrals of Birkhoff, Debreu, Dunford, Gould type.

References

[1] H. Agahi, R. Mesiar, *On Pettis–Choquet expectation of Banach-valued functions: a counterexample*, International Journal of Uncertinity, Fuzziness and Knowledge-Based Systems 26.02 (2018), 255-259.

[2] R.J. Aumann, *Integrals of set-valued maps*, J. Math. Anal. Appl. 12 (1965), 1-12.

[3] A. Boccuto, D. Candeloro, A.R. Sambucini, *Vitali-type theorems for filter convergence related to vector lattice-valued modulars and applications to stochastic processes*, J. Math. Anal. Appl. 419 (2014), 818-838.

[4] T. Calvo, M. Komornikova, A. Kolesarova, *Aggregation operators. New trends and applications*, (2002).

[5] D. Candeloro, L. Di Piazza, K. Musial, A.R. Sambucini, *Some new results on integration for multifunction*, Ric. Mat. (2018), 67, 361-372.

[6] D. Candeloro, A. Croitoru, A. Gavriluţ, A.R. Sambucini, *An extension of the Birkhoff integrability for multifunctions*, Mediterranean J. Math. 13, Issue 5 (2016), 2551-2575. DOI:10.1007/s0009-015-0639-7.

[7] D. Candeloro, A. Croitoru, A. Gavriluţ, A. Iosif, A.R. Sambucini, *Properties of the Riemann-Lebesgue integrability in the non-additive case*, Rendiconti del Circolo Matematico di Palermo Series 2 (2019), 1-13. https://doi.org/10.1007/s12215-019-00419-y.

[8] B. Cascales, V. Kadels, J. Rodriguez, *The Pettis integral for multivalued functions via single-valued ones*, J. Math. Anal. Appl. 332 (2007), 1-10.

[9] C. Castaing, M. Valadier, *Convex Analysis and Measurable Multifunctions*, Springer-Verlag, Lecture Notes in Math. 580, (1977).

[10] G. Choquet, *Theory of capacities*, Annales de l'Institut Fourier 5, (1953), 131-295.

[11] A. Croitoru, *Fuzzy integral of measurable multifunctions*, Iranian Journal of Fuzzy Systems 9 (2012), 133-140.

[12] A. Croitoru, *Strong integral of multifunctions relative to a fuzzy measure*, Fuzzy Sets and Systems 244 (2014), 20-33.

[13] A. Croitoru, *An integral for multifunctions with respect to a multimeasure*, An. Şt. Univ. "Al. I. Cuza" Iaşi 49 (2003), 95-106.

[14] A. Croitoru, C. Godet-Thobie, *Set-Valued Integration in Seminorm I* An. Sci. Univ. Ovidius Constanta, 13, 55-66 (2005).

[15] A. Croitoru, C. Godet-Thobie, *Set-Valued Integration in Seminorm II* Annals of University of Craiova, Math. Comp. Sci. Ser., 33, 16-25 (2006).

[16] L. Di Piazza, K. Musial, *Set-valued Kurzweil-Henstock-Pettis integral*, Set-valued Analysis 13 (2) (2005), 167-179.

[17] L. Di Piazza, V. Marraffa, *Pettis integrability of fuzzy mappings with values in arbitrary Banach spaces*, Mathematica Slovaca 67 (6) (2017), 1359-1370.

[18] L. Drewnowski, *Topological rings of sets, continuous set functions. Integration,* I, II, III, Bull. Acad. Polon. Sci. Sér. Math. Astron. Phys. 20 (1972), 269-286.

[19] A. Gavriluţ, *The general Gould type integral with respect to a multisubmeasure*, Math. Slovaca 58 (2008), 43-62.

[20] L.C. Jang, B.M. Kil, Y.K. Kim, J.S. Kwan, *Some properties of Choquet integrals of set-valued function*, Fuzzy Sets and Systems 91 (1997), 95-98.

[21] E.P. Klement, D. Ralescu, *Nonlinearity of the Fuzzy Integral*, Fuzzy Sets and Systems 11 (1983), pp. 309-315.

[22] W.L. Liu, X.Q. Song, S.B. Zhang, *(T)-Fuzzy integral of multi-dimensional function with respect to multi-valued measure*, Iranian Journal of Fuzzy Systems, vol.9, No. 3 (2012), 111-126.

[23] R. Mesiar, B. De Baets, *New construction methods for agregation operators*, Proceedings of the IPMU'2000, Madrid, (2000), pp. 701-707.

[24] Y. Narukawa, V. Torra, *Multidimensional generalized fuzzy integral*, Fuzzy Sets and Systems 169 (2009), 802-815.

[25] E. Pap, *Pseudo-additive measures and their applications*, In: Pap E. (ed) Handbook of Measure Theory, vol.II, Elsevier, (2002), pp. 1403-1465.

[26] E. Pap, *An integral generated by decomposable measure* Univ. u Novom Sadu Zh. Rad. Prirod.-Mat. Fak. Ser. Mat. 20 (1990) 135-144.

[27] E. Pap, A. Iosif, A. Gavriluţ, *Integrability of an interval-valued multifunction with respect to an interval-valued set multifunction*, Iranian Journal of Fuzzy Systems 15, No. 3 (2018), 47-63.

[28] M. Potyrala, *Some remarks about Birkhoff and Riemann–Lebesgue integrability of vector valued functions*, Tatra Mt. Math. Publ. 35 (2007), 97-106.

[29] C.K. Park, *Set-valued Choquet-Pettis integrals*, Korean J. Math., 20 (2012), No. 4, 381-393.

[30] B. Pettis, *On integration in vector spaces* Trans. Amer. Math. Soc. 57(1940), 277-304.

[31] V. Postolică, *Vectorial optimization programs with multifunctions and duality*, Ann. Sci. Math. Quebec 10 (1986), No. 1, 85-102.

[32] A.M. Precupanu, A. Gavrilut, A. Croitoru, *A fuzzy Gould type integral*, Fuzzy Sets and Systems 161 (2010), 661-680.

[33] A.R. Sambucini *A survey on multivalued integration*, Atti Semis. Mat. Fis. Univ. Modena 50 (2002), 53-63.

[34] A.R. Sambucini, *The Choquet integral with respect to fuzzy measures and applications*, Math. Slovaca 67 (2017), No. 6, 1427-1450. DOI: 10.1515/ms-2017-0049.

[35] C. Stamate, *A general Pettis–Choquet type integral*, Zilele Academice Ieşene, 2019.

[36] C. Stamate, *Vector fuzzy integral*, Recent Advances in Neural Network, Fuzzy Systsmes and Evolutionary Computing, 2010, 221-224.

[37] C. Stamate, *About fuzzy integrals for vector valued multifunctions*, Recent Researchers in Neural Networks, Fuzzy Systems, Evolutionary Computing and Automation, 2011, 110-113.

[38] C. Stamate, *About the weak efficiencies in vector optimization problems*, Recent Researches in Computational Techniques, Nonlinear Systems Control, Proceedings of NOLASC'11, Iaşi, Romania, July 1-3. 2011.

[39] C. Stamate, A. Croitoru, *Nonlinear integrals, properties and relationships*, Recent Advances in Telecommunications, Signals and Systems (Proceedings of NOLASC 13), WSEAS Press, 2013, pp. 118-123.

[40] C. Stamate, A. Croitoru, *Aumann–Pettis–Sugeno integral for vector multifunctions relative to vector multisubmeasures*, Fuzzy sets and Systems, 444 (2022), 172-181.

[41] C. Stamate, A. Croitoru, *The general Pettis–Sugeno integral for vector multifunctions relative to vector multisubmeasures*, Fuzzy sets and Systems, 327 (2017), 123-136.

[42] C. Stamate, A. Croitoru, $\mathcal{G}$-*Choquet integral of vector multifunctions relative to vector multisubmeasures*, submitted for publication.

[43] M. Sugeno, *Theory of fuzzy integrals and its applications*, Ph.D. Thesis, Tokyo Institute of Technology, (1971).

[44] D. Zhang, Z. Wang, *Fuzzy integrals of set-valued mappings and fuzzy mappings*, Fuzzy sets and Systems, 75 (1995), 103-109.

[45] D. Zhang, C. Guo, C., *Generalized fuzzy integrals of set-valued functions*, Fuzzy Sets abd Systems 76, (1995), pp. 365-373.

[46] D. Zhang, C. Guo, D. Liu, *Set-valued Choquet integrals revisited* Fuzzy Sets and Systems, 147 (2004), 475-485.

[47] Z. Wang, G.J. Klir, *Fuzzy Measure Theory*, Plenum Press, New York, (1992).

[48] C. Wu, T. Mamadou, *An extension of Sugeno integral*, Fuzzy Sets and Systems 138 (2003), 537-550.

© 2025 World Scientific Publishing Company
https://doi.org/10.1142/9789819812202_0006

Chapter 6

Convergence of Riemann Integrable Functions over Banach Spaces on Time Scales

Hemen Bharali

Department of Mathematics, Assam Don Bosco University, Sonapur, Assam 782402, India
Email: hemen.bharali@dbuniversity.ac.in

Vikuozonuo Sekhose

Department of Mathematics, Assam Don Bosco University, Sonapur, Assam 782402, India
Email: vikuosekhose4@gmail.com

Hemanta Kalita

Mathematics Division, VIT Bhopal University, Kothrikalan, Bhopal-Indore Highway, Madhya Pradesh-466114, India
Email: hemanta30kalita@gmail.com

In this chapter, we study the convergence of sequences and series of Riemann integrable functions over Banach spaces on time scales. We discuss several classical convergence theorems in the context of Riemann integrable functions over Banach spaces on time scales.

We formulate a few results concerning the notion of uniform convergence for sequences and series of Riemann integrable functions over Banach spaces on time scales.

Keywords and Phrases: Riemann Δ-integral, Riemann ∇-integral, time scales, Banach space.

AMS subject classification (2020): 26A42, 40A10, 40A30, 46B25, 46G10.

1. Introduction

German mathematician Stefan Hilger in his Ph.D. thesis, which was later published as [13], "Ein Maßkettenkalkül mit Anwendung auf Zentrumsmannigfaltigkeiten", 1988, supervised by Bernd Aulbach, introduced the theory of time scale calculus.

As can be observed from his work, Hilger's primary motivation was the analogy between discrete and continuous analysis and the aim to unify them. A direct quote from him [13], "*These analogies which are described in the relevant literature lead to the idea to develop some higher ranging calculus which in special cases covers those two concepts*", beautifully portrays this objective.

Hilger, in [13], states a system of axioms to develop the foundational framework of the theory of time scale, and introduces the delta derivative, and also a descriptive sense of the integral (named the Cauchy integral). More than a decade after the so-called delta derivative was formulated, another derivative called the nabla derivative was introduced by Atici and Guseinov [2], which was previously hinted in the works of Calvin and Bohner [1], who introduced a so-called alpha derivative which consisted both the delta and nabla derivative as special cases.

For an excellent introduction to this subject with theoretical developmental summary and rich history, the reader is referred to the following [5, 6, 13–15].

In literature, a number of integration concepts are treated in their constructive sense, including the Riemann integral for Banach space-valued functions on time scales, formulated by B. Aulbach and L. Neidhart [3].

In this paper, we discuss the convergence of sequences and series of Riemann integrable functions over Banach spaces on time scales.

We formulate a few results concerning the notion of uniform convergence for sequences and series of Riemann integrable functions over Banach spaces on time scales.

2. Preliminaries

In this section, we recall a few definitions, results and notations on the classical Banach space theory and on the theory of time scale calculus, which will be used throughout the paper.

Definition 1. [16] Metric Space: A metric space is a set A along with a metric or distance function d : $A \times A \to \mathbb{R}$ such that the following three conditions are satisfied, $\forall\ x, y, z \in A$:

(1) $d(x, y) \geq 0$, and $d(x, y) = 0$ if and only if $x = y$;
(2) $d(x, y) = d(y, x)$;
(3) $d(x, z) \leq d(x, y) + d(y, z)$ (the triangle inequality).

Definition 2. [16] Normed Space: Let $\mathfrak{Y}$ be a vector space. A norm on $\mathfrak{Y}$ is a real-valued function $\|\cdot\|$ on $\mathfrak{Y}$ such that the following conditions are satisfied by all members x and y of $\mathfrak{Y}$ and each scalar α:

(1) $\|x\| \geq 0$, and $\|x\| = 0$ if and only if $x = 0$;

(2) $\|\alpha x\| = |\alpha| \|x\|$;

(3) $\|x + y\| \leq \|x\| + \|y\|$ (the triangle inequality).

The ordered pair $(\mathfrak{Y}, \|\cdot\|)$ is called a normed space/normed vector space/normed linear space.

Notation: $\mathbb{K}$ will denote the fields $\mathbb{R}$ or $\mathbb{C}$.

Definition 3. [16] Banach Space: A Banach space $\mathfrak{X}$ is a vector space over $\mathbb{K}$ equipped with the norm $\|\cdot\|$ such that $\mathfrak{X}$ is a complete metric space with respect to the metric

$$\mathrm{d}(x_1, x_2) = \|x_1 - x_2\|, \quad x_1, x_2 \in \mathfrak{X}.$$

Throughout this paper $\mathfrak{X}$ will denote a Banach space.

For more insight on the classical Banach space theory one may refer [16].

We now provide some definitions and results on the theory of time scales (one may refer [6, 13] for more insight).

A time scale $\mathbf{T}$ is any non-empty closed subset of $\mathbb{R}$.

Definition 4. [13] Forward Jump Operator: The forward jump operator denoted by σ is a mapping, $\sigma : \mathbf{T} \to \mathbf{T}$ defined by $\sigma(t) = \inf \{r \in \mathbf{T} : r > t\}$.

Definition 5. [13] Backward Jump Operator: The backward jump operator denoted by ρ is a mapping, $\rho : \mathbf{T} \to \mathbf{T}$ defined by $\rho(t) = \sup \{r \in \mathbf{T} : r < t\}$.

Assuming $p \leq q$, intervals in $\mathbf{T}$ are defined as [5]:

$[p, q]_{\mathbf{T}} = \{t \in \mathbf{T} : p \leq t \leq q\}$; $(p, q)_{\mathbf{T}} = \{t \in \mathbf{T} : p < t < q\}$;

$[p, q)_{\mathbf{T}} = \{t \in \mathbf{T} : p \leq t < q\}$; $(p, q]_{\mathbf{T}} = \{t \in \mathbf{T} : p < t \leq q\}$.

Let $[p, q]_{\mathbf{T}}$ be a closed interval on $\mathbf{T}$ such that $p < q$. Let $\mathfrak{P}$ be the collection of all possible partitions of $[p, q]_{\mathbf{T}}$.

Before proceeding we first establish a few preliminary information required for the definition. For the sake of clarity, $\mathcal{V} \in \mathfrak{P}$ will denote the partition for the Δ-integral and $\mathcal{W} \in \mathfrak{P}$ will denote the partition for the ∇-integral.

Let $\mathcal{V} \in \mathfrak{P}$, $\mathcal{V} = \{p = t_0 < t_1 < \ldots < t_n = q\}$, with t_0, t_1, ..., t_n being the finite points of division. We consider subintervals of the form $[t_{h-1}, t_h)_{\mathbf{T}}$, for $1 \leq h \leq n$, and from each subinterval we choose ϑ_h arbitrarily, defined as $\vartheta_h \in [t_{h-1}, t_h)_{\mathbf{T}}$, and call it the tag point of the respective subinterval. For $\mathcal{V} \in \mathfrak{P}$, we define a point-interval collection as $\check{\mathcal{V}} = \{(\vartheta_h, [t_{h-1}, t_h)_{\mathbf{T}})\}_{h=1}^{n}$, and call it the tagged partition. We define the mesh of $\mathcal{V}$ as, mesh-$(\mathcal{V}) = \max_{1 \leq h \leq n}(t_h - t_{h-1}) > 0$. For some $\delta > 0$, $\mathcal{V}_\delta$ will represent a partition of $[p, q]_{\mathbf{T}}$ with mesh δ satisfying the property: For each $h = 1, 2, \ldots, n$ we have either- $(t_h - t_{h-1}) \leq \delta$ or $(t_h - t_{h-1}) > \delta \wedge \rho(t_h) = t_{h-1}$. Hence, $\check{\mathcal{V}}_\delta$ will mean a tagged partition with mesh δ satisfying the above property.

We proceed to give the definition of Riemann Δ-integral according to Aulbach and Neidhart [3] for Banach space-valued functions.

We form Riemann Δ-sum, $(\overline{R})(f; \check{\mathcal{V}}_\delta)$, of the Banach valued function, f, evaluated at the tags as,

$$(\overline{R})(f; \check{\mathcal{V}}_\delta) = \sum_{h=1}^{n} (t_h - t_{h-1}) \cdot f(\vartheta_h).$$

Definition 6. [3] A bounded function $f : [p, q]_{\mathbf{T}} \to \mathfrak{X}$ is Riemann Δ-integrable if there exists a number $\overline{I} \in \mathfrak{X}$ such that, for any $\epsilon > 0$ there exists a $\delta > 0$ such that for any tagged partition $\check{\mathcal{V}}_\delta$ we have $\left\| (\overline{R})(f; \check{\mathcal{V}}_\delta) - \overline{I} \right\| < \epsilon$. Here $\overline{I} = (\overline{R}) \int_p^q f(t) \Delta t$ and $(\overline{R})(f; \check{\mathcal{V}}_\delta) = \sum_{h=1}^{n} (t_h - t_{h-1}) \cdot f(\vartheta_h)$.

The set of all Riemann Δ-Banach space-valued integrable function on $[p, q]_{\mathbf{T}}$ will be denoted by $(\mathfrak{R})_\Delta ([p, q]_{\mathbf{T}}, \mathfrak{X})$.

Now let $\mathcal{W} \in \mathfrak{P}$, $\mathcal{W} = \{p = t_0 < t_1 < \ldots < t_n = q\}$, with t_0, t_1, ..., t_n being the finite points of division. We consider subintervals of the form $(t_{h-1}, t_h]_{\mathbf{T}}$, for $1 \leq h \leq n$, and from each subinterval we choose ξ_h arbitrarily, defined as $\xi_h \in (t_{h-1}, t_h]_{\mathbf{T}}$, and call it the tag point of the respective subinterval. For $\mathcal{W} \in \mathfrak{P}$, we define a point-interval collection as $\check{\mathcal{W}} = \{(\xi_h, (t_{h-1}, t_h]_{\mathbf{T}})\}_{h=1}^{n}$, and call it the tagged partition. We define the mesh of $\mathcal{W}$ as, mesh-$(\mathcal{W}) = |\mathcal{W}| = \max_{1 \leq h \leq n}(t_h - t_{h-1}) > 0$. For some $\delta > 0$, $\mathcal{W}_\delta$ will represent a partition of $[p, q]_{\mathbf{T}}$ with mesh δ satisfying the property: For each $h = 1, 2, \ldots, n$ we have either- $(t_h - t_{h-1}) \leq \delta$ or $(t_h - t_{h-1}) > \delta \wedge \sigma(t_{h-1}) = t_h$. Hence, $\check{\mathcal{W}}_\delta$ will mean a tagged partition with mesh δ satisfying the above property.

Convergence of Riemann Integrable Functions over Banach Spaces

We proceed to give the definition of Riemann ∇-integral for Banach space-valued functions.

We form Riemann ∇-sum, $(\underline{R})(f; \breve{W}_\delta)$, of the Banach valued function, f, evaluated at the tags as,

$$(\underline{R})(f; \breve{W}_\delta) = \sum_{h=1}^{n} (t_h - t_{h-1}) \cdot f(\xi_h).$$

Definition 7. [19] A bounded function $f : [p, q]_\mathbf{T} \to \mathfrak{X}$ is Riemann ∇-integrable if there exists a number $\underline{I} \in \mathfrak{X}$ such that, for any $\epsilon > 0$ there exists a $\delta > 0$ such that for any tagged partition $\breve{W}_\delta$ we have $\left\| (\underline{R})(f; \breve{W}_\delta) - \underline{I} \right\| < \epsilon$. Here $\underline{I} = (\underline{R}) \int_p^q f(t) \nabla t$ and $(\underline{R})(f; \breve{W}_\delta) = \sum_{h=1}^{n} (t_h - t_{h-1}) \cdot f(\xi_h)$.

The set of all Riemann ∇-Banach space-valued integrable function on $[p, q]_\mathbf{T}$ will be denoted by $(\mathfrak{R})_\nabla ([p, q]_\mathbf{T}, \mathfrak{X})$.

Below we recall a few definitions on sequences and series in the classical sense.

Definition 8. [4] Sequence of functions: Let A be a set and suppose that for each $r \in \mathbb{N}$ there is a function $f_r : A \to \mathfrak{X}$, then we say $\{f_r\}$ is a sequence of functions on A to $\mathfrak{X}$.

Definition 9. [4] Pointwise Convergence of sequences: Let $\{f_r\}$ be a sequence of functions on A to $\mathfrak{X}$ and let $f : A \to \mathfrak{X}$. We say that the sequence $\{f_r\}$ converges pointwise on A to a function f, if for every $\epsilon > 0$, there exists $N \in \mathbb{N}$ such that $\left\| f_r(z) - f(z) \right\| < \epsilon$ for each $z \in A$ and all $r \geq N$.

Definition 10. [4] Uniform Convergence of sequences: Let $\{f_r\}$ be a sequence of functions on A to $\mathfrak{X}$ and let $f : A \to \mathfrak{X}$. We say that the sequence $\{f_r\}$ converges uniformly on A to a function f, if for every $\epsilon > 0$, there exists $N \in \mathbb{N}$ such that $\left\| f_r(z) - f(z) \right\| < \epsilon$ for all $z \in A$ and all $r \geq N$.

Definition 11. [4] Cauchy sequence: Let $\{f_r\}$ be a sequence of functions on A to $\mathfrak{X}$, then $\{f_r\}$ is said to be Cauchy if for every $\epsilon > 0$, there exists $N \in \mathbb{N}$ such that $\left\| f_r(z) - f_m(z) \right\| < \epsilon$ for all $z \in A$ and all $r, m \geq N$.

Definition 12. [4] Contractive sequence: Let $\{f_r\}$ be a sequence of functions on A to $\mathfrak{X}$, then $\{f_r\}$ is said to be contactive if there exists a constant C, $0 < C < 1$, such that $\left\| f_{r+2}(z) - f_{r+1}(z) \right\| \leq C \left\| f_{r+1}(z) - f_r(z) \right\|$ for all $z \in A$ and all $r \in \mathbb{N}$. The number C is called the constant of the contractive sequence.

Definition 13. [4] Pointwise Bounded: A sequence $\{f_r\}$ is said to be pointwise bounded on A if for every $z \in A$ there exists a finite-valued function ϕ defined on A such that $\|f_r(z)\| < \phi(z)$.

Definition 14. [4] Uniform Bounded: A sequence $\{f_r\}$ is said to be uniformly bounded on A if for all $z \in A$ there exists a number D such that $\|f_r(z)\| < D$.

Definition 15. [4] Series of functions: If $\{f_r\}$ is a sequence of functions defined on a set A with values in $\mathfrak{X}$, the sequence of partial sums $\{s_r\}$ of the series $\sum f_r$ is defined for z in A by

$$s_1(z) := f_1(z)$$

$$s_2(z) := s_1(z) + f_2(z)$$

$$\cdots$$

$$s_{r+1}(z) := s_r(z) + f_{r+1}(z)$$

$$\cdots\cdot$$

In case the sequence $\{s_r\}$ of functions converges on A to a function f, we say that the series of functions $\sum f_r$ converges to f on A. We will often write

$$\sum f_r \ \ or \ \ \sum_{r=1}^{\infty} f_r$$

to denote either the series of the limit of the function, when it exists.

Definition 16. [4] Cauchy series: If $\{f_r\}$ is a sequence of functions defined on a set A with values in $\mathfrak{X}$, the sequence of partial sums $\{s_r\}$ of the series $\sum f_r$ is said to be Cauchy if for every $\epsilon > 0$, there exists $N \in \mathbb{N}$ such that $\|s_r(z) - s_m(z)\| < \epsilon$ for all $z \in A$ and all $r, m \geq N$.

Definition 17. [4] Pointwise Convergence of series: The series $\sum f_r$ converges pointwise on A to f if the sequence $\{s_r\}$ of partial sums, defined by $\sum_{i=1}^{r} f_i(z) = s_r(z)$, converges pointwisely on A to f.

Definition 18. [4] Uniform Convergence of series: The series $\sum f_r$ converges uniformly on A to f if the sequence $\{s_r\}$ of partial sums, defined by $\sum_{i=1}^{r} f_i(z) = s_r(z)$, converges uniformly on A to f.

For more insight one may refer [4] and [17].

Convergence of Riemann Integrable Functions over Banach Spaces   129

In the following section, we discuss the convergence theorems for sequences and series of Riemann integrable functions over Banach spaces on time scales.

3. Convergence Theorems

In this section, we discuss the convergence theorems involving the notion of uniform convergence for sequences of Riemann Δ-integrable Banach space-valued functions on time scales. The case of the ∇-integral can be obtained in a similar manner using the above ∇-integral definition (Def. 7).

3.1. *Convergence of sequences*

We first establish the convergence theorem involving the notion of uniform convergence for sequences and establish a few results related to the uniform convergence theorem.

Theorem 1. *If the sequence $\{f_r\}$ of Riemann Δ-integrable functions defined on $[p, q]_{\mathbf{T}}$ converges uniformly to f on $[p, q]_{\mathbf{T}}$, then f is also Riemann Δ-integrable on $[p, q]_{\mathbf{T}}$ and*

$$(\overline{R}) \int_p^q f(t)\Delta t = \lim_{r \to \infty} (\overline{R}) \int_p^q f_r(t)\Delta t.$$

Proof. Given the sequence $\{f_r\}$ of Riemann Δ-integrable functions defined on $[p, q]_{\mathbf{T}}$ converges uniformly to f on $[p, q]_{\mathbf{T}}$, implies for any $\epsilon > 0$, there exists $N \in \mathbb{N}$ such that for all $r \geq N$ and any $t \in [p, q)_{\mathbf{T}}$, we have $\left\| f_r(t) - f(t) \right\| < \epsilon$.
Consequently, if $r, m \geq N$, then

$$\left\| f_r(t) - f_m(t) \right\| < \left\| f_r(t) - f(t) \right\| + \left\| f(t) - f_m(t) \right\| < \epsilon + \epsilon = 2\epsilon$$

for any $t \in [p, q)_{\mathbf{T}}$.
Hence $\left\| f_r(t) - f_m(t) \right\| \leq \sup_{t \in [p, q)_{\mathbf{T}}} \left\| f_r(t) - f_m(t) \right\| < 2\epsilon$, implying $-2\epsilon < f_r(t) - f_m(t) < 2\epsilon$ for any $t \in [p, q)_{\mathbf{T}}$. Thus,

$$-2\epsilon(q - p) < (\overline{R}) \int_p^q f_r(t)\Delta t - (\overline{R}) \int_p^q f_m(t)\Delta t < 2\epsilon(q - p).$$

Since $\epsilon > 0$ is arbitrary, hence the sequence $\left((\overline{R}) \int_p^q f_r(t)\Delta t \right)$ is a Cauchy sequence in $\mathfrak{X}$ and therefore converges to some number, say $\overline{I} \in \mathfrak{X}$. We will now show that $f \in (\mathfrak{R})_\Delta \big([p,q]_\mathbf{T}, \mathfrak{X} \big)$ with integral $\overline{I}$.

If $\breve{\mathcal{V}} = \left\{ \left(\vartheta_h, [t_{h-1}, t_h)_\mathbf{T} \right) \right\}_{h=1}^n$ be any tagged partition of $[p,q]_\mathbf{T}$ and if $r \geq N$, then

$$\left\| \overline{R}(f_r; \breve{\mathcal{V}}_\delta) - \overline{R}(f; \breve{\mathcal{V}}_\delta) \right\| \leq \sum_{h=1}^n \left\| f_r(\vartheta_h) - f(\vartheta_h) \right\| (t_h - t_{h-1})$$

$$\leq \sum_{h=1}^n \epsilon(t_h - t_{h-1}) = \epsilon(q - p).$$

We now choose $a \geq N$ such that $\left\| (\overline{R}) \int_p^q f_a(t)\Delta t - \overline{I} \right\| < \epsilon$, and we let $\delta_{a,\epsilon} > 0$ be such that $\left\| (\overline{R}) \int_p^q f_a(t)\Delta t - \overline{R}(f_a; \breve{\mathcal{V}}_{\delta_{a,\epsilon}}) \right\| < \epsilon$, whenever length of the subintervals in the partition is less than $\delta_{a,\epsilon}$. Then we have,

$$\left\| \overline{R}(f; \breve{\mathcal{V}}_{\delta_{a,\epsilon}}) - \overline{I} \right\| \leq \left\| \overline{R}(f; \breve{\mathcal{V}}_{\delta_{a,\epsilon}}) - \overline{R}(f_a; \breve{\mathcal{V}}_{\delta_{a,\epsilon}}) \right\|$$

$$+ \left\| \overline{R}(f_a; \breve{\mathcal{V}}_{\delta_{a,\epsilon}}) - (\overline{R}) \int_p^q f_a(t)\Delta t \right\|$$

$$+ \left\| (\overline{R}) \int_p^q f_a(t)\Delta t - \overline{I} \right\|$$

$$\leq \epsilon(q - p) + \epsilon + \epsilon = \epsilon(q - p + 2).$$

But since $\epsilon > 0$ is arbitrary, it follows that $f \in (\mathfrak{R})_\Delta \big([p,q]_\mathbf{T}, \mathfrak{X} \big)$ and $(\overline{R}) \int_p^q f(t)\Delta t = \overline{I}$, and for all $r \geq N$ it follows that $(\overline{R}) \int_p^q f(t)\Delta t = \lim_{r \to \infty} (\overline{R}) \int_p^q f_r(t)\Delta t$. $\qquad \square$

Theorem 1 is called the uniform convergence theorem, here the sequence of functions under consideration are Riemann Δ-integrable functions over Banach spaces.

Below we state a similar theorem, but considering the sequence of functions to be Riemann ∇-integrable functions over Banach spaces, the proof of which is similar to Theorem 1 hence omitted.

Theorem 2. *If the sequence $\{f_r\}$ of Riemann ∇-integrable functions defined on $[p,q]_\mathbf{T}$ converges uniformly to f on $[p,q]_\mathbf{T}$, then f is also Riemann ∇-integrable on $[p,q]_\mathbf{T}$ and*

$$(\underline{R}) \int_p^q f(t)\nabla t = \lim_{r \to \infty} (\underline{R}) \int_p^q f_r(t)\nabla t.$$

Convergence of Riemann Integrable Functions over Banach Spaces 131

We now proceed to state and proof a few theorems and corollaries which hold true, given a sequence of functions converges uniformly. We present here the statements for the Riemann Δ-integration which will hold similarly for the Riemann ∇-integration hence not explicitly stated here.

Corollary 1. *Let $\{f_r\}$ be a sequence of Riemann Δ-integrable functions defined on $[p,q]_\mathbf{T}$ that converges pointwise to a function f on $[p,q]_\mathbf{T}$. If given for any $\epsilon > 0$, there exists $N \in \mathbb{N}$ such that $\sup_{t \in [p,q)_\mathbf{T}} \left\| f_r(t) - f(t) \right\| < \epsilon$, for all $t \in [p,q)_\mathbf{T}$ and all $r \geq N$, then the given function f is also Riemann Δ-integrable on $[p,q]_\mathbf{T}$ and, $(\overline{R}) \int_p^q f(t)\Delta t = \lim_{r \to \infty} (\overline{R}) \int_p^q f_r(t)\Delta t$.*

Below we prove the uniqueness of the limit function f to which the sequence of Riemann Δ-integrable functions converge to.

Theorem 3. *Let $\{f_r\}$ be a sequence of Riemann Δ-integrable functions defined on $[p,q]_\mathbf{T}$ that converges uniformly to a Riemann Δ-integrable function f on $[p,q]_\mathbf{T}$. Then the function f is unique.*

Proof. Suppose that f' and f'' are both limits of sequence $\{f_r\}$.
For every $\epsilon > 0$ there exists $N' \in \mathbb{N}$ such that for all $r \geq N'$ and all $t \in [p,q)_\mathbf{T}$, we have $\left\| f_r(t) - \left(f(t) \right)' \right\| < \frac{\epsilon}{2}$, and there exists $N'' \in \mathbb{N}$ such that for all $r \geq N''$ and all $t \in [p,q)_\mathbf{T}$, we have $\left\| f_r(t) - \left(f(t) \right)'' \right\| < \frac{\epsilon}{2}$.
Now we choose $N = \min\{N', N''\}$ and all $t \in [p,q)_\mathbf{T}$ we apply the triangle inequality, we get

$$\left\| \left(f(t) \right)' - \left(f(t) \right)'' \right\| \leq \left\| \left(f(t) \right)' - f_r(t) \right\| + \left\| f_r(t) - \left(f(t) \right)'' \right\|$$
$$< \frac{\epsilon}{2} + \frac{\epsilon}{2} = \epsilon.$$

Since $\epsilon > 0$ is arbitrary, we conclude that $\left(f(t) \right)' - \left(f(t) \right)'' = 0$, for all $t \in [p,q)_\mathbf{T}$. $\qquad\square$

Theorem 4. *Let $\{f_r\}$ be a sequence of Riemann Δ-integrable functions defined on $[p,q]_\mathbf{T}$ that converges pointwise to a function f on $[p,q]_\mathbf{T}$. If given $\lim_{r \to \infty} \sup_{t \in [p,q_\mathbf{T})} \left\| f_r(t) - f(t) \right\| = 0$, then the given function f is also Riemann Δ-integrable on $[p,q]_\mathbf{T}$ and, $(\overline{R}) \int_p^q f(t)\Delta t = \lim_{r \to \infty} (\overline{R}) \int_p^q f_r(t)\Delta t$.*

Proof. Given $\{f_r\}$ be a sequence of Riemann Δ-integrable functions defined on $[p,q]_\mathbf{T}$ that converges pointwise to f on $[p,q]_\mathbf{T}$, implies for any $\epsilon > 0$, there exists $N \in \mathbb{N}$ such that

$$\left\| f_r(t) - f(t) \right\| < \epsilon \tag{1}$$

for each $t \in [p, q)_\mathbf{T}$ and all $r \geq N$. Also given,

$$\lim_{r \to \infty} \sup_{t \in [p,q)_\mathbf{T}} \left\| f_r(t) - f(t) \right\| = 0 \tag{2}$$

for all $t \in [p, q)_\mathbf{T}$.

From Eq. 1 and Eq. 2 for sequence $\{f_r\}$, given any $\epsilon > 0$, there exists $N \in \mathbb{N}$ such that $\left\| f_r(t) - f(t) \right\| < \epsilon$ for all $t \in [p, q)_\mathbf{T}$ for all $r \geq N$. Therefore $\{f_r\}$ converges uniformly on $[p, q]_\mathbf{T}$ to f.

Thus, the given function f is also Riemann Δ-integrable on $[p, q]_\mathbf{T}$ and, $(\overline{R}) \int_p^q f(t) \Delta t = \lim_{r \to \infty} (\overline{R}) \int_p^q f_r(t) \Delta t$, following Theorem 1. $\qquad\square$

Theorem 5. *Let $\{f_r\}$ be a sequence of Riemann Δ-integrable functions defined on $[p, q]_\mathbf{T}$ that converges pointwise to a function f on $[p, q]_\mathbf{T}$, such that for any $\epsilon > 0$, there exists $N \in \mathbb{N}$ such that $\sup_{t \in [p,q)_\mathbf{T}} \left\| f_r(t) - f_m(t) \right\| < \epsilon$, for all $t \in [p, q)_\mathbf{T}$ and all $r, m \geq N$, then the given function f is also Riemann Δ-integrable on $[p, q]_\mathbf{T}$ and, $(\overline{R}) \int_p^q f(t) \Delta t = \lim_{r \to \infty} (\overline{R}) \int_p^q f_r(t) \Delta t$.*

Proof. For all $t \in [p, q)_\mathbf{T}$ we have,

$$\left\| f_r(t) - f_m(t) \right\| \leq \sup_{t \in [p,q)_\mathbf{T}} \left\| f_r(t) - f_m(t) \right\| < \epsilon \quad \forall \quad r, m \geq N(\epsilon),$$

implying that the sequence $\{f_r\}$ is a Cauchy sequence, hence it is convergent.

Also, given $\{f_r\}$ converges pointwise to f on $[p, q]_\mathbf{T}$, thus for each t, $f_r(t)$ converges to $f(t)$ as $r \to \infty$. Now, by passing $m \to \infty$ above,

$$\left\| f_r(t) - f(t) \right\| \leq \sup_{t \in [p,q)_\mathbf{T}} \left\| f_r(t) - f(t) \right\| < \epsilon$$

for all $t \in [p, q)_\mathbf{T}$ and all $r \geq N$. Hence, sequence $\{f_r\}$ converges uniformly on $[p, q]_\mathbf{T}$ to f.

Thus, the given function f is also Riemann Δ-integrable on $[p, q]_\mathbf{T}$ and, $(\overline{R}) \int_p^q f(t) \Delta t = \lim_{r \to \infty} (\overline{R}) \int_p^q f_r(t) \Delta t$, following Theorem 1. $\qquad\square$

Corollary 2. *Let $\{f_r\}$ be a Cauchy sequence of Riemann Δ-integrable functions defined on $[p, q]_\mathbf{T}$ that converges pointwise to a function f on $[p, q]_\mathbf{T}$, then the given function f is also Riemann Δ-integrable on $[p, q]_\mathbf{T}$ and, $(\overline{R}) \int_p^q f(t) \Delta t = \lim_{r \to \infty} (\overline{R}) \int_p^q f_r(t) \Delta t$.*

Proof. Proof follows directly from Theorem 5. $\qquad\square$

Convergence of Riemann Integrable Functions over Banach Spaces   133

Theorem 6. *Let $\{f_r\}$ and $\{g_r\}$ be sequences of Riemann Δ-integrable functions defined on $[p,q]_{\mathbf{T}}$. Sequence $\{g_r\}$ is defined as $g_r = f_{m+r}$, given $m \in \mathbb{N}$. If $\{f_r\}$ converges uniformly to function f on $[p,q]_{\mathbf{T}}$ implies $\{g_r\}$ also converges uniformly to f on $[p,q]_{\mathbf{T}}$ and vice versa.*

Theorem 7. *Let $\{f_r\}$ be a sequence of Riemann Δ-integrable functions defined on $[p,q]_{\mathbf{T}}$ that converges uniformly to function f on $[p,q]_{\mathbf{T}}$. Let $\tilde{r} \in \mathbb{K}$ then the sequence $\{\tilde{r}f_r\}$ is also a sequence of Riemann Δ-integrable functions and converges uniformly to Riemann Δ-integrable function $\tilde{r}f$ on $[p,q]_{\mathbf{T}}$ and*

$$(\overline{R}) \int_p^q \tilde{r}f(t)\Delta t = \lim_{r \to \infty} (\overline{R}) \int_p^q \tilde{r}f_r(t)\Delta t.$$

Proof. Given $\{f_r\}$ be a sequence of Riemann Δ-integrable functions defined on $[p,q]_{\mathbf{T}}$ that converges uniformly to function f on $[p,q]_{\mathbf{T}}$, implies that for $\tilde{r} \in \mathbb{K}$, sequence $\{\tilde{r}f_r\}$ converges uniformly to function $\tilde{r}f$ (Theorem 3.2.3 [4, pp. 64]).

Also for any $r \in \mathbb{N}$, since f_r is given to be Riemann Δ-integrable, implies $\tilde{r}f_r$ is also Riemann Δ-integrable for any $\tilde{r} \in \mathbb{K}$.

Hence the sequence $\{\tilde{r}f_r\}$ of Riemann Δ-integrable functions converges uniformly to Riemann Δ-integrable function $\tilde{r}f$, thus from Theorem 1 we conclude.

$$\square$$

Theorem 8. *Let $\{f_r\}$ and $\{\tilde{f}_r\}$ be sequences of Riemann Δ-integrable functions defined on $[p,q]_{\mathbf{T}}$ that converges uniformly to function f and $\tilde{f}$ on $[p,q]_{\mathbf{T}}$ respectively. Then the sequence $\{f_r + \tilde{f}_r\}$ is also a sequence of Riemann Δ-integrable functions and converges uniformly to Riemann Δ-integrable function $(f + \tilde{f})$ on $[p,q]_{\mathbf{T}}$ and*

$$(\overline{R}) \int_p^q (f + \tilde{f})(t)\Delta t = \lim_{r \to \infty} (\overline{R}) \int_p^q (f_r + \tilde{f}_r)(t)\Delta t.$$

Proof. Given $\{f_r\}$ and $\{\tilde{f}_r\}$ be sequences of Riemann Δ-integrable functions defined on $[p,q]_{\mathbf{T}}$ that converges uniformly to function f and $\tilde{f}$ on $[p,q]_{\mathbf{T}}$, implies that sequence $\{f_r + \tilde{f}_r\}$ converges uniformly to function $(f + \tilde{f})$ (Theorem 3.2.3 [4, pp. 64]).

Also for any $r \in \mathbb{N}$, since f_r and $\tilde{f}_r$ are given to be Riemann Δ-integrable, implies $f_r + \tilde{f}_r$ is also Riemann Δ-integrable.

Hence the sequence $\{f_r + \tilde{f}_r\}$ of Riemann Δ-integrable functions converges uniformly to Riemann Δ-integrable function $f_r + \tilde{f}_r$, thus from Theorem 1 we conclude. $\qquad\square$

Corollary 3. *Let* $\{f_r\}$ *and* $\{\tilde{f}_r\}$ *be sequences of Riemann Δ-integrable functions defined on* $[p,q]_{\mathbf{T}}$ *that converges uniformly to function f and $\tilde{f}$ on* $[p,q]_{\mathbf{T}}$ *respectively. Then the sequence* $\{f_r - \tilde{f}_r\}$ *is also a sequence of Riemann Δ-integrable functions and converges uniformly to Riemann Δ-integrable function* $(f - \tilde{f})$ *on* $[p,q]_{\mathbf{T}}$ *and*

$$(\overline{R}) \int_p^q (f - \tilde{f})(t)\Delta t = \lim_{r \to \infty} (\overline{R}) \int_p^q (f_r - \tilde{f}_r)(t)\Delta t.$$

Theorem 9. *Let* $\{f_r\}$ *and* $\{\tilde{f}_r\}$ *be sequences of Riemann Δ-integrable functions defined on* $[p,q]_{\mathbf{T}}$ *that converges uniformly to function f and $\tilde{f}$ on* $[p,q]_{\mathbf{T}}$ *respectively. Given* $\{g_r\}$ *be also sequences of Riemann Δ-integrable functions, if* $f_r \leq g_r \leq \tilde{f}_r$ *for all* $n \in \mathbb{N}$, *and if* $f = \tilde{f}$, *then the sequence* $\{g_r\}$ *also converges uniformly to Riemann Δ-integrable function f on* $[p,q]_{\mathbf{T}}$.

Theorem 10. *Let* $\{f_r\}$ *be a contractive sequence of Riemann Δ-integrable functions defined on* $[p,q]_{\mathbf{T}}$ *that converges pointwise to a function f on* $[p,q]_{\mathbf{T}}$, *then the given function f is also Riemann Δ-integrable on* $[p,q]_{\mathbf{T}}$ *and,* $(\overline{R}) \int_p^q f(t)\Delta t = \lim_{r \to \infty} (\overline{R}) \int_p^q f_r(t)\Delta t$.

Proof. Given $\{f_r\}$ is contractive, we get

$$\left\| f_{r+2} - f_{r+1} \right\| \leq C \left\| f_{r+1} - f_r \right\| \leq C^2 \left\| f_r - f_{r-1} \right\| \leq \ldots \leq C^r \left\| f_2 - f_1 \right\|.$$

For $r, m \in \mathbb{N}$ such that $m > r$, applying the triangle inequality we get,

$$\left\| f_m - f_r \right\| \leq \left\| f_m - f_{m-1} \right\| + \left\| f_{m-1} - f_{m-2} \right\| + \ldots + \left\| f_{r+1} - f_r \right\|$$

$$\leq (C^{m-2} + C^{m-3} + \ldots + C^{r-1}) \left\| f_2 - f_1 \right\|$$

$$= C^{r-1} \left(\frac{1 - C^{m-r}}{1 - C} \right) \left\| f_2 - f_1 \right\| \leq C^{r-1} \left(\frac{1}{1 - C} \right) \left\| f_2 - f_1 \right\|.$$

Since $0 < C < 1$, we know $\lim_{r \to \infty}(C^r) = 0$, implying that $\{f_r\}$ is a Cauchy sequence. Hence from Corollary 6, we conclude. $\qquad\square$

Theorem 11. *Let* $\{f_r\}$ *be a uniformly bounded sequence of Riemann Δ-integrable functions defined on* $[p,q]_{\mathbf{T}}$ *that converges pointwise to a function f on* $[p,q]_{\mathbf{T}}$. *If* $\{f_r\}$ *converges uniformly to f on each subinterval*

Convergence of Riemann Integrable Functions over Banach Spaces   135

$[t_{h-1}, t_h)_\mathbf{T}$ of $[p, q]_\mathbf{T}$, then f is Riemann Δ-integrable and $(\overline{R}) \int_p^q f(t)\Delta t = \lim_{r \to \infty} (\overline{R}) \int_p^q f_r(t)\Delta t$.

Proof. Let D be a uniform bound for the sequence $\{f_r\}$ on $[p, q]_\mathbf{T}$, note D also bounds function f on $[p, q]_\mathbf{T}$.

Given $\{f_r\}$ converges uniformly to f on each subinterval of the form $[t_{h-1}, t_h)_\mathbf{T}$ of $[p, q]_\mathbf{T}$, hence by Theorem 1 function f is Riemann Δ-integrable on each subinterval of the form $[t_{h-1}, t_h)_\mathbf{T}$ of $[p, q]_\mathbf{T}$, implying f is Riemann Δ-integrable on $[p, q]_\mathbf{T}$.

Let $\epsilon > 0$. Choose points $a, b \in (p, q)_\mathbf{T}$ such that $a < b$, $a - p < \frac{\epsilon}{2D}$, and $q - b < \frac{\epsilon}{2D}$. Since $\{f_r\}$ converges uniformly to f on $[a, b]_\mathbf{T}$, by Theorem 1 there exists a positive integer N such that $\left\| (\overline{R}) \int_a^b f_r(t)\Delta t - (\overline{R}) \int_a^b f(t)\Delta t \right\| < \epsilon$ for all $t \in [a, b)_\mathbf{T}$ and all $r \geq N$. Then

$$\left\| (\overline{R}) \int_p^q f_r(t)\Delta t - (\overline{R}) \int_p^q f(t)\Delta t \right\| \leq \left\| (\overline{R}) \int_p^a \{f_r(t) - f(t)\}\Delta t \right\|$$
$$+ \left\| (\overline{R}) \int_a^b f_r(t)\Delta t - (\overline{R}) \int_a^b f(t)\Delta t \right\|$$
$$+ \left\| (\overline{R}) \int_b^q \{f_r(t) - f(t)\}\Delta t \right\|$$
$$< 2D(a - p) + \epsilon + 2D(q - b) < \epsilon + \epsilon + \epsilon = 3\epsilon$$

for all $t \in [p, q)_\mathbf{T}$ and all $r \geq N$, and it follows that $(\overline{R}) \int_p^q f(t)\Delta t = \lim_{r \to \infty} (\overline{R}) \int_p^q f_r(t)\Delta t$, given ϵ is arbitrary. $\square$

Theorem 12. *Let $\{f_r\}$ be a sequence of Riemann Δ-integrable functions defined on $[p, q]_\mathbf{T}$ that converges uniformly to a function f on $[p, q]_\mathbf{T}$. Given the functions in the sequence $\{f_r\}$ on $[p, q]_\mathbf{T}$ are continuous implies the Riemann Δ-integrable function f is also continuous on $[p, q]_\mathbf{T}$.*

Proof. Given for every $\epsilon > 0$ there exists $N := N(\frac{\epsilon}{3}) \in \mathbb{N}$ such that if $r \geq N(\frac{\epsilon}{3})$ then $\left\| f_r(t) - f(t) \right\| < \frac{\epsilon}{3}$ for all $t \in [p, q)_\mathbf{T}$.

Let $t_0 \in [p, q)_\mathbf{T}$ be arbitrary. As f_r is continuous, there exists $\delta > 0$ such that,

$$\left\| f_r(t) - f_r(t_0) \right\| < \frac{\epsilon}{3} \quad if \quad |t - t_0| < \delta, \quad t \in [p, q).$$

Therefore, for $t \in [p, q]_{\mathbf{T}}$, $|t - t_0| < \delta$,

$$\left\|f(t) - f(t_0)\right\| \leq \left\|f(t) - f_r(t)\right\| + \left\|f_r(t) - f_r(t_0)\right\| + \left\|f_r(t_0) - f(t_0)\right\|$$
$$< \frac{\epsilon}{3} + \frac{\epsilon}{3} + \frac{\epsilon}{3} = \epsilon,$$

which shows that f is continuous at t_0. Since t_0 is arbitrary, it follows that f is continuous on $[p, q]_{\mathbf{T}}$. $\qquad\square$

Corollary 4. Let $\{f_r\}$ be a sequence of continuous functions that converges uniformly to a function f on $[p, q]_{\mathbf{T}}$, implies that f is also continuous on $[p, q]_{\mathbf{T}}$ and Riemann Δ-integrable on $[p, q]_{\mathbf{T}}$.

Theorem 12 can also be expressed as Theorem 13.

Theorem 13. Let $\{f_r\}$ be a sequence of Riemann Δ-integrable functions defined on $[p, q]_{\mathbf{T}}$ that converges uniformly to a function f on $[p, q]_{\mathbf{T}}$. Given the functions in the sequence $\{f_r\}$ on $[p, q]_{\mathbf{T}}$ are continuous implies the below equation holds-

$$\lim_{t \to t_0} \lim_{r \to \infty} f_r(t) = \lim_{r \to \infty} \lim_{t \to t_0} f_r(t). \tag{3}$$

Proof. Following Theorem 12, the left hand side of Eq. 3 is $\lim_{t \to t_0} f(t)$ which, by the continuity of f is equal to $f(t_0)$.
On the other hand, the right hand side of Eq. 3 is equal to $\lim_{r \to \infty} f_r(t_0)$, which by the uniform convergence of f_r, is also equal to $f(t_0)$. $\qquad\square$

Theorem 14. Suppose that $\{f_r\}$ is a monotone sequence of continuous Riemann Δ-integrable functions defined on $[p, q]_{\mathbf{T}}$ that converges pointwise to a continuous function f on interval $[p, q]_{\mathbf{T}}$, then f is also Riemann Δ-integrable on $[p, q]_{\mathbf{T}}$ and $(\overline{R}) \int_p^q f(t) \Delta t = \lim_{r \to \infty} (\overline{R}) \int_p^q f_r(t) \Delta t$.

Proof. We suppose that the sequence $\{f_r\}$ is decreasing and let $g_r := f_r - f$. Then $\{g_r\}$ is a decreasing sequence of continuous functions converging on $[p, q]_{\mathbf{T}}$ to the 0-function. We will show that the convergence is uniform on $[p, q]_{\mathbf{T}}$.
By contradiction, assuming g_r does not converge to 0 uniformly, by definition there exists some ϵ_0 such that $\sup_{t \in [p,q)_{\mathbf{T}}} \left\|f_r(t) - f(t)\right\| \geq 2\epsilon_0 > 0$ for infinitely many r's. For simplicity let us assume it is so for all r's.

Then we find, for each r, a point t_r in $[p,q]_{\mathbf{T}}$ such that $\left\|f_r(t_r) - f(t_r)\right\| \geq \epsilon_0$. By Bolzano-Weierstrass theorem, $\{t_r\}$ contains a subsequence $\{t_{r_j}\}$ convergent to some t^*. As each f_r is decreasing,

$$\left\|f_m(t) - f(t)\right\| \geq \left\|f_r(t) - f(t)\right\|, \quad \forall m, r, \quad m \leq r.$$

Taking $m = r_k$ and $r = r_j$, $j \geq k$, we obtain

$$\left\|f_{r_k}(t_{r_j}) - f(t_{r_j})\right\| \geq \left\|f_{r_j}(t_{r_j}) - f(t_{r_j})\right\| \geq \epsilon_0.$$

Now fix r_k and let r_j go to infinity, we get

$$\left\|f_{r_k}(t^*) - f(t^*)\right\| \geq \epsilon_0,$$

for all r_k, but this is not possible as $g_r = f_r - f$ tends to 0 pointwisely. Hence $\{f_r\}$ converges uniformly to f on $[p,q]_{\mathbf{T}}$.

Thus, the given function f is also Riemann Δ-integrable on $[p,q]_{\mathbf{T}}$ and, $(\overline{R}) \int_p^q f(t)\Delta t = \lim_{r \to \infty} (\overline{R}) \int_p^q f_r(t)\Delta t$, following Theorem 1. $\qquad \square$

3.2. Convergence of series

We now establish the convergence theorem involving the notion of uniform convergence for series and establish a few results.

Theorem 15. *Suppose $\{f_r\}$ is a sequence of Riemann Δ-integrable functions defined on $[p,q]_{\mathbf{T}}$. If the series $\sum_{r=1}^{\infty} (f_r)$ converges uniformly to f on $[p,q]_{\mathbf{T}}$, then f is also Riemann Δ-integrable on $[p,q]_{\mathbf{T}}$ and*

$$(\overline{R}) \int_p^q f(t)\Delta t = \sum_{r=1}^{\infty} (\overline{R}) \int_p^q f_r(t)\Delta t.$$

Proof. Given the sequence $\{f_r\}$ of Riemann Δ-integrable functions defined on $[p,q]_{\mathbf{T}}$ implies the sequence of partial sums $\{s_r\}$ of $\{f_r\}$'s are also Riemann Δ-integrable functions.

Also, given the series $\sum_{r=1}^{\infty} \{f_r\}$ converges uniformly to f on $[p,q]_{\mathbf{T}}$, implies that the sequence of partial sums $\{s_r\}$ of $\{f_r\}$'s converges uniformly to f.

Now, for every $\epsilon > 0$, there exists $N \in \mathbb{N}$ such that for all $r \geq N$ and all $t \in [p,q)_{\mathbf{T}}$, we have $\left\|s_r(t) - f(t)\right\| = \left\|s_r(t) - f(t)\right\| < \epsilon$.

Consequently, if $r, m \geq N$, then

$$\left\|s_r(t) - s_m(t)\right\| < \left\|s_r(t) - f(t)\right\| + \left\|f(t) - s_m(t)\right\| < \epsilon + \epsilon = 2\epsilon$$

for all $t \in [p, q]_{\mathbf{T}}$.

Hence, $\left\|s_r(t) - s_m(t)\right\| \leq \sup_{t \in [p,q]_{\mathbf{T}}} \left\|s_r(t) - s_m(t)\right\| < 2\epsilon$, implying $-2\epsilon < s_r(t) - s_m(t) < 2\epsilon$ for all $t \in [p, q]_{\mathbf{T}}$. Thus,

$$-2\epsilon(q - p) < (\overline{R}) \int_p^q s_r(t)\Delta t - (\overline{R}) \int_p^q s_m(t)\Delta t < 2\epsilon(q - p).$$

Since $\epsilon > 0$ is arbitrary, hence the sequence $\left((\overline{R}) \int_p^q s_r(t)\Delta t\right)$ is a Cauchy sequence in $\mathfrak{X}$ and therefore converges to some $\overline{I} \in \mathfrak{X}$.

We will now show that $f \in (\mathfrak{R})_\Delta\left([p, q]_{\mathbf{T}}, \mathfrak{X}\right)$ with integral $\overline{I}$.

If $\check{\mathcal{V}} = \left\{\left(\vartheta_h, [t_{h-1}, t_h)_{\mathbf{T}}\right)\right\}_{h=1}^n$ be any tagged partition of $[p, q]_{\mathbf{T}}$ and if $r \geq N$, then

$$\left\|(\overline{R})(s_r; \check{\mathcal{V}}_\delta) - (\overline{R})(f; \check{\mathcal{V}}_\delta)\right\| \leq \sum_{h=1}^n \left\|s_r(\vartheta_h) - f(\vartheta_h)\right\|(t_h - t_{h-1})$$

$$\leq \sum_{h=1}^n \epsilon(t_h - t_{h-1}) = \epsilon(q - p).$$

We now choose $a \geq N$, such that $\left\|(\overline{R}) \int_p^q s_a(t)\Delta t - \overline{I}\right\| < \epsilon$ and we let $\delta_{a,\epsilon} > 0$ be such that $\left\|(\overline{R}) \int_p^q s_a(t)\Delta t - (\overline{R})(s_a; \check{\mathcal{V}}_{\delta_{a,\epsilon}})\right\| < \epsilon$, whenever length of the subintervals in the partition is less than $\delta_{a,\epsilon}$. Then we have,

$$\left\|(\overline{R})(f; \check{\mathcal{V}}_{\delta_{a,\epsilon}}) - \overline{I}\right\| \leq \left\|(\overline{R})(f; \check{\mathcal{V}}_{\delta_{a,\epsilon}}) - (\overline{R})(s_a; \check{\mathcal{V}}_{\delta_{a,\epsilon}})\right\| + \left\|(\overline{R})(s_a; \check{\mathcal{V}}_{\delta_{a,\epsilon}}) - (\overline{R}) \int_p^q s_a(t)\Delta t\right\| + \left\|(\overline{R}) \int_p^q s_a(t)\Delta t - \overline{I}\right\| \leq \epsilon(q - p) + \epsilon + \epsilon = \epsilon(q - p + 2).$$

But since $\epsilon > 0$ is arbitrary, it follows that $f \in (\mathfrak{R})_\Delta\left([p, q]_{\mathbf{T}}, \mathfrak{X}\right)$ and $(\overline{R}) \int_p^q f(t)\Delta t = \overline{I}$, and for all $r \geq N$ it follows that $(\overline{R}) \int_p^q f(t)\Delta t = \lim_{r \to \infty} (\overline{R}) \int_p^q s_r(t)\Delta t$. $\qquad \square$

Theorem 15 is called the uniform convergence theorem for series, here the sequence of functions under consideration are Riemann Δ-integrable functions over Banach spaces.

Below we state a similar theorem, but considering the sequence of functions to be Riemann ∇-integrable functions over Banach spaces, the proof of which is is omitted as it is similar to that of Theorem 15.

Convergence of Riemann Integrable Functions over Banach Spaces

Theorem 16. *Suppose $\{f_r\}$ is a sequence of Riemann ∇-integrable functions defined on $[p,q]_{\mathbf{T}}$. If the series $\sum_{r=1}^{\infty}(f_r)$ converges uniformly to f on $[p,q]_{\mathbf{T}}$, then f is also Riemann ∇-integrable on $[p,q]_{\mathbf{T}}$ and*

$$(\underline{R}) \int_p^q f(t)\nabla t = \sum_{r=1}^{\infty} (\underline{R}) \int_p^q f_r(t)\nabla t.$$

We now proceed to state and prove a few theorems and corollaries which hold true, given a series of functions converges uniformly. We present here the statements for the Riemann Δ-integration which will hold similarly for the Riemann ∇-integration hence not explicitly stated here.

Corollary 5. *Suppose $\{f_r\}$ is a sequence of Riemann Δ-integrable functions defined on $[p,q]_{\mathbf{T}}$. If the series $\sum_{r=1}^{\infty}(f_r)$ converges pointwise to a function f on $[p,q]_{\mathbf{T}}$, and given for any $\epsilon > 0$, there exists $N \in \mathbb{N}$ such that $\sup_{t \in [p,q)_{\mathbf{T}}} \left\| s_r(t) - f(t) \right\| < \epsilon$, for all $t \in [p,q)_{\mathbf{T}}$ and all $r \geq N$, then the given function f is also Riemann Δ-integrable on $[p,q]_{\mathbf{T}}$ and, $(\overline{R}) \int_p^q f(t)\Delta t = \sum_{r=1}^{\infty} (\overline{R}) \int_p^q f_r(t)\Delta t.$*

Below we prove the uniqueness of the limit function f to which the series of Riemann Δ-integrable functions converge to.

Theorem 17. *Suppose $\{f_r\}$ is a sequence of Riemann Δ-integrable functions defined on $[p,q]_{\mathbf{T}}$. If the series $\sum_{r=1}^{\infty}(f_r)$ converges uniformly to a Riemann Δ-integrable function f on $[p,q]_{\mathbf{T}}$. Then the function f is unique.*

Proof. Suppose that f' and f'' are both limits of the series $\sum_{r=1}^{\infty}(f_r)$. For every $\epsilon > 0$ there exists $N' \in \mathbb{N}$ such that for all $r \geq N'$ and all $t \in [p,q)_{\mathbf{T}}$, we have $\left\| s_r(t) - (f(t))' \right\| < \frac{\epsilon}{2}$, and there exists $N'' \in \mathbb{N}$ such that for all $r \geq N''$ and all $t \in [p,q)_{\mathbf{T}}$, we have $\left\| s_r(t) - (f(t))'' \right\| < \frac{\epsilon}{2}$. Now we choose $N = \min\{N', N''\}$ and all $t \in [p,q)_{\mathbf{T}}$ we apply the triangle inequality, we get

$$\left\| (f(t))' - (f(t))'' \right\| \leq \left\| (f(t))' - s_r(t) \right\| + \left\| s_r(t) - (f(t))'' \right\|$$
$$< \frac{\epsilon}{2} + \frac{\epsilon}{2} = \epsilon.$$

Since $\epsilon > 0$ is arbitrary, we conclude that $(f(t))' - (f(t))'' = 0$, for all $t \in [p,q)_{\mathbf{T}}$. $\qquad\square$

Theorem 18. *Suppose $\{f_r\}$ is a sequence of Riemann Δ-integrable functions defined on $[p,q]_{\mathbf{T}}$. If the series $\sum_{r=1}^{\infty}(f_r)$ converges pointwise to a function f on $[p,q]_{\mathbf{T}}$, such that for any $\epsilon > 0$, there exists $N \in \mathbb{N}$ such that $\sup_{t\in[p,q)_{\mathbf{T}}}\left\|s_r(t) - s_m(t)\right\| < \epsilon$, for all $t \in [p,q)_{\mathbf{T}}$ and all $r,m \geq N$, then the given function f is also Riemann Δ-integrable on $[p,q]_{\mathbf{T}}$ and, $(\overline{R})\int_p^q f(t)\Delta t = \sum_{r=1}^{\infty}(\overline{R})\int_p^q f_r(t)\Delta t.$*

Proof. For all $t \in [p,q)_{\mathbf{T}}$ we have,

$$\left\|s_r(t) - s_m(t)\right\| \leq \sup_{t\in[p,q)_{\mathbf{T}}}\left\|s_r(t) - s_m(t)\right\| < \epsilon \quad \forall\, r,m \geq N(\epsilon),$$

implying that the series $\sum_{r=1}^{\infty}(f_r)$ is a Cauchy series, and hence convergent. Also, given $\{s_r\}$ converges pointwise to f on $[p,q]$, thus for each t, $s_r(t)$ converges to $f(t)$ as $r \to \infty$. Now, by passing $m \to \infty$ above,

$$\left\|s_r(t) - f(t)\right\| \leq \sup_{t\in[p,q)}\left\|s_r(t) - f(t)\right\| < \epsilon$$

for all $t \in [p,q)_{\mathbf{T}}$ and all $r \geq N$.
Hence, sequence of partial sums $\{s_r\}$ converges uniformly on $[p,q]_{\mathbf{T}}$ to f. Thus, the given function f is also Riemann Δ-integrable on $[p,q]_{\mathbf{T}}$ and, $(\overline{R})\int_p^q f(t)\Delta t = \sum_{r=1}^{\infty}(\overline{R})\int_p^q f_r(t)\Delta t$, following Theorem 15. $\qquad\square$

Corollary 6. *Suppose $\{f_r\}$ is a sequence of Riemann Δ-integrable functions defined on $[p,q]_{\mathbf{T}}$. If the series $\sum_{r=1}^{\infty}(f_r)$ is given to be Cauchy and it converges pointwise to a function f on $[p,q]_{\mathbf{T}}$, then the given function f is also Riemann Δ-integrable on $[p,q]_{\mathbf{T}}$ and, $(\overline{R})\int_p^q f(t)\Delta t = \sum_{r=1}^{\infty}(\overline{R})\int_p^q f_r(t)\Delta t.$*

Proof. Proof follows directly from Theorem 18. $\qquad\square$

Theorem 19. *Suppose $\{f_r\}$ is a sequence of Riemann Δ-integrable functions defined on $[p,q]_{\mathbf{T}}$. If the series $\sum_{r=1}^{\infty}(f_r)$ converges pointwise to a function f on $[p,q]_{\mathbf{T}}$, and there exists a convergent series $\sum M_r$ of positive numbers such that for all $r \in \mathbb{N}$ and all $t \in [p,q)_{\mathbf{T}}$ we have $\left\|f_r(t)\right\| \leq M_r$, then the given function f is also Riemann Δ-integrable on $[p,q]_{\mathbf{T}}$ and, $(\overline{R})\int_p^q f(t)\Delta t = \sum_{r=1}^{\infty}(\overline{R})\int_p^q f_r(t)\Delta t.$*

Proof. Let $\epsilon > 0$. Since $\sum M_r$ is given to be convergent, hence there exists a $N \in \mathbb{N}$ such that

$$\left\|M_{r+1} + M_{r+2} + \ldots + M_{r+n}\right\| < \epsilon,$$

for all $r \geq N$ and all $n \geq 0$.

Convergence of Riemann Integrable Functions over Banach Spaces 141

Hence we see that, for all $t \in [p, q)$ and all $r \geq N$ and all $n \geq 0$,

$$\left\| f_{r+1}(t) + \ldots + f_{r+n}(t) \right\| \leq \left\| M_{r+1} + \ldots + M_{r+n} \right\| < \epsilon.$$

Hence, series $\sum_{r=1}^{\infty} (f_r)$ is uniformly convergent in $[p, q]_{\mathbf{T}}$. Thus by Theorem 15 we conclude. $\qquad\square$

Conclusion

In this chapter, we study the convergence of sequences and series of Riemann integrable functions over Banach spaces on time scales. We formulated a few results concerning the notion of uniform convergence for sequences and series of Riemann integrable functions over Banach spaces on time scales.

References

[1] Ahlbrandt, C.D., Bohner, M.: *Hamiltonian Systems on Time Scales.* Journal of Mathematical Analysis and Applications **250**, (2000), 561-578.

[2] Atici, F.M., Guseinov, G.Sh.: *On Green's functions and positive solutions for boundary value problems on time scales.* Journal of Computational and Applied Mathematics **141**, (2002), 75-99.

[3] Aulbach, B., Neidhart, L.: *Integration on Measure Chains*, in: Proc. Sixth Int. Conf. Difference Equations, CRC, Boca Raton, FL, (2004), 239-252.

[4] Bartle, R.G., Sherbert, D.R.: *Introduction to real analysis*, 4th Edition; Wiley India Edition, 2011.

[5] Bohner, M., Georgiev, S.G.: *Multivariable Dynamic Calculus on Time Scales*; Springer International Publishing Switzerland, 2016.

[6] Bohner, M., Peterson, A.: *Advances in Dynamic Equations on Time Scales*; Birkhäuser Boston, 2003.

[7] Cichoń, M.: *On integrals of vector-valued functions on time scales.* Communications in Mathematical Analysis, **111**, (2011), 94-110.

[8] Gordan, R.: *Riemann Integration in Banach Spaces.* Rocky Mountain, Journal of Mathematics, **213**, (1927), 163-177.

[9] Gordan, R.A.: *The Integrals of Lebesgue, Denjoy, Perron, and Henstock*; American Mathematical Society, (1995).

[10] Graves, L.M.: *Riemann Integration and Taylor's Theorem in general analysis**. American Mathematical Society, **291**, (1991), 923-949.

[11] Guseinov, G.Sh., Kaymakçalan, B.: *Basics of Riemann Delta and Nabla Integration on Time Scales.* Journal of Difference Equations and Applications, **811**, (2002), 1001-1017.

[12] Guseinov, G.Sh.: *Integration on time scales.* Journal of Mathematical Analysis and Applications, **285**, (2003), 107-127.

142   H. Bharali, V. Sekhose, H. Kalita

[13] Hilger, S.: *Analysis on Measure Chains- A unified approach to continuous and discrete calculus.* Results in Mathematics, **18**, (1990).

[14] Hilger, S.: *Differential and Difference Calculus- Unified!.* Nonlinear Analysis, Theory, Methods & Applications, **30**5, (1997), 2683-2694.

[15] Lakshmikantham, V., Sivasundaram, S., Kaymakcalan, B.: *Dynamic Systems on Measure Chains*; Springer-Science+ Business Media, B.Y., 1996.

[16] Megginson, R.E.: *An Introduction to Banach Space Theory.* Springer-Verlag New York, Inc., 1998.

[17] Rubin, W.: *Principles of Mathematical Analysis*; International Series in Pure and Applied Mathematics, 3rd Edition, McGraw-Hill, Inc., 1976.

[18] Schwabik, Š., Guojo, Y.: *Topics in Banach Space Integration*; Series in Real Analysis- Vol. 10, World Scientific, (2005).

[19] Sethose, V., Bharali, H., Kalita, H., *Extensibility of Banach-valued Riemann type intgrals on time scale*; Journal of Nonlinear and Convex Analysis, **24**5, (2023), 2037-2057.

© 2025 World Scientific Publishing Company
https://doi.org/10.1142/9789819812202_0007

Chapter 7

Comparative Results among Different Types of Generalized Integrals

Hemanta Kalita

Mathematics Division, VIT Bhopal University, Kothrikalan, Bhopal-Indore Highway, Madhya Pradesh-466114, India
Email: hemanta30kalita@gmail.com

Anca Croitoru

Faculty of Mathematics, University "Alexandru Ioan Cuza", Bd. Carol I, No. 11, 700506 Iasi, Romania
Email: croitoru@uaic.ro

We present some comparative results concerning set-valued integrals: a Dunford type integral of multifunctions with respect to a multimeasure, a Gould type integral of real functions in relation to a multimeasure, and a Gould type integral of multifunctions relative to a non-negative set function.

1. Introduction

The domain of generalized integrals experienced rapid development due to numerous applications in a large number of fields, such as probabilities, economics, game theory, statistics, artificial intelligence. Many authors have studied different pure and applied aspects of generalized integrability (for instance, [1]-[15]).

The subject of set-valued integrals was intensively studied after 1965, when Aumann defined such an integral for multifunctions, motivated by many important applications in economic mathematics and control theory. There are several approaches in defining a set-valued integral:

I. The integral of a multifunction relative to a measure (MF-M). Different problems in economic mathematics, such as the theory of economic equilibrium or the competitive market theory, led to the need to define an integral for multifunctions. These concepts were introduced and studied in [16]-[30].

143

II. The integral of a function with respect to a multimeasure (F-MM). The need to introduce such an integral appeared in economic mathematics, along with the study of Vind [31] in the theory of economic equilibrium. Some types of this set-valued integral were defined in [32]-[40].

III. The integral of a multifunction in relation to a multimeasure (MF-MM), whose definition was motivated by different problems regarding aggregation operators, processes of subjective evaluation, decision-making theory, diagnosing an illness, accident statistics or efficiency theory. These types of set-valued integrals can be found in [41]-[53].

In this work, we present some comparative results concerning set-valued integrals:

- a Dunford type integral of multifunctions with respect to a multimeasure (MF-MM),

- a Gould type integral of real functions in relation to a multimeasure (F-MM),

- a Gould type integral of multifunctions relative to a non-negative set function (MF-M).

The structure of the work is as follows: Section 2 is dedicated to some preliminaries and basic definitions. In Section 3, we present some comparative results among different types of generalized set-valued integrals. In the final Section 4, some conclusions are exposed.

2. Preliminaries and basic definitions

Everywhere in this work, T is a nonempty set and $\mathcal{C}$ is a σ-algebra of subsets of T. If $(X, \| \cdot \|)$ is a real Banach space, let $KC(X)$ be the family of all nonempty compact convex subsets of X and let h be the Hausdorff metric. For every $E \in KC(X)$, denote $\|E\| = h(E, \{0\}) = \sup_{x \in E} \|x\|$. If $G : T \to KC(X)$ is a multifunction, denote by $\|G\|$ the function defined for every $t \in T$ by $\|G\|(t) = \|G(t)\|$. Denote $\mathbb{N}^* = \{1, 2, 3, \ldots\}$.

Suppose $(X, \| \cdot \|)$ is a real Banach algebra with the identity $\theta \neq 0$. $C(X)$ is the set of all nonempty closed subsets of X, $B(X)$ is the set of all nonempty bounded subsets of X, $BC(X)$ is the family of all nonempty closed bounded subsets of X and $WKC(X)$ is the space of all nonempty weakly compact convex subsets of X. For an arbitrary subset E of X, denote by $\overline{E}$ the closure of E in the topology induced by the norm of X and $cE = X \setminus E$. The operations of addition, product and multiplication with scalars of the sets are usually defined: $A + B = \{a + b \,|\, a \in A, b \in B\}$,

Comparative Results among Different Types of Generalized Integrals 145

$AB = \{ab \mid a \in A, b \in B\}$, $\alpha A = \{\alpha a \mid a \in A\}$, for every $A \subset X$, $B \subset X$, $\alpha \in \mathbb{R}$.

Consider a semigroup $S \subset KC(X)$ with identity $\{\theta\}$ relative to the product of the sets, satisfying the conditions:

- $A, B \in S \Rightarrow AB \in S$,
- $A, B, C \in S \Rightarrow A(B + C) = AB + AC$,
- $A \in S, \alpha \in [0, \infty) \Rightarrow \alpha A \in S$,
- $\{\theta\} \in S$.

Example 1. A semigroup satisfying the conditions above is

$$S = \{E \mid E \in KC(\mathbb{R}), E \subset [0, \infty)\}. \tag{1}$$

Definition 1. A finite family of pairwise disjoint subsets of T, $P = \{B_k\}_{k=1}^{n} \subset \mathcal{C}$, is called a partition of T if $\bigcup_{k=1}^{n} B_k = T$. If $P_1 = \{B_k\}_{k=1}^{n}$ and $P_2 = \{C_j\}_{j=1}^{m}$ are two partitions of T, it is said that P_2 is finer than P_1, denoted $P_1 \leq P_2$ or $P_2 \geq P_1$ if for every $j \in \{1, \ldots, m\}$, there is $k_j \in \{1, \ldots, n\}$ such that $C_j \subset B_{k_j}$.

Definition 2. ([54]) A non-negative set function $\nu : \mathcal{C} \to [0, +\infty)$ is called:

(2.3.a) finitely additive if $\nu(\emptyset) = 0$ and for every disjoint sets $A, B \in \mathcal{C}$, it holds $\nu(A \cup B) = \nu(A) + \nu(B)$;

(2.3.b) monotone if for every $A, B \in \mathcal{C}$, $A \subset B \Rightarrow \nu(A) \leq \nu(B)$;

(2.3.c) σ-null-null-additive if for every $\{A_n\}_{n \in \mathbb{N}} \subset \mathcal{C}$, $\nu(A_n) = 0$, $\forall n \in \mathbb{N} \Rightarrow \nu\left(\bigcup_{n=0}^{\infty} A_n\right) = 0$.

Definition 3. For the set function $\nu : \mathcal{C} \to [0, +\infty]$, its semivariation $\nu^* : \mathcal{P}(T) \to [0, +\infty]$ is defined as $\nu^*(E) = \inf\{\nu(A) \mid E \subset A, E \in \mathcal{C}\}$, $\forall E \in \mathcal{P}(T)$.

Definition 4. Let $\nu : \mathcal{C} \to [0, +\infty)$ be a non-negative set function. A set $E \in \mathcal{C}$ is called an atom of ν if $\nu(E) > 0$ and for every $A \in \mathcal{C}$,

$$A \subset E \Rightarrow \{\nu(A) = 0 \text{ or } \nu(E \setminus A) = 0\}.$$

Example 2. If $T = \{x, y, z\}$, $\mathcal{C} = \mathcal{P}(T)$ and

$$\nu(A) = \begin{cases} 2, & A = T, \\ 1, & A = \{x, y\} \text{ or } A = \{z\}, , \\ 0, & \text{otherwise}, \end{cases}$$

then the set $E = \{x, y\}$ is an atom of ν.

Definition 5. A set multifunction $\Gamma : \mathcal{C} \to KC(X)$ is called a multimeasure if it satisfies the conditions:

(2.7.a) $\Gamma(\emptyset) = \{0\}$,

(2.7.b) $\Gamma(A \cup B) = \Gamma(A) + \Gamma(B)$, $\forall A, B \in KC(X)$, $A \cap B = \emptyset$.

Example 3. ([55]) If $\nu_1, \nu_2 : \mathcal{C} \to [0, +\infty)$ are finitely additive set functions, with $\nu_1(\emptyset) = \nu_2(\emptyset) = 0$, then the set multifunction $\Gamma(A) = [-\nu_1(A), \nu_2(A)]$, $\forall A \in \mathcal{C}$, is a multimeasure.

Definition 6. For an arbitrary set multifunction $\Gamma : \mathcal{C} \to KC(X)$, its variation is defined to be the set function $\overline{\Gamma} : \mathcal{C} \to [0, +\infty]$,

$$\overline{\Gamma}(A) = \sup \left\{ \sum_{k=1}^{n} \| \Gamma(B_k) \|; \, (B_k)_{k=1}^{n} \subset \mathcal{C}, \, B_k \cap B_j = \emptyset, \, (k \neq j), \, \bigcup_{k=1}^{n} B_k = A \right\}$$

for every $A \in \mathcal{C}$.

Remark 1. If Γ is a multimeasure, then $\overline{\Gamma}$ is finitely additive.

Definition 7. Suppose $X = \mathbb{R}$ and S is that of (1). For a set multifunction $\Gamma : \mathcal{C} \to S$, we define

$$S(\Gamma) = \{\nu : \mathcal{C} \to [0, +\infty) \, |$$
$$\nu \text{ is a finitely additive measure and } \nu(A) \in \Gamma(A), \forall A \in \mathcal{C}\}.$$

3. Comparative results among different types of generalized integrals

MF-MM

First, we present some comparisons regarding a Dunford type integral of multifunctions relative to a multimeasure, defined in [42]. The multifunction and the multimeasure take their values in the space of all nonvoid compact convex subsets of a real Banach algebra.

In the sequel, $\Gamma : \mathcal{C} \to S$ is a multimeasure with finite variation.

Definition 8. A multifunction $G : T \to S$ is called simple if it is of the following form:

$$G = \sum_{i=1}^{n} E_i \cdot \mathcal{X}_{B_i},$$

where $E_i \in S$, $i \in \{1, \dots, n\}$, $\{B_i\}_{i=1}^{n} \subset \mathcal{C}$ is a finite partition of T and χ_{B_i}

is the characteristic function of B_i. Its integral on every set $A \in \mathcal{C}$ relative to Γ is defined by $\int_A G d\Gamma = \sum_{i=1}^{n} E_i \Gamma(B_i \cap A)$.

Remark 2. The integral $\int_A G d\Gamma$ is an element of S, for every set $A \in \mathcal{C}$.

We introduced in [42] the following definition of a Dunford type integral of multifunctions relative to Γ.

Definition 9. ([42]) A multifunction $G : T \to S$ is called Γ-integrable (on T) if there exists a sequence of simple multifunctions (G_n), $G_n : T \to S$, $n \in \mathbb{N}$, satisfying the conditions:

(3.3.i) $\forall n \in \mathbb{N}$, $h(G_n, G)$ is $\overline{\Gamma}$-measurable (according to [56]),

(3.3.ii) $h(G_n, G) \xrightarrow{\overline{\Gamma}} 0$ (according to [56]),

(3.3.iii) $\lim_{n,m \to \infty} (D) \int_T h(G_n, G_m) d\overline{\Gamma} = 0$.

The integral of (3.3.iii) is of Dunford type ([56]), simply denoted by $\int_T h(G_n, G_m) d\overline{\Gamma}$.

The sequence (G_n) is called a defining sequence for G. The integral of G (on T) relative to Γ is defined by

$$\int_T G d\Gamma = \lim_{n \to \infty} \int_T G_n d\Gamma. \tag{2}$$

The Γ-integrability on every $A \in \mathcal{C}$ is defined in the usual way.

Denote by $D(A, \Gamma)$ the space of all multifunctions that are Γ-integrable on A, $\forall A \in \mathcal{C}$, according to Definition 3.3.

Definition 10. A real function $g : T \to \mathbb{R}$ is called Γ-simple if $g = \sum_{i=1}^{n} a_i \cdot \chi_{B_i}$, where $a_i \in \mathbb{R}$, $\forall i \in \{1, \ldots, n\}$ and $\{B_i\}_{i=1}^{n} \subset \mathcal{C}$ is a finite partition of T, satisfying the condition: $a_i \neq 0 \Rightarrow \overline{\Gamma}(B_i) < +\infty$.

The integral of g with respect to Γ on an arbitrary set $A \in \mathcal{C}$ is defined by $\int_A g d\Gamma = \sum_{i=1}^{n} a_i \Gamma(B_i \cap A)$.

Definition 11. ([32]) A non-negative real function $g : T \to [0, +\infty)$ is called Brooks Γ-integrable (on T) if there exists a sequence (g_n) of Γ-simple functions, $g_n : T \to [0, +\infty)$, $n \in \mathbb{N}$, such that:

(3.5.i) $g_n \xrightarrow{\overline{\Gamma}^*} g$,

(3.5.ii) $\displaystyle\lim_{n,m\to\infty}\int_T |g_n - g_m|\,d\overline{\Gamma} = 0.$

The sequence (g_n) is called a defining sequence of g. The Brooks integral of g on $A \in \mathcal{C}$ relative to Γ is defined by $(B)\displaystyle\int_A g\,d\Gamma = \lim_{n\to\infty}\int_A g_n\,d\Gamma$ (it shows that the integral does not depend on the defining sequence).

Definition 12. ([32]) A real function $g : T \to \mathbb{R}$ is called Brooks Γ-integrable if g^+ and g^- are both Brooks Γ-integrable, where $g^+ = \max\{f,0\}$ and $g^- = \max\{-f,0\}$. In this case, the Brooks integral of g on $A \in \mathcal{C}$ relative to Γ is defined by:

$$(B)\int_A g\,d\Gamma = (B)\int_A g^+\,d\Gamma - (B)\int_A g^-\,d\Gamma,$$

where $(B)\displaystyle\int_A g^+\,d\Gamma - (B)\int_A g^-\,d\Gamma = \{x - y \mid x \in (B)\int_A g^+\,d\Gamma, y \in (B)\int_A g^-\,d\Gamma\}$.

Denote by $B(A,\Gamma)$ the space of all functions that are Brooks Γ-integrable on A, $\forall A \in \mathcal{C}$.

Remark 3. It shows that the integral of (2) is well defined, since it does not depend on the defining sequence.

Example 4. In some particular cases, the integral of Definition 3.3 is reduced to some classical integrals.

I. If $X = \mathbb{R}$, $S = \{\{y\}; y \in \mathbb{R}\}$, $G = \{g\}$, where $g : T \to \mathbb{R}$ is a real function and G is Γ-integrable, then g is Brooks Γ-integrable and $\displaystyle\int_T G\,d\Gamma = (B)\int_T g\,d\Gamma.$

II. For X, S and G as before, if $\Gamma = \{\nu\}$, where $\nu : \mathcal{C} \to [0,\infty)$ is a finitely additive measure and G is Γ-integrable, then $\displaystyle\int_T G\,d\Gamma = \left\{(D)\int_T g\,d\nu\right\}.$

III. If $X = \mathbb{R}$, $\Gamma = \{\nu\}$, where ν is a finitely additive measure, and G is Γ-integrable, then $\displaystyle\int_T G\,d\Gamma = (MS)\int_T G\,d\nu$ (the integral of Martellotti-Sambucini [20]).

Denote by $MS(A,\nu)$ the space of all multifunctions that are ν-integrable on A, $\forall A \in \mathcal{C}$.

Example 5. Suppose $X = \mathbb{R}$, S is given by (1), $\Gamma = [\nu_1, \nu_2]$, where $\nu_1, \nu_2 : \mathcal{C} \to [0, \infty)$ are finitely additive measures, such that ν_2 is bounded and $\nu_1 \leq \nu_2$, and $G = [g_1, g_2]$, such that $g_1 \leq g_2$ and there exist (g'_n), (g''_n) defining sequences for g_1, g_2 respectively, such that $0 \leq g'_n \leq g''_n$, $\forall n \in \mathbb{N}$. Then G is Γ-integrable and

$$\int_T Gd\Gamma = \left[\int_T g_1 d\nu_1, \int_T g_2 d\nu_2 \right].$$

We begin by presenting a comparison with the MS-integral.

Consider the following hypothesis: (H) $X = \mathbb{R}$ and S is that of (1).

Theorem 1. *The following inclusions hold:*

(3.10.i) $D(A, \Gamma) \subset MS(A, \overline{\Gamma})$, $\forall A \in \mathcal{C}$.

In addition, in hypothesis (H), $(MS) \int_A Gd\overline{\Gamma} \subset \int_A Gd\Gamma$, $\forall A \in \mathcal{C}$.

(3.10.ii) Suppose $\mathcal{S}(\Gamma) \neq \emptyset$ and let $\nu \in \mathcal{S}(\Gamma)$. Then $D(A, \Gamma) \subset MS(A, \overline{\nu})$.

In addition, in hypothesis (H), $(MS) \int_A Gd\overline{\nu} \subset \int_A Gd\Gamma$, $\forall A \in \mathcal{C}$.

Now, we give a comparative result with the Brink-Maritz integral ([41]). We recall the following definition.

Definition 13. ([41]) Let $G : T \to S$ be a multifunction such that $\mathcal{S}(\Gamma) \neq \emptyset$. The Brink-Maritz integral of G relative to Γ is defined for every $A \in \mathcal{C}$ by

$$(BM) \int_A Gd\Gamma = \left\{ \int_A gd\nu \,|\, \nu \in \mathcal{S}(\Gamma), g \in \mathcal{S}^1(G) \right\},$$

where $\mathcal{S}^1(G) = \{g \in L^1(\nu); g(t) \in G(t) \; \nu - a.e. \; t \in A\} \neq \emptyset$.

G is called Brink-Maritz Γ-integrable on $A \in \mathcal{C}$ if $(BM) \int_A Gd\Gamma \neq \emptyset$.

Denote by $BM(A, \Gamma)$ the space all Brink-Maritz Γ-integrable multifunctions on A.

Theorem 2. *Suppose $\mathcal{S}(\Gamma) \neq \emptyset$, the hypothesis (H) is fulfilled and let $G : T \to S$ be a multifunction such that $\mathcal{S}^1(G) \neq \emptyset$. Then for every $A \in \mathcal{C}$ it holds $D(A, \Gamma) \subset BM(A, \Gamma)$ and $(BM) \int_A Gd\Gamma \subset \int_A Gd\Gamma$.*

The next theorem gives a comparison with the Brooks integral.

Theorem 3. *If $G \in D(T, \Gamma)$, then $\|G\| \in B(T, \Gamma)$. Besides, in hypothesis (H), $(B) \int_A \|G\| d\Gamma \subset \int_A Gd\Gamma$, $\forall A \in \mathcal{C}$.*

F-MM

Secondly, we present some comparative results concerning a Gould type integral of real functions relative to a multimeasure.

We recall the following definitions.

Definition 14. ([40]) Suppose $\Gamma : \mathcal{C} \to BC(X)$ is monotone and of finite variation. A bounded real function $g : T \to \mathbb{R}$ is called Gould Γ-integrable (on T) if there exists a set $U \in BC(X)$ satisfying the properties: for every $\varepsilon > 0$, there is P_ε a partition of T such that for every $P = \{A_i\}_{i=1}^n$, a partition of T, with $P \geq P_\varepsilon$ and every $t_i \in A_i$, $i \in \{1, \ldots, n\}$, it holds $H(\sigma(P), U) < \varepsilon$, where $\sigma(P) = \sum_{i=1}^n g(t_i)\Gamma(A_i)$. The set U is called the integral of G relative to Γ (on T) and is denoted by $U = (G) \int_T g d\Gamma$. The Gould Γ-integrability on every set $A \in \mathcal{C}$ is usually defined. For every $A \in \mathcal{C}$, denote by $G(A, \Gamma)$ the space of all Gould Γ-integrable multifunctions on A.

Definition 15. ([38]) Suppose $\Gamma : \mathcal{C} \to C(X)$. For every $A \in \mathcal{C}$ and a real function $g : S \to \mathbb{R}$, $g \in G(A, \nu)$, denote $\mathcal{S}_g(A, \Gamma) = \{\nu : \mathcal{C} \to X; \ \nu$ is finitely additive, $\nu(B) \in \Gamma(B)$ for every $B \in \mathcal{C}\}$, where $G(A, \nu)$ means the space of all functions that are Gould ν-integrable on A ([57]).

A real function $g : T \to \mathbb{R}$ is called Aumann-Gould (AG) Γ-integrable (on T) if the set $\left\{ (G) \int_T g d\nu; \ \nu \in \mathcal{S}_g(T, \Gamma) \right\}$, denoted by $(AG) \int_T g d\Gamma$, is nonvoid.

The AG Γ-integrability on every $A \in \mathcal{C}$ is usually defined and denote by $AG(A, \Gamma)$ the space of all AG Γ-integrable functions on A.

Theorem 4. ([38]) Suppose $(T, \mathcal{C}, \overline{\Gamma})$ is complete and X has the Radon-Nikodym Property. Let $\Gamma : \mathcal{C} \to WKC(X)$ be a multimeasure. Then, for every $A \in \mathcal{C}$ it holds $G(A, \Gamma) \subset AG(A, \Gamma)$ and $(G) \int_A g d\Gamma = (AG) \int_A g d\Gamma$.

Theorem 5. Suppose $\Gamma : \mathcal{C} \to KC(X)$ is a multimeasure of total variation. Let $g : T \to [0, +\infty)$ be a non-negative function, $g \in B(T, \Gamma)$, such that there exists (g_n) a defining sequence of g and there is $\alpha \in [0, +\infty)$ such that $0 \leq g_n(t) \leq \alpha$, for every $t \in T$ and every $n \in \mathbb{N}$, satisfying the property:

Comparative Results among Different Types of Generalized Integrals 151

(3.17.i) $\forall \varepsilon > 0$, $\exists \delta > 0$ *s.t.* $\forall n \in \mathbb{N}^*$, $\forall a_i, b_i \in \mathbb{R}$, $|a_i - b_i| < \delta$, $i \in \{1, \ldots, n\}$ *and* $\forall \{A_i\}_{i=1}^n \subset \mathcal{C}$, $\sum_{i=1}^n |a_i - b_i| \cdot |G(A_i)| < \varepsilon$.

Then $g \in D(T, \Gamma)$ *and* $(G) \int_T g d\Gamma = (B) \int_T g d\Gamma$.

Proof. Consider $\varepsilon > 0$. By the hypothesis, it holds $(B) \int_T g d\Gamma = \lim_{n \to \infty} \int_T g_n d\Gamma$, so there exists $n_1 \in \mathbb{N}$ so that $\forall n \geq n_1$, $h\left(\int_T g_n d\Gamma, (B) \int_T g d\Gamma\right) < \frac{\varepsilon}{3}$. From the hypothesis, there is a subsequence of (g_n), denoted also by (g_n), with $g_n \xrightarrow{au} g$. So, there is $B \in \mathcal{C}$, with $\overline{\Gamma}(B) < \frac{\varepsilon}{6\alpha}$, so that $g_n \xrightarrow[cB]{u} g \Rightarrow \exists n_2 \in \mathbb{N}$ s.t. $\forall n \geq n_2$, $|g_n(t) - g(t)| < \delta$, $\forall t \in cB$. Let $n_0 = \max\{n_1, n_2\}$.

Suppose $g_{n_0} = \sum_{i=1}^m b_i \cdot \mathcal{X}_{A_i}$, where $b_i \in [0, +\infty)$, $\forall i \in \{1, \ldots, m\}$ and $P' = \{A_i\}_{i=1}^m$ is a partition of T. Let $P_0 = P' \wedge \{B, cB\}$, an arbitrary partition of T, $P = \{E_i\}_{j=1}^n$, with $P \geq P_0$ and $t_j \in E_j$, $\forall j \in \{1, \ldots, m\}$. From the above inequalities, it results:

$$h\left(\sum_{j=1}^n g(t_j) \cdot \Gamma(E_j), (B) \int_T g d\Gamma\right) \leq h\left(\sum_{j=1}^n g(t_j) \cdot \Gamma(E_j), \int_T g_{n_0} d\Gamma\right) +$$

$$h\left(\int_T g_{n_0} d\Gamma, (B) \int_T g d\Gamma\right) < h\left(\sum_{j=1}^n g(t_j)\Gamma(E_j), \sum_{i=1}^m g_{n_0}(t_i) \cdot \Gamma(A_i)\right) + \frac{\varepsilon}{3} \leq$$

$$h\left(\sum_{\substack{E_i \subset B \\ \cup E_i = B}} g(t_i)\Gamma(E_i), \sum_{\substack{E_i \subset B \\ \cup E_i = B}} g_{n_0}(t_i)\Gamma(E_i)\right) +$$

$$h\left(\sum_{\substack{E_i \subset cB \\ \cup E_i = cB}} g(t_i)\Gamma(E_i), \sum_{\substack{E_i \subset cB \\ \cup E_i = cB}} g_{n_0}(t_i)\Gamma(E_i)\right) + \frac{\varepsilon}{3} \leq \sum_{\substack{E_i \subset B \\ \cup E_i = B}} |g(t_i)| \cdot |\Gamma(E_i)| +$$

$$\sum_{\substack{E_i \subset B \\ \cup E_i = B}} |g_{n_0}(t_i)| \cdot |\Gamma(E_i)| + \sum_{\substack{E_i \subset cB \\ \cup E_i = cB}} |g(t_i) - g_{n_0}(t_i)| \cdot |\Gamma(E_i)| + \frac{\varepsilon}{3} \leq$$

$$2\alpha \sum_{\substack{E_i \subset B \\ \cup E_i = B}} \overline{\Gamma}(E_i) + \frac{\varepsilon}{3} + \frac{\varepsilon}{3} = 2\alpha\overline{\Gamma}(B) + \frac{\varepsilon}{3} + \frac{\varepsilon}{3} < \frac{\varepsilon}{3} + \frac{\varepsilon}{3} + \frac{\varepsilon}{3} = \varepsilon. \text{ So, } g \in D(T, \Gamma)$$

and $(G) \int_T g d\Gamma = (B) \int_T g d\Gamma$. $\qquad\square$

MF-M

Lastly, we highlight some comparison theorems between Gould λ-integrable multifunctions and Birkhoff simple λ-integrable multifunctions. In the sequel, λ is a non-negative set function, $\lambda : \mathcal{C} \to [0, +\infty)$, such that $\lambda(\emptyset) = 0$.

Definition 16. ([26]) A multifunction $G : T \to B(X)$ is called Gould λ-integrable (on T) if there exists a set $U \in BC(X)$ satisfying the property: $\forall \varepsilon > 0, \exists P_\varepsilon$ a finite partition of T s.t. $\forall P = \{C_i\}_{i=1}^n$ a finite partition of T, $P \geq P_\varepsilon$ and $\forall t_i \in C_i, i \in \{1, \ldots, n\}$, $h\left(\sum_{i=1}^n G(t_i)\lambda(C_i), U\right) < \varepsilon.$

The set U is called the Gould integral of G on T relative to λ and is denoted: $U = (G) \int_T Gd\lambda$. The Gould λ-integrability on every $A \in \mathcal{C}$ is defined in the usual way and denote by $G(A, \lambda)$ the space of all multifunctions that are Gould λ-integrable on A.

Definition 17. ([26]) A multifunction $G : T \to B(X)$ is called Birkhoff simple λ-integrable (on T) if there exists a set $V \in BC(X)$ satisfying the property:
$\forall \varepsilon > 0, \exists P_\varepsilon$ a countable partition of T s.t. $\forall P = \{C_n\}_{n \in \mathbb{N}}$ a countable partition of T, $P \geq P_\varepsilon$ and $\forall t_n \in C_n, n \in \mathbb{N}$, $\limsup_n h\left(\sum_{k=1}^n G(t_k)\lambda(C_k), V\right) < \varepsilon.$

The set V is called the Birkhoff simple integral of G on T relative to λ and is denoted: $V = (Bs) \int_T Gd\lambda$. The Birkhoff simple λ-integrability on every $A \in \mathcal{C}$ is usually defined and denote by $Bs(A, \lambda)$ the space of all multifunctions that are Birkhoff simple λ-integrable on A.

Theorem 6. ([26]) *Consider* $(T, \mathcal{C}, \lambda)$ *is a measure space and let* $G : T \to B(X)$ *be a bounded multifunction. Then* $G \in G(T, \lambda) \Leftrightarrow G \in Bs(T, \lambda)$ *and* $(G) \int_T Gd\lambda = (Bs) \int_T Gd\lambda.$

Theorem 7. ([26]) *Consider* $\lambda : \mathcal{C} \to [0, +\infty)$ *to be a* σ-*null-null additive monotone set function. Let* $B \in \mathcal{C}$ *be an atom of* λ *and* $G : T \to B(X)$ *is a multifunction. Then* $G \in G(A, \lambda) \Leftrightarrow G \in Bs(A, \lambda)$ *and* $(G) \int_A Gd\lambda = (Bs) \int_A Gd\lambda.$

4. Conclusions

We present some comparisons regarding different types of set-valued integrals:

- a Dunford type integral for multifunctions relative to a multimeasure;
- a Gould type integral of real functions with respect to a multimeasure;
- comparative results between a Gould type integral of multifunctions and a Birkhoff simple integral of multifunctions in relation to a non-negative set function.

In future work, we look for comparison results with other integrals of Choquet type or Henstock-Kurzweil type.

References

[1] Precupanu A. M. - Some applications of the regular multimeasures, An. St. Univ. "Al. I. Cuza" Iasi, 31 (1985), 5-15.

[2] Chiţescu, I. - Finitely purely atomic measures: coincidence and rigidity properties, Rendiconti del Circolo Matematico di Palermo, Serie II, Tomo L (2001), 455-476.

[3] Sambucini A. R. - A survey on multivalued integration, Atti Sem. Mat. Fis. Univ. Modena 50 (2002), 53-63.

[4] Zhang D., Guo C., Liu D. - Set-valued Choquet integrals revisited, Fuzzy Sets Syst. 147 (2004) 475-485.

[5] Mahmoud, Muhammad M.A.S., Moraru S. - Accident rates estimation modeling based on human factors using fuzzy c-means clustering algorithm, World Academy of Science, Engineering and Technology (WASET), 2012, 64, 1209-1219.

[6] Stamate C., Croitoru A. - Non-linear integrals, properties and relationships, in: Recent Advances in Telecommunications, Signals and Systems (Proceedings of NOLASC 13), WSEAS Press, 2013, pp. 118-123.

[7] Klement E. P., Li J., Mesiar R., Pap E. - Integrals based on monotone set functions, Fuzzy Sets and Systems, 281 (2015), 88-102. DOI: 10.1016/j.fss.2015.07.010.

[8] Pap E. - Multivalued functions integration: from additive to arbitrary non-negative set function, in: S. Saminger-Platz, R. Mesiar (Eds.), On Logical, Algebraic and Probabilistic Aspects of Fuzzy Set Theory, Springer, 2016, pp. 257-274.

[9] Patriche, M., Minimax theorems for set-valued maps without continuity assumptions, Optimization 65 (5) (2016), 957-976.
Doi: 10.1080/02331934.2015.1091822.

[10] Candeloro D., Croitoru A., Gavriluţ A., Sambucini A. R. - Atomicity related to non-additive integrability, Rend. Circolo Matem. Palermo, Volume 65 (3) (2016), 435-449. DOI: 10.1007/s12215-016-0244-z.

[11] Sambucini A. R. - The Choquet integral with respect to fuzzy measures and applications, Math. Slov., 67 (6) (2017), 1427-1450. DOI: 10.1515/ms-2017-0049.

[12] Candeloro D., Mesiar R., Sambucini A.R., A special class of fuzzy measures: Choquet integral and applications, Fuzzy Sets and Syst. 355 (2019), 83-99. DOI: 10.1016/j.fss.2018.04.008.

[13] Kalita H., Hazarika B. - A convergence theorem for ap-Henstock-Kurzweil integral and its relation to topology, Filomat, 6(20) (2022), 6831-6839.

[14] Kalita H., Hazarika B., Becerra T. P. - On AP-Henstock-Kurzweil integrals and non-atomic Radon measure, Mathematics 11 (6), 1552, (2023), 1-17.

[15] Di Piazza L., Musial K., Marraffa V., Sambucini A. R. - Convergence for varying measures, J. Math. Anal. Appl. 518 (2023) Doi: 10.1016/j.jmaa.2022.126782.

[16] Aumann, R.J. - Integrals of set-valued maps, J. Math. Anal. Appl. 12 (1965), 1-12.

[17] Debreu G. - Integration of correspondences, in: 5-th Berkely Symposium on Math. Stat. Prob. II, Part I, 1967, pp. 351-372.

[18] Hukuhara M. - Integration des applications mesurables dont la valeur est un compact convexe, Funkc. Ekvac., 10 (1967), 205-223.

[19] Arstein Z., Byrns J. - Integration of compact set-valued functions, Pac. J. Math. 58 (1975), 297-307.

[20] Martellotti A., Sambucini A. R. - A Radon-Nikodym theorem for multimeasures, Atti Sem. Mat. Fis. Univ. Modena, 42 (1994), 579-599.

[21] Jang L. C., Kil B. M., Kim Y. K., Kwan J. S. - Some properties of Choquet integrals of set-valued function, Fuzzy Sets Syst. 91 (1997), 95-98.

[22] Di Piazza L., Musial K. - Set-valued Kurzweil-Henstock-Pettis integral, Set-Valued Anal. 13 (2) (2005), 167-179.

[23] Croitoru A. - Fuzzy integral of measurable multifunctions, Iran. J. Fuzzy Syst. 9 (4) (2012), 133-140

[24] Croitoru A. - Strong integral of multifunctions relative to a fuzzy measure, Fuzzy Sets Syst. 244 (2014), 20-33.

[25] Cascales B., Rodriguez J. - Birkhoff integral for multi-valued functions, J. Math. Anal. Appl. 297 (2004), 540-560.

[26] Candeloro D., Croitoru A., Gavriluţ A., Sambucini A. R. - An extension of the Birkhöff integrability for multifunctions, Mediterr. J. Math. 13 (5) (2016), 2551-2575. http://dx.doi.org/10.1007/s0009-015-0639-7.

[27] Cascales B., Kadets V., Rodriguez J. - The Pettis integral for multivalued functions via single-valued ones, J. Math. Anal. Appl. 332 (2007), 1-10.

[28] Cichon K., Cichon M., Satco B. - Differential inclusions and multivalued integrals, Discuss. Math., Differ. Incl. Control Optim. 39 (2013), 171-191.

[29] Boccuto A., Sambucini A.R. - A note on comparison between Birkhoff and McShane-type integrals for multifunctions, Real Anal. Exch. 37 (2) (2011/2012), 315-324.

[30] Stamate C. - About fuzzy integrals for vector valued multifunctions, in: Recent Researchers in Neural Networks, Fuzzy Systems, Evolutionary Computing and Automation, 2011, pp. 110-113.

[31] Vind K. - Edgeworth allocations in an exchange economy with many traders, Econ. Rev. 5 (1964), 165-177.

[32] Brooks, J.K. - An integration theory for set-valued measures, I/II, Bull. Soc. Roy. Sci. de Liege, 37 (1968), 312-319/375-380.

[33] Godet-Thobie C. - Multimesures et multimesures de transition, These de Doctorat, 1975, Univ. des Sci. et Techn. du Languedoc, Montpellier.

[34] Helsel R. G., Pu H. W. - A set-valued integral, Bull. Soc. Roy. Sci. Liege, 33 (1964), 272-286.

[35] Kandilakis D. A. - On the extension of multimeasures and integration with respect to a multimeasure, Proc. Amer. Math. Soc. 116 (1992), 85-92.

[36] Precupanu A. M., Croitoru A. - A Gould type integral with respect to a multimeasure I, Analele Stiintifice ale Univ. "Al. I. Cuza" Iasi, 48 (2002), 165-200.

[37] Precupanu A. M., Croitoru A. - A Gould type integral with respect to a multimeasure II, Analele Stiintifice ale Univ. "Al. I. Cuza" Iasi, 49 (2003), 183-207.

[38] Precupanu A., Satco B. - The Aumann-Gould integral, Mediterr. J. Math. 5 (2008), 429-441.

[39] Gavriluţ A. - The general Gould type integral with respect to a multisubmeasure, Math. Slovaca 58 (2008), 43-62.

[40] Precupanu A. M., Gavriluţ A., Croitoru A. - A fuzzy Gould type integral, Fuzzy Sets and Systems, 161 (2010), 661-680.

[41] Brink, H. E., Maritz, P. - Integration of multifunctions with respect to a multimeasure, Glasnik Math. 35 (2000), 313-334.

[42] Croitoru A. - A set-valued integral, Analele Stiintifice ale Univ. "Al. I. Cuza" Iasi, 44 (1998), 101-112.

[43] Croitoru A. - A Radon-Nikodym theorem for multimeasures, Analele Stiintifice ale Univ. "Al. I. Cuza" Iasi, 44 (1988), 395-402.

[44] Croitoru, A. - An integral for multifunctions with respect to a multimeasure, Analele Stiintifice ale Univ. "Al. I. Cuza" Iasi, 49 (2003), 95-106.

[45] Croitoru, A. - On a set-valued integral, Carpathian J. Math. 19 (2003), 41-50.

[46] Croitoru, A., Godet-Thobie, C. - Set-valued integration in seminorm. I, Annals of Univ. of Craiova, Math. Comp. Sci. Ser. 33 (2006), 16-25.

[47] Croitoru, A., Godet-Thobie, C. - Set-valued integration in seminorm. II, Analele Stiintifice ale Univ. Ovidius Constanta, 13 (2005), 55-66.

[48] Croitoru, A. - Multivalued version of Radon-Nikodym theorem, Carpathian J. Math. 21 (2005), 27-38.

[49] Croitoru A. - Lebesgue type convergence theorems, Studia Univ. Babes-Bolyai, Mathematica, 52 (2007), 39-50.

[50] Stamate C., Croitoru A. - The general Pettis-Sugeno integral of vector multifunctions relative to a vector fuzzy multimeasure, Fuzzy Sets Syst. 327 (2017), 123-136.

[51] Iosif A., Gavriluţ A. - Integrability in interval-valued (set) multifunctions setting, Bul. Inst. Politehnic Iasi, Secţia MATEMATICĂ. MECANICĂ TEORETICĂ. FIZICĂ, 63 (67), No 1, (2017), 65-79.

[52] Pap E., Iosif A., Gavriluţ A. - Integrability of an Interval-valued Multifunction with respect to an Interval-valued Set Multifunction, Iranian Journal of Fuzzy Systems, 15 (3), (2018), 47-63.

[53] Costarelli D., Croitoru A., Gavriluţ A., Iosif A., Sambucini A. R. - The Riemann-Lebesgue integral of interval-valued multifunctions, Mathematics 8 (12), 2250 (2020), 1-17.

[54] Croitoru A., Gavriluţ A. - Comparison Between Birkhoff Integral and Gould Integral, Mediterr. J. Math. 12 (2015), 329-347. DOI: 10.1007/s00009-014-0410-5.

[55] Gavriluţ A. - Regular set multifunctions, Publishing House PIM, Iasi, 2012.

[56] Dunford N., Schwartz J. - Linear Operators, I. General Theory, Interscience, New York, 1958.

[57] Gould G. G. - Integration over vector-valued measures, Proc. London Math. Soc. 15 (1965), 193-225.

© 2025 World Scientific Publishing Company
https://doi.org/10.1142/9789819812202_0008

Chapter 8

The Heat Equation with the L^p Primitive Integral

Erik Talvila

Department of Mathematics & Statistics,
University of the Fraser Valley,
Abbotsford, BC Canada V2S 7M8
Email: Erik.Talvila@ufv.ca

For each $1 \leq p < \infty$ a Banach space of integrable Schwartz distributions is defined by taking the distributional derivative of all functions in $L^p(\mathbb{R})$. Such distributions can be integrated when multiplied by a function that is the integral of a function in $L^q(\mathbb{R})$, where q is the conjugate exponent of p. The heat equation on the real line is solved in this space of distributions. The initial data is taken to be the distributional derivative of an $L^p(\mathbb{R})$ function. The solutions are shown to be smooth functions. Initial conditions are taken on in norm. Sharp estimates of solutions are obtained and a uniqueness theorem is proved.

1. Introduction

In this paper, we study solutions of the heat equation on the real line with initial value data that is the distributional derivative of an L^p function $(1 \leq p < \infty)$. The solution is given by convolution of the Gauss–Weierstrass heat kernel with the initial data. This produces smooth solutions to the heat equation. We introduce a norm that is isometric to the L^p norm and show that solutions tend to the initial condition within this norm as $t \to 0^+$. Since the initial data is the distributional derivative of an L^p function it can fail to have pointwise values everywhere, can be zero almost everywhere, and can be the difference of translated Dirac distributions. And yet, the initial conditions are taken on in the strong (norm) sense rather than in the weak (distributional) sense.

For $u : \mathbb{R} \times (0, \infty) \to \mathbb{R}$ write $u_t(x) = u(x, t)$.

The classical problem of the heat equation on the real line is, given a function $F \in L^p$ for some $1 \leq p \leq \infty$, find a function $u : \mathbb{R} \times (0, \infty) \to \mathbb{R}$

157

158   E. Talvila

such that $u_t \in C^2(\mathbb{R})$ for each $t > 0$, $u(x, \cdot) \in C^1((0, \infty))$ for each $x \in \mathbb{R}$ and

$$\frac{\partial^2 u(x,t)}{\partial x^2} - \frac{\partial u(x,t)}{\partial t} = 0 \text{ for each } (x,t) \in \mathbb{R} \times (0, \infty) \tag{1}$$

$$\lim_{t \to 0^+} \|u_t - F\|_p = 0. \tag{2}$$

If $p = \infty$, then F is also assumed to be continuous. Under suitable growth conditions on u, the unique solution is given by $u(x,t) = f * \Theta_t(x)$, where the Gauss–Weierstrass heat kernel is $\Theta_t(x) = \exp(-x^2/(4t))/(2\sqrt{\pi t})$.

Instead of the initial data being an L^p function, we take it to be the distributional derivative of such a function. A well-defined integration process for these functions and distributions appears in [16]. The space of derivatives of L^p functions is denoted L'^p. It is a Banach space isometrically isomorphic to L^p. In L'^p, many properties of L^p functions continue to hold, such as convolution with functions in Lebesgue spaces, a type of Hölder inequality, continuity of translations, etc. Some of the distributions in L'^p can be identified with signed measures, some can be identified with smooth functions and some have no pointwise values.

See [3] and [19] for classical results on the heat equation, including extensive bibliographies. For initial data in L^p spaces, see [9] and [10]. Distributional solutions of the heat equation are considered in [4] and [14], where the initial data is taken on in the distributional sense. For initial data in the Sobolev space H^s, see [12]. Another paper studying the heat equation with distributions in Banach spaces is [17], in which the initial data is taken to be the distributional derivative of a continuous function.

For $1 \leq p < \infty$, the Lebesgue space on the real line is L^p, which is the set of measurable functions $f : \mathbb{R} \to \mathbb{R}$ such that $\int_{-\infty}^{\infty} |f(x)|^p \, dx < \infty$. To distinguish between other types of integrals introduced later, Lebesgue integrals will always explicitly show the integration variable and differential as above.

The Schwartz space (test functions), of rapidly decreasing smooth functions, is denoted $\mathcal{S}$ and the tempered distributions $\mathcal{S}'$. See, for example, [7] or [8].

The outline of the paper is as follows.

The heat equation is considered in L'^p in Section 2. With initial data $f \in L'^p$ it is proved that convolution with f and the Gauss–Weierstrass heat kernel is a smooth function in L'^s for each $p \leq s < \infty$ but need not be in L'^s if $1 \leq s < p$ (Theorem 1). Sharp estimates are given in $\|\cdot\|'_r$, with $1/p + 1/q = 1 + 1/r$. This convolution gives the unique solution to the heat equation (Corollary 1). Existence is proven using a type of Hardy space. In Proposition 1, continuity is proved with respect to the initial conditions.

In Section 3, various sharp L^p and pointwise estimates are given. It is shown that these solutions cannot be of one sign, have limit 0 at infinity, and have a vanishing integral over the real line.

An appendix contains a lemma on pointwise differentiation of convolutions and proof of some sharp estimates for solutions of (1) and (2).

2. Solutions of the heat equation in L'^p

For $1 \leq p < \infty$, let $L'^p = \{f \in \mathcal{S}' \mid f = F'$ for some $F \in L^p\}$. As explained in [16], if $f \in L'^p$ then it has a unique primitive in L^p. In this case, $||f||'_p = ||F||_p$ and which makes L'^p and L^p isometrically isomorphic Banach spaces. As in [16], define $I^p = \{G : \mathbb{R} \to \mathbb{R} \mid G(x) = \int_0^x g(t)\, dt$ for some $g \in L^p\}$.

If $f \in L'^p$, then $f * \Theta_t$ is a smooth solution of the heat equation. The initial value f is taken on in the $||\cdot||'_p$ norm and norm boundedness ensures uniqueness. Solutions are shown to be in L'^r for each $p \leq r < \infty$ and sharp norm estimates are given using the results of Theorem 4 in the Appendix. If a solution v_t of the heat equation has bounded norms $||v_t||'_p$ for $t > 0$, then there exists a distribution $f \in L'^p$ such that $v_t = \Theta_t * f$. This is proved using an analogue of a harmonic Hardy space. There is continuity with respect to initial conditions. The example of initial data that is the difference to two Dirac measures is considered.

Theorem 1. *Let $1 \leq p < \infty$. Let $f \in L'^p$ with primitive $F \in L^p$. Let $t > 0$ and define $v(x,t) = v_t(x) = f * \Theta_t(x)$.*
(a) For each $x \in \mathbb{R}$ we have

$$v_t(x) = \int_{-\infty}^{\infty} f(y)\Theta_t(x - y)\, dy = F * \Theta_t'(x) = \Theta_t' * F(x) = (F * \Theta_t)'(x).$$

The function v_t is uniformly continuous on $\mathbb{R}$.
(b) For each $n \geq 1$ we have $v_t^{(n)}(x) = f * \Theta_t^{(n)}(x)$ and $\partial^n v(x,t)/\partial t^n = f * \partial^n \Theta(x,t)/\partial t^n$. Hence, $v \in C^\infty(\mathbb{R}) \times C^\infty((0,\infty))$.
(c) If $p \leq s < \infty$, then $f * \Theta_t \in L'^s$.
(d) Let $q, r \in [1, \infty]$ such that $1/p + 1/q = 1 + 1/r$. There is a constant $K_{p,q}$ such that $||f * \Theta_t||'_r \leq K_{p,q}||f||'_p\, t^{-(1-1/q)/2}$ for all $t > 0$. The estimate is sharp in the sense that if $\psi : (0,\infty) \to (0,\infty)$ such that $\psi(t) = o(t^{-(1-1/q)/2})$ as $t \to 0^+$ or $t \to \infty$ then there is $g \in L'^p$ such that $||g * \Theta_t||'_r/\psi(t)$ is not bounded. The constant $K_{p,q}$ cannot be replaced

160   *E. Talvila*

by any smaller number. In particular, $||f * \Theta_t||'_p \leq ||f||'_p$ is a sharp estimate.

(e) If $1 \leq s < p$ then, $f * \Theta_t$ need not be in L'^s.

Note that if $1/p + 1/q = 1 + 1/r$ then, for suitable q, r can take on any of the values in $[p, \infty]$, with $r = p$ corresponding to $q = 1$ and $r = \infty$ corresponding to p and q being conjugates. And, for suitable r, q can take on any of the values in $[1, p/(p-1)]$, with $q = 1$ corresponding to $r = p$ and $q = p/(p-1)$ corresponding to $r = \infty$. If $p = 1$, then q can take on all values in $[1, \infty]$.

The constants $K_{p,q}$ in Theorem 4(b) and Part (d) above are the same. When $r = p$ and $q = 1$, the inequality in Part (d) reads $||f * \Theta_t||'_p \leq ||f||'_p$.

Proof. (a) Since Θ_t and its derivatives are in each L^q space, the equalities follow by Theorem 5.1 in [16], Lemma 1 in the Appendix, and usual properties of convolution for Lebesgue integrals. To prove uniform continuity, note that if q is the conjugate of p then

$$|v_t(x) - v_t(y)| = |F * [\Theta'_t(x) - \Theta'_t(y)]| \leq ||F||_p ||\Theta'_t(x - \cdot) - \Theta'_t(y - \cdot)||_q \to 0$$

as $y \to x$. The last line uses the Hölder inequality and continuity in the L^q norm. (When $q = \infty$, this requires boundedness and uniform continuity of Θ'_t.)

(b) Differentiation in x follows from Lemma 1. For differentiation in t, use Part (a) to write

$$v_t(x) = F * \Theta'_t(x) = -\frac{1}{4\sqrt{\pi} t^{3/2}} \int_{-\infty}^{\infty} F(x - y) y e^{-y^2/(4t)} \, dy.$$

Then

$$\frac{\partial v(x,t)}{\partial t} = \frac{3}{8\sqrt{\pi} t^{5/2}} \int_{-\infty}^{\infty} F(x - y) y e^{-y^2/(4t)} \, dy$$

$$- \frac{1}{16\sqrt{\pi} t^{7/2}} \int_{-\infty}^{\infty} F(x - y) y^3 e^{-y^2/(4t)} \, dy.$$

Differentiation under the integral sign is justified by dominated convergence since, if $t \leq t_0$, then $|y^3 e^{-y^2/(4t)}| \leq |y|^3 e^{-y^2/(4t_0)}$.

(c), (d) From Theorems 3.6 and 5.1 in [16] we have

$$||f * \Theta_t||'_r = ||F * \Theta'_t||'_r = ||(F * \Theta_t)'||'_r = ||F * \Theta_t||_r \leq C_{p,q} ||F||_p ||\Theta_t||_q.$$

The rest of the proof is identical to Theorem 4(b).

The Heat Equation with the L^p Primitive Integral 161

(e) Since $v_t = (F * \Theta_t)' \in L'^s$ if and only if $F * \Theta_t \in L^s$, this result is equivalent to Theorem 4(c). □

To prove uniqueness for the initial value problem we use a classical theorem, which follows from Theorem 6.1 in [19].

Theorem 2. *Let* $u : \mathbb{R} \times (0, \infty) \to \mathbb{R}$ *such that* $u_t \in C^2(\mathbb{R})$ *for each* $t > 0$, $u(x, \cdot) \in C^1((0, \infty))$ *for each* $x \in \mathbb{R}$, $u \in C(\mathbb{R} \times [0, \infty))$, $\partial^2 u(x, t)/\partial x^2 - \partial u(x, t)/\partial t = 0$ *for* $(x, t) \in \mathbb{R} \times (0, \infty)$, u *is bounded*, $u(x, 0) = f(x)$ *for a bounded continuous function* $f : \mathbb{R} \to \mathbb{R}$. *Then the unique solution is given by* $u_t(x) = f * \Theta_t(x)$.

Corollary 1. *Let* $1 \leq p < \infty$.
(a) Let $f \in L'^p$. *The unique solution of the problem to find* $v : \mathbb{R} \times (0, \infty) \to \mathbb{R}$ *such that*

$$v_t \in L'^p \text{ for each } t > 0 \tag{3}$$

$$v_t \in C^2(\mathbb{R}) \text{ for each } t > 0, v(x, \cdot) \in C^1((0, \infty)) \text{ for each } x \in \mathbb{R} \tag{4}$$

$$\frac{\partial^2 v(x, t)}{\partial x^2} - \frac{\partial v(x, t)}{\partial t} = 0 \text{ for each } (x, t) \in \mathbb{R} \times (0, \infty) \tag{5}$$

$$||v_t||'_p \text{ is bounded} \tag{6}$$

$$\lim_{t \to 0^+} ||v_t - f||'_p = 0 \tag{7}$$

is given by $v_t = f * \Theta_t$.
(b) If $f \in L'^p$ *and* $v_t = f * \Theta_t$, *then* $\lim_{t \to 0^+} ||v_t||'_p = ||f||'_p$.
(c) Let $1 < p < \infty$ *and let* h'^p *be the set of functions satisfying* (3), (4), (5) *and* (6). *If* $v \in h'^p$, *define* $||v||_p^{'h} = \sup_{t > 0} ||v_t||'_p$. *Let* $T : L'^p \to h'^p$ *be given by* $T[f] = \Theta_t * f$. *Then* T *is a linear isometric isomorphism.*
(d) Let $1 < p < \infty$. *If* (3), (4), (5) *and* (6) *hold for some* $v \in h'^p$, *then there is a unique distribution* $f \in L'^p$ *such that* $v_t = \Theta_t * f$ *and* (7) *holds.*

The proof uses methods from [10, Theorem 9.2, p. 195] and [1, pp. 115-120]. Part (c) implies h'^p is a Banach space.

Proof. (a) By Part (b) of the theorem we can differentiate under the integral sign and this shows $v_t = f * \Theta_t$ is a smooth solution of the heat equation. By Part (c) of the theorem, $v_t \in L'^p$. To see that the initial conditions are taken on in the $||\cdot||'_p$ norm, note that $||v_t - f||'_p = ||F * \Theta_t - F||_p$. Since $\int_{-\infty}^{\infty} \Theta_t(y) \, dy = 1$ we have

$$||F * \Theta_t - F||_p = \left(\int_{-\infty}^{\infty} \left| \int_{-\infty}^{\infty} [F(x - y) - F(x)] \Theta_t(y) \, dy \right|^p dx \right)^{1/p}.$$

By the Minkowski inequality for integrals, for example [7, p. 194],

$$\lim_{t \to 0^+} ||F * \Theta_t - F||_p \leq \lim_{t \to 0^+} \int_{-\infty}^{\infty} ||F(\cdot - y) - F(\cdot)||_p \Theta_t(y)\, dy$$
$$= \lim_{y \to 0} ||F(\cdot - y) - F(\cdot)||_p = 0.$$

The last line follows by the fact that Θ_t is an approximate identity (delta sequence) and by continuity in the L^p norm.

To prove the solution is unique, suppose there were two solutions v and w. Let $z = v - w$. Then z satisfies (3), (4), (5) and (6). And,

$$||z_t||_p' \leq ||v_t - f||_p' + ||w_t - f||_p' \to 0 \text{ as } t \to 0^+.$$

Now define $K : \mathbb{R} \to \mathbb{R}$ by $K(x) = 1 - |x|$ for $|x| \leq 1$ and $K(x) = 0$ for $|x| \geq 1$, and let $K_h(x) = (1/h)K(x/h)$ for $h > 0$. Then K_h is absolutely continuous with compact support. Note that $\int_{-\infty}^{\infty} K(y)\, dy = \int_{-\infty}^{\infty} K_h(y)\, dy = 1$. We have

$$K_h'(x) = \left\{ \begin{array}{ll} 0, & |x| > h \\ -\text{sgn}(x)/h^2, & |x| < h. \end{array} \right.$$

Now define $\psi_h(x,t) = K_h * z_t(x) = \int_{-h}^{h} K_h(y)z_t(x - y)\, dy$. Since $\partial z_t(x)/\partial t$ and $z_t''(x)$ are continuous on compact intervals in $\mathbb{R} \times (0, \infty)$ we can use dominated convergence to differentiate under the integral sign and this shows ψ_h is a solution of the heat equation. By usual properties of convolutions, $\psi_h \in C^2(\mathbb{R}) \times C^1((0, \infty))$. From the Hölder inequality [16, Theorem 3.6], $|\psi_h(x,t)| \leq ||z_t||_p' ||K_h'||_q$, where q is the conjugate of p. And, $||K_h'||_q = 2^{1-1/p}/h^{1+1/p}$. It follows that ψ_h is bounded on $\mathbb{R} \times (0, \infty)$ and $\lim_{t \to 0^+} \psi_h(x,t) = 0$. Define $\psi_h(x,0) = 0$. The above estimate now shows $\psi_h \in C(\mathbb{R} \times [0, \infty))$. By Theorem 2, $\psi_h = 0$. Since z_t is continuous, for $(x,t) \in \mathbb{R} \times (0, \infty)$ we get $\lim_{h \to 0^+} \psi_h(x,t) = 0 = \lim_{h \to 0^+} K_h * z_t(x) = z_t(x) \lim_{h \to 0^+} \int_{-\infty}^{\infty} K_h(y)\, dy = z_t(x)$.

(b) This follows from (7) and the triangle inequality.

(c) Clearly, T is linear.

Show T is an isometry. From Theorem 1(d), $||T[f]||_p'^h \leq \sup_{t>0} ||f||_p' = ||f||_p'$. And, if $f \in L'^p$ then

$$||f||_p' \leq ||f - f * \Theta_t||_p' + ||f * \Theta_t||_p' \leq ||f - f * \Theta_t||_p' + ||T[f]||_p'^h.$$

Letting $t \to 0^+$ and using (7) shows $||f||_p' \leq ||T[f]||_p'^h$. Hence, T is an isometry and necessarily one-to-one.

Now show T is onto h'^p. Let $v \in h'^p$ and find $f \in L'^p$ such that $T[f] = v$. For $h > 0$ use notation from the proof of (a) and define $w_h = K_h * v_t$. Let

$\delta > 0$. Then the function $(x,t) \mapsto w_h(x, t+\delta)$ satisfies (3), (4), (5) and (6) and is continuous and bounded on $\mathbb{R} \times [0, \infty)$. By Theorem 2, we have $w_h(x, t+\delta) = \int_{-\infty}^{\infty} \Theta_t(x-y)w_h(y,\delta)\,dy$. And, $||K_h * v_\delta - v_\delta||_p' \to 0$ as $h \to 0^+$ by [16, Theorem 5.3(g)]. (The second sentence in this part of the theorem should read, Let $F \in L^1$.) If q is the conjugate of p ([16, Theorem 3.3(c)]), then $I^{q*} = L'^p$ so, by weak* convergence, for each $G \in I^q$ we have $\int_{-\infty}^{\infty}(K_h * v_\delta - v_\delta)G \to 0$. In particular, take $G = \Theta_t$ and note Definition 3.5 in [16]. Then, also using the continuity of v,

$$\lim_{h \to 0^+} w_h(x, t+\delta) = v_{t+\delta}(x) = \lim_{h \to 0^+} \int_{-\infty}^{\infty} \Theta_t(x-y)w_h(y,\delta)\,dy$$
$$= \int_{-\infty}^{\infty} \Theta_t(x-y)v_\delta(y)\,dy. \tag{8}$$

By definition, $v \in h'^p$ is norm-bounded in $L'^p = I^{q*}$ where q is the conjugate of p. Every norm-bounded sequence in a separable Banach space contains a weak* convergent subsequence. Since I^q is separable (since $q < \infty$) there is a sequence δ_i that decreases to 0 and a distribution $f \in L'^p$ such that $\int_{-\infty}^{\infty} v_{\delta_i} G \to \int_{-\infty}^{\infty} fG$ for each $G \in I^q$. In particular, take $G = \Theta_t$. Then $\lim_{i \to \infty} \int_{-\infty}^{\infty} v_{\delta_i}(y)\Theta_t(x-y)\,dy = f * \Theta_t(x)$. Replacing δ by δ_i in (8) and letting $i \to \infty$ now shows $v_t = T[f]$.

(d) This follows from (c). $\qquad\square$

The following proposition shows continuity with respect to initial conditions.

Proposition 1. *Let $1 \le p < \infty$. Let $f, g \in L'^p$.*
(a) Let $q, r \in [1, \infty]$ such that $1/p + 1/q = 1 + 1/r$. Let $K_{p,q}$ be the constant in the proof of Theorem 1. Then

$$||f * \Theta_t - g * \Theta_t||_r' \le K_{p,q}||f - g||_p' t^{-(1-1/q)/2}.$$

When $r = p$, $q = 1$, this reads $||f \cdot \Theta_t - g * \Theta_t||_p' \le ||f - g||_r'$.
(b) Let $v, w : \mathbb{R} \times (0, \infty) \to \mathbb{R}$ such that for each $t > 0$ we have v_t and w_t in L'^p. Let $\epsilon > 0$. Suppose $||v_t - f||_p' \to 0$ and $||w_t - g||_p' \to 0$ as $t \to 0^+$. If $f, g \in L'^p$ such that $||f - g||_p' < \epsilon$, then for small enough $t > 0$ we have $||v_t - w_t||_p' < 2\epsilon$.

Proof. (a) This follows from Theorem 1(d) since $||f * \Theta_t - g * \Theta_t||_r' = ||(F - G) * \Theta_t||_r$, where F and G are the respective primitives of f, g in L^p.
(b) This follows from the triangle inequality. $\qquad\square$

We remark the initial conditions are also taken on in the weak (distributional) sense.

Remark 1. If $f \in L'^p$ with primitive $F \in L^p$ for some $1 \leq p < \infty$, then we can define $v(x,t) = v_t(x) = f * \Theta_t(x)$. As shown in Corollary 1, v satisfies the heat equation and the initial condition is taken on in $||\cdot||'_p$ as $t \to 0^+$. Such a solution also takes on the initial condition in the weak distributional sense. To see this let $\phi \in \mathcal{S}$. Then

$$\langle v_t - f, \phi \rangle = \langle (F * \Theta_t - F)', \phi \rangle = -\langle F * \Theta_t - F, \phi' \rangle$$

$$= -\int_{-\infty}^{\infty} [F * \Theta_t(x) - F(x)] \, \phi'(x) \, dx.$$

We know $||F * \Theta_t - F||_p \to 0$ so $\langle F * \Theta_t - F, \psi \rangle \to 0$ for each $\psi \in L^q$ where q is conjugate to p. In particular, this holds when $\psi = \phi'$. Therefore, the initial condition is taken on in the weak distributional sense.

Several distributions in L'^p are given in [16, Section 4], for which (7) of course holds but (2) does not hold in any Lebesgue space. We now consider the example of the difference of two translated Dirac distributions.

Example 1. Denote the Dirac distribution supported at $a \in \mathbb{R}$ by δ_a. Let $F = \chi_{[a,b]}$, then F is in every L^p space. Therefore, $F' = \delta_a - \delta_b$ is in every L'^p space. Hence, L'^p is not a subset of any of the Lebesgue spaces. Since step functions are dense in L^p, linear combinations of such differences of Dirac distributions are dense in L'^p.

Now let $f = \delta_{-a} - \delta_a$ for fixed $a > 0$. Then $v_t(x) = f * \Theta_t(x) = \Theta_t(x + a) - \Theta_t(x - a)$. Note that v is a solution of the heat equation in $\mathbb{R} \times (0, \infty)$. According to Corollary 1, the initial condition is taken on as $||v_t - f||'_p \to 0$ as $t \to 0^+$.

One might wonder in what other sense the initial conditions are realised. We've shown above that they are taken on in the weak distributional sense. For no $1 \leq r \leq \infty$ is $||v_t - f||_r$ defined so (2) is meaningless. As a signed Borel measure, we have $d\mu(x) = [\Theta_t(x+a) - \Theta_t(x-a)]dx - d\delta_{-a}(x) + d\delta_a(x)$. If the Jordan decomposition of μ is $\mu = \mu_+ - \mu_-$, then μ converges to 0 in variation if $V(\mu) = \mu_+(\mathbb{R}) + \mu_-(\mathbb{R}) \to 0$ as $t \to 0^+$. For any measurable set E we have

$$V(\mu) \geq \mu_+(E) + \mu_-(E) \geq |\mu(E)|$$
$$= \left| \frac{1}{2\sqrt{\pi t}} \int_E \left(e^{-(x+a)^2/(4t)} - e^{-(x-a)^2/(4t)} \right) dx - \chi_E(-a) + \chi_E(a) \right|.$$

Take $E = (-\infty, -a)$. Then

$$V(\mu) \geq \frac{1}{2\sqrt{\pi t}} \int_{-\infty}^{-a} \left(e^{-(x+a)^2/(4t)} - e^{-(x-a)^2/(4t)} \right) dx$$

$$= \frac{1}{2\sqrt{\pi t}} \int_0^{2a} e^{-x^2/(4t)} \, dx = \frac{1}{\sqrt{\pi}} \int_0^{a/\sqrt{t}} e^{-y^2} \, dy \to \frac{1}{2} \text{ as } t \to 0^+.$$

Hence, v_t does not converge to f in variation.

This also shows v_t does not converge to f in the Alexiewicz norm. See [15].

3. Estimates

In the previous section, we looked at estimates of $f * \Theta_t$ in L'^p. Here we look at estimates in the spaces L^r where $1/p + 1/q = 1 + 1/r$. If $f \in L'^p$, then $f * \Theta_t \in L^r$ and Young's inequality gives estimates in terms of $||f||'_p$. It is also shown that $f * \Theta_t(x)$ cannot be of one sign, has limit 0 as $|x| \to \infty$, and has a vanishing integral over the real line. Pointwise estimates are given when f has compact support.

Let $1 \leq q \leq \infty$. Then

$$||\Theta'_t||_q = \frac{\delta_q}{t^{(2-1/q)/2}} \text{ where } \delta_q = \begin{cases} \frac{1}{\sqrt{\pi}}, & q = 1, \\ \frac{\Gamma^{1/q}((q+1)/2)}{2^{1-1/q}\sqrt{\pi}q^{(1+1/q)/2)}}, & 1 < q < \infty, \\ \frac{1}{2^{3/2}\sqrt{\pi e}}, & q = \infty. \end{cases} \quad (9)$$

When $q = 1$, the integral is elementary. When $1 < q < \infty$ the calculation uses the definition of the gamma function, $\Gamma(s) = \int_0^\infty e^{-x} x^{s-1} \, dx$. When $q = \infty$, this is a calculus minimisation problem.

Theorem 3. *Let $1 \leq p < \infty$. Let $f \in L'^p$ with primitive $F \in L^p$. Let $t > 0$ and define $v(x,t) = v_t(x) = f * \Theta_t(x)$.*
(a) Let $q, r \in [1, \infty]$ such that $1/p + 1/q = 1 + 1/r$. There is a constant $L_{p,q}$ such that $||f * \Theta_t||_r \leq L_{p,q}||f||'_p t^{-(2-1/q)/2}$ for all $t > 0$. The estimate is sharp in the sense that if $\psi : (0, \infty) \to (0, \infty)$ such that $\psi(t) = o(t^{-(2-1/q)/2})$ as $t \to 0^+$ or $t \to \infty$, then there is $g \in L'^p$ such that $||g * \Theta_t||_r/\psi(t)$ is not bounded as $t \to 0^+$ or $t \to \infty$.
(b) Fix $t > 0$. Then v_t cannot be of one sign.
(c) For each $t > 0$, $\int_{-\infty}^\infty v_t(x) \, dx = 0$.
(d) For each $t > 0$, $\lim_{|x| \to \infty} v_t(x) = 0$.

(e) Suppose f has compact support in $[-R, R]$. Then $v_t(x) = O(|x|^{1/p} e^{-(|x|-R)^2/(4t)})$ for fixed $t > 0$ as $|x| \to \infty$. Or,

$$|v_t(x)| \leq \begin{cases} M_p \|f\|_1' |x| e^{-(|x|-R)^2/(4t)} t^{-3/2}, & p = 1, \quad |x| \geq R + \sqrt{2t}, \\ M_p \|f\|_p' |x|^{1/p} e^{-x^2/(16t)} t^{-(1/2+1/p)}, & 1 < p < \infty, \quad |x| \geq 2R, \end{cases}$$

where

$$M_p = \begin{cases} \dfrac{1}{4\sqrt{\pi}}, & p = 1, \\ \dfrac{3^{1/p}(p-1)^{1-1/p}}{2^{1+2/p}\sqrt{\pi}p^{1-1/p}}, & 1 < p < \infty. \end{cases}$$

In Part (a), remarks about how possible values of q and r are the same as those following Theorem 1.

When $r = p$ and $q = 1$, the inequality in Part (a) reads $\|f * \Theta_t\|_p \leq \|F\|_p/\sqrt{\pi t}$. When $r = \infty$, then p and q are conjugates and the inequality in Part (a) reads $\|f * \Theta_t\|_\infty \leq \delta_q \|F\|_p t^{-(1+1/p)/2}$. When $p = 1$, then $q = r$ and the sharp constant in (a) is $L_{p,q} = \delta_q$. This can be proved similarly to Part (b) of Theorem 4 using $F = \Theta_t^\beta$ in the limit $\beta \to \infty$ (effectively, $f = \delta'$). It is not known in the other cases if the constant $L_{p,q}$ is sharp or not.

Part (c) appears for $f \in L^1(\mathbb{R})$ as Exercise 8.1.5 in [5].

Positive solutions of the heat equation are known to be convolutions of the Gauss–Weierstrass heat kernel with an increasing function. See [19].

An example for Part (d) is $F(x) = x^{-1/p}\sin(x)$ for $x > 1$ and $F(x) = 0$ for $x \leq 1$. Then $\int_{-\infty}^\infty F(y)\,dy$ exists. And, $F \in L^s$ if and only if $s > p$.

Proof. (a) By Young's inequality,

$$\|v_t\|_r = \|F * \Theta_t'\|_r \leq C_{p,q}\|F\|_p\|\Theta_t'\|_q = \frac{C_{p,q}\|f\|_p'\delta_q}{t^{(2-1/q)/2}}.$$

The constant $C_{p,q}$ is given in the proof of Theorem 4. We then take $L_{p,q} = C_{p,q}\delta_q$.

As in the proof of Theorem 4, define $S_t : L'^p \to L^r$ by $S_t[f](x) = f * \Theta_t(x)/\psi(t)$. The estimate $\|S_t[f]\|_r \leq C_{p,q}\delta_q\|f\|_p' t^{-(2-1/q)/2}/\psi(t)$ shows that, for each $t > 0$, S_t is a bounded linear operator. Let $f_t = \Theta_t'$. Then

$$\frac{\|S_t[f_t]\|_r}{\|f_t\|_p'} = \frac{\|\Theta_{2t}'\|_r}{\psi(t)\|\Theta_t\|_p} = \frac{\delta_r}{\alpha_p 2^{(2-1/r)/2}\psi(t)t^{(2-1/q)/2}}.$$

This is not bounded in the limit $t \to 0^+$. Hence, S_t is not uniformly bounded. By the Uniform Bounded Principle, it is not pointwise bounded. Therefore, there is a distribution $g \in L'^p$ such that $\|g * \Theta_t\|_r \neq O(\psi(t))$ as $t \to 0^+$. And, the growth estimate $\|v_t\|_r = O(t^{-(2-1/q)/2})$ as $t \to 0^+$ is sharp. Similarly for sharpness as $t \to \infty$.

The Heat Equation with the L^p Primitive Integral   167

(b) Note that $v_t(x) = f * \Theta_t(x) = (F * \Theta_t)'(x)$ for each $x \in \mathbb{R}$ by Theorem 1(a). If $v_t \geq 0$, then $F * \Theta_t$ is increasing. But $F * \Theta_t \in L^p$ by Theorem 4(a), so $F * \Theta_t = 0$ and hence $v_t = 0$.

(c) From Example 1, we have that linear combinations of differences of Dirac distributions are dense in L'^p. Given $\epsilon > 0$ there are $-\infty < b_0 < b_1 < \cdots < b_N < \infty$ and $a_n \in \mathbb{R}$ such that if $g = \sum_{n=1}^{N} a_n(\tau_{b_{n-1}}\delta - \tau_{b_n}\delta)$ then $||f - g||_p' < \epsilon$. Then, by Theorem 3(a),

$$\left| \int_{-\infty}^{\infty} f * \Theta_t(x)\, dx - \int_{-\infty}^{\infty} g * \Theta_t(x)\, dx \right| \leq ||f - g||_p' t^{-(2-1/q)/2},$$

where q is conjugate to p. It suffices then to show $\int_{-\infty}^{\infty} g * \Theta_t(x)\, dx = 0$. We have

$$\int_{-\infty}^{\infty} g * \Theta_t(x)\, dx = \int_{-\infty}^{\infty} \sum_{n=1}^{N} a_n(\tau_{b_{n-1}}\delta * \Theta_t(x) - \tau_{b_n}\delta * \Theta_t(x))\, dx$$

$$= \sum_{n=1}^{N} a_n \int_{-\infty}^{\infty} (\Theta_t(x - b_{n-1}) - \Theta_t(x - b_n))\, dx = 0.$$

(d) Let $R > 0$. Since $F \in L^1_{loc}$ and Θ_t' is bounded, dominated convergence shows

$$\lim_{|x| \to \infty} \int_{-R}^{R} F(y)\Theta_t'(x - y)\, dy = \int_{-R}^{R} F(y) \lim_{|x| \to \infty} \Theta_t'(x - y)\, dy = 0.$$

Given $\epsilon > 0$ now take $R > 0$ such that $||F\chi_{(R,\infty)}||_p < \epsilon$. By the Hölder inequality,

$$\left| \int_{R}^{\infty} F(y)\Theta_t'(x - y)\, dy \right| \leq ||F\chi_{(R,\infty)}||_p ||\Theta_t'||_q.$$

Similarly, for integration over $(-\infty, -R)$.

(d) From the Hölder inequality, $|v_t(x)| \leq ||F||_p ||\chi_{(x-R,x+R)}\Theta_t'||_q$. When $p = 1$ and $q = \infty$ we compute that if $|x| \geq R + \sqrt{2t}$ then the maximum of $|\Theta_t'|$ occurs at $|x| - R$. And, $\Theta_t'(|x| - R) = (|x| - R)e^{-(|x|-R)^2/(4t)}/[4\sqrt{\pi}t^{3/2}]$.

Now let $1 < p < \infty$ and q be its conjugate. Suppose $x > 2R$. Then

$$||\chi_{(x-R,x+R)}\Theta_t'||_q^q = \frac{1}{(4\sqrt{\pi}t^{3/2})^q} \int_{x-R}^{x+R} y^q e^{-qy^2/(4t)}\, dy$$

$$\leq \frac{(x+R)^{q-1}2t}{(4\sqrt{\pi}t^{3/2})^q q} \left[e^{-q(x-R)^2/(4t)} - e^{-q(x+R)^2/(4t)} \right]$$

$$\leq \frac{3^{q-1}x^{q-1}t e^{-qx^2/(16t)}}{2^{q-2}(4\sqrt{\pi}t^{3/2})^q q}.$$

$\square$

168   E. Talvila

4. Appendix

The Appendix contains a proof of sharp L^p estimates for the heat equation and a lemma on differentiation of convolutions.

The solution $u(x,t)$ that is in $C^2(\mathbb{R})$ with respect to x and in $C^1((0,\infty))$ with respect to t of (1) and (2) is given by the convolution $u_t(x) = F * \Theta_t(x) = \int_{-\infty}^{\infty} F(x-y)\Theta_t(y)\,dy$ where the Gauss–Weierstrass heat kernel is $\Theta_t(x) = \exp(-x^2/(4t))/(2\sqrt{\pi t})$. For example, see [6].

The heat kernel has the following properties. Let $t > 0$ and let $s \neq 0$ such that $1/s + 1/t > 0$. Then

$$\Theta_t * \Theta_s = \Theta_{t+s} \tag{10}$$

$$\|\Theta_t\|_q = \frac{\alpha_q}{t^{(1-1/q)/2}} \text{ where } \alpha_q = \begin{cases} 1, & q = 1, \\ \frac{1}{(2\sqrt{\pi})^{1-1/q}\,q^{1/(2q)}}, & 1 < q < \infty, \\ \frac{1}{2\sqrt{\pi}}, & q = \infty. \end{cases} \tag{11}$$

The last of these follows from the probability integral $\int_{-\infty}^{\infty} e^{-x^2}\,dx = \sqrt{\pi}$.

We now present some estimates for $F * \Theta_t$ when $F \in L^p$, as they are used in Theorem 1 with data in L'^p. These estimates are well-known, for example, [9] when $r = \infty$. And, [11, Proposition 3.1]. However, we have not been able to find a proof in the literature that the estimates are sharp.

Theorem 4. Let $1 \leq p \leq \infty$ and $F \in L^p$.
(a) If $p \leq s \leq \infty$, then $F * \Theta_t \in L^s$.
(b) Let $q,r \in [1,\infty]$ such that $1/p + 1/q = 1 + 1/r$. There is a constant $K_{p,q}$ such that $\|F * \Theta_t\|_r \leq K_{p,q}\|F\|_p t^{-(1-1/q)/2}$ for all $t > 0$. The estimate is sharp in the sense that if $\psi : (0,\infty) \to (0,\infty)$ such that $\psi(t) = o(t^{-(1-1/q)/2})$ as $t \to 0^+$ or $t \to \infty$ then there is $G \in L^p$ such that $\|G * \Theta_t\|_r/\psi(t)$ is not bounded as $t \to 0^+$ or $t \to \infty$. The constant $K_{p,q} = (c_p c_q/c_r)^{1/2}\alpha_q$, where $c_p = p^{1/p}/(p')^{1/p'}$ with p, p' being conjugate exponents. It cannot be replaced with any smaller number.
(c) If $1 \leq s < p$, then $F * \Theta_t$ need not be in L^s.

When $r = p$ and $q = 1$ the inequality in Part (b) reads $\|F * \Theta_t\|_p \leq \|F\|_p$. When $r = \infty$ then p and q are conjugates and the inequality in Part (b) reads $\|F * \Theta_t\|_\infty \leq \|F\|_p t^{-1/(2p)}$.

The condition for sharpness in Young's inequality is that both functions be Gaussians. This fact is exploited in the proof of Part (b). See [13, p. 99], [2] and [18].

Proof. (a), (b) Young's inequality gives

$$||F * \Theta_t||_r \leq C_{p,q}||F||_p||\Theta_t||_q = \frac{C_{p,q}||F||_p\alpha_q}{t^{(1-1/q)/2}}, \tag{12}$$

where α_q is given in (11). The sharp constant, given in [13, p. 99], is $C_{p,q} = (c_pc_q/c_r)^{1/2}$ where $c_p = p^{1/p}/(p')^{1/p'}$ with p, p' being conjugate exponents. Note that $c_1 = c_\infty = 1$. Also, $0 < C_{p,q} \leq 1$. We then take $K_{p,q} = C_{p,q}\alpha_q$.

To show the estimate $||u_t||_r = O(t^{-(1-1/q)/2})$ is sharp as $t \to 0^+$ and $t \to \infty$, let ψ be as in the statement of the theorem. Fix $p \leq r \leq \infty$. Define the family of linear operators $S_t : L^p \to L^r$ by $S_t[F](x) = F * \Theta_t(x)/\psi(t)$. The estimate $||S_t[F]||_r \leq K_{p,q}||F||_pt^{-(1-1/q)/2}/\psi(t)$ shows that, for each $t > 0$, S_t is a bounded linear operator. Let $F_t = \Theta_t$. Then, from (10) and (11),

$$\frac{||S_t[F_t]||_r}{||F_t||_p} = \frac{||\Theta_t * \Theta_t||_r}{\psi(t)||\Theta_t||_p} = \frac{||\Theta_{2t}||_r}{\psi(t)||\Theta_t||_p} = \frac{\alpha_r}{\alpha_p2^{(1-1/r)/2}\psi(t)t^{(1-1/q)/2}}.$$

This is not bounded in the limit $t \to 0^+$. Hence, S_t is not uniformly bounded. By the Uniform Bounded Principle it is not pointwise bounded. Therefore, there is a function $F \in L^p$ such that $||F * \Theta_t||_r \neq O(\psi(t))$ as $t \to 0^+$. And, the growth estimate $||F * \Theta_t||_r = O(t^{-(1-1/q)/2})$ as $t \to 0^+$ is sharp. Similarly for sharpness as $t \to \infty$.

Now we show that the constant $K_{p,q}$ cannot be reduced. A calculation shows that there is equality in (12) when $F = \Theta_t^\beta$ and β is given by the equation

$$\frac{\beta^{1-1/q}}{(\beta+1)^{1-1/r}} = \frac{c_pc_q}{c_r}\left(\frac{\alpha_p\alpha_q}{\alpha_r}\right)^2$$

$$= \left(1 - \frac{1}{p}\right)^{1-1/p}\left(1 - \frac{1}{q}\right)^{1-1/q}\left(1 - \frac{1}{r}\right)^{-(1-1/r)}. \tag{13}$$

First consider the case $p \neq 1$ and $q \neq 1$. Notice that $1 - 1/r = (1 - 1/q) + (1 - 1/p) > 1 - 1/q$. Let $g(x) = x^A(x+1)^{-B}$ with $B > A > 0$. Then g is strictly increasing on $(0, A/(B - A))$ and strictly decreasing for $x > A/(B - A)$ so there is a unique maximum for g at $A/(B - A)$. Put $A = 1 - 1/q$ and $B = 1 - 1/r$. Then

$$g\left(\frac{A}{B - A}\right) = \frac{\beta^{1-1/q}}{(\beta+1)^{1-1/r}}$$

$$= \left(1 - \frac{1}{p}\right)^{1-1/p}\left(1 - \frac{1}{q}\right)^{1-1/q}\left(1 - \frac{1}{r}\right)^{-(1-1/r)}.$$

Hence, (13) has a unique positive solution for β given by $\beta = (1-1/q)/(1-1/p)$.

If $p = 1$, then $q = r$. In this case, (13) reduces to $(1+1/\beta)^{1-1/q} = 1$ and the solution is given in the limit $\beta \to \infty$. Sharpness of (12) is then given in this limit. It can also be seen that taking F to be the Dirac distribution gives equality.

If $q = 1$, then $p = r$. Now, (13) reduces to $(\beta + 1)^{1-1/p} = 1$ and $\beta = 0$. There is equality in (12) when $F = 1$. This must be done in the limit $\beta \to 0^+$.

If $p = q = r = 1$, then there is equality in (12) for each $\beta > 0$.

Hence, the constant in (12) is sharp.

(c) Suppose $F \geq 0$ and F is decreasing on $[c, \infty)$ for some $c \in \mathbb{R}$. Let $x > c$. Then

$$F * \Theta_t(x) \geq \int_c^x F(y)\Theta_t(x - y)\, dy \geq F(x) \int_c^x \Theta_t(x - y)\, dy$$
$$= \frac{F(x)}{\sqrt{\pi}} \int_0^{(x-c)/(2\sqrt{t})} e^{-y^2}\, dy \sim F(x)/2 \quad \text{as } x \to \infty.$$

Now put $F(x) = 1/[x^{1/p} \log^2(x)]$ for $x \geq e$ and $F(x) = 0$, otherwise. For $p = \infty$ replace $x^{1/p}$ by 1. $\qquad\square$

We now prove a lemma on pointwise differentiation of L^p convolutions. A similar result for $p = 1$ and functions with bounded derivatives appears in [7, Proposition 8.10].

Lemma 1 (Differentiation of Convolutions). *Let $F \in L^p$ for some $1 \leq p \leq \infty$. Let q be the conjugate of p and suppose $\psi \in L^q$.*
(a) Suppose ψ is a differentiable function such that ψ' is absolutely continuous and $\psi'' \in L^q$. Then $F * \psi$ is differentiable and $(F * \psi)'(x) = F * \psi'(x)$ for each $x \in \mathbb{R}$.
(b) Suppose ψ is a differentiable function such that ψ is absolutely continuous, ψ' is bounded on each compact interval and $\psi' \in L^q$. Then $F * \psi$ is differentiable and $(F * \psi)'(x) = F * \psi'(x)$ for each $x \in \mathbb{R}$.
(c) Suppose ψ is a C^∞ function such that $\psi^{(n)} \in L^q$ for each $n \geq 0$. Then $F * \psi \in C^\infty(\mathbb{R})$ and $(F * \psi)^{(n)}(x) = F * \psi^{(n)}(x)$ for each $n \geq 1$ and each $x \in \mathbb{R}$.

The Heat Equation with the L^p Primitive Integral

Proof. Let $w = F * \psi$. (a) Without loss of generality, we can take $h > 0$ and consider

$$\frac{w(x+h) - w(x)}{h} = \int_{-\infty}^{\infty} F(y) \left[\frac{\psi(x+h-y) - \psi(x-y)}{h} \right] dy$$
$$= F * \psi'(x) - R(h).$$

The remainder from Taylor's theorem is

$$R(h) = \frac{1}{h} \int_{-\infty}^{\infty} F(y) \int_{x-y}^{x+h-y} (z - x - h + y)\psi''(z) \, dz \, dy.$$

And, by the Hölder inequality, Jensen's inequality and the Fubini–Tonelli theorem,

$$|R(h)| \leq ||F||_p \left(\int_{-\infty}^{\infty} \int_{x-y}^{x+h-y} |z - x - h + y|^q |\psi''(z)|^q \frac{dz}{h} \right)^{1/q} dy$$
$$= ||F||_p \left(\int_{-\infty}^{\infty} |\psi''(z)|^q \int_{x-z}^{x+h-z} |z - x - h + y|^q \frac{dy}{h} \, dz \right)^{1/q}$$
$$= \frac{||F||_p ||\psi''||_q |h|}{(q+1)^{1/q}} \to 0 \text{ as } h \to 0.$$

Minor changes are needed when $p = 1$. Therefore, $(F * \psi)'(x) = F * \psi'(x)$.

(b) Let $R > 0$. Write

$$F * \psi(x) = \int_{-R}^{R} F(y)\psi(x-y) \, dy + \int_{|y|>R} F(y)\psi(x-y) \, dy.$$

Since $F \in L^1_{loc}$, and ψ' is bounded on compact intervals, dominated convergence shows

$$\frac{d}{dx} \int_{-R}^{R} F(y)\psi(x-y) \, dy = \int_{-R}^{R} F(y)\psi'(x-y) \, dy.$$

Now, let $h > 0$ and consider

$$\left| \int_{R}^{\infty} F(y) \left[\frac{\psi(x+h-y) - \psi(x-y)}{h} \right] dy \right| \leq$$
$$||F\chi_{(R,\infty)}||_p \left(\int_{R}^{\infty} \left| \frac{\psi(x+h-y) - \psi(x-y)}{h} \right|^q dy \right)^{1/q}.$$

Given $\epsilon > 0$, take R large enough so that $||F\chi_{(R,\infty)}||_p < \epsilon$. Using the fundamental theorem of calculus, Jensen's inequality and then the Fubini–Tonelli theorem, we have

$$
\int_R^\infty \left| \frac{\psi(x+h-y) - \psi(x-y)}{h} \right|^q dy = \int_R^\infty \left| \int_{x-y}^{x+h-y} \psi'(z) \frac{dz}{h} \right|^q dy
$$
$$
\leq \int_R^\infty \int_{x-y}^{x+h-y} |\psi'(z)|^q \frac{dz}{h} dy
$$
$$
= \int_{-\infty}^\infty |\psi'(z)|^q \int_{x-z}^{x+h-z} \frac{dy}{h} dz
$$
$$
= ||\psi'||_q^q.
$$

Therefore,

$$
\left| \int_R^\infty F(y) \frac{\psi(x+h-y) - \psi(x-y)}{h} dy \right| \leq \epsilon ||\psi'||_q.
$$

The case $p = 1$ is similar.

(c) Follows from (a) or (b). $\qquad\square$

References

[1] S. Axler, P. Bourdon and W. Ramey, *Harmonic function theory*, New York, Springer-Verlag, 2001.

[2] W. Beckner, *Inequalities in Fourier analysis*, Ann. of Math. (2) **102**(1975), 159–182.

[3] J.R. Cannon, *The one-dimensional heat equation*, Menlo Park, Addison–Wesley, 1984.

[4] R. Dautray and J.-L. Lions, *Mathematical analysis and numerical methods for science and technology, vol. 5* (trans. A. Craig), Berlin, Springer-Verlag, 1992.

[5] B. Epstein, *Partial differential equations*, New York, McGraw-Hill, 1962.

[6] G.B. Folland, *Introduction to partial differential equations*, Princeton, Princeton University Press, 1995.

[7] G.B. Folland, *Real analysis*, New York, Wiley, 1999.

[8] F.G. Friedlander and M. Joshi, *Introduction to the theory of distributions*, Cambridge, Cambridge University Press, 1999.

[9] K.E. Gustafson, *Introduction to partial differential equations and Hilbert space methods*, New York, Dover, 1999.

[10] I.I. Hirschman and D.V. Widder, *The convolution transform*, Princeton, Princeton University Press, 1955.

[11] T. Iwabuchi, T. Matsuyama and K. Taniguchi, *Boundedness of spectral multipliers for Schrödinger operators on open sets*, Rev. Mat. Iberoam. **34**(2018), 1277–1322.

[12] R.J. Iorio, Jr. and V. de Magãlhaes Iorio, *Fourier analysis and partial differential equations*, New York, Cambridge University Press, 2001.

[13] E.H. Lieb and M. Loss, *Analysis*, Providence, American Mathematical Society, 2001.

[14] Z. Szmydt, *Fourier transforms and linear differential equations* (trans. M. Kuczma), Dordrecht, D. Reidel, 1977.

[15] E. Talvila, *The regulated primitive integral*, Illinois J. Math. **53**(2009), 1187–1219.

[16] E. Talvila, *The L^p primitive integral*, Math. Slovaca **64**(2014), 1497–1524.

[17] E. Talvila, *The one-dimensional heat equation in the Alexiewicz norm*, Adv. Pure Appl. Math. **6**(2015), 13 37.

[18] G. Toscani, *Heat equation and the sharp Young's inequality*, arXiv:1204.2086 (2012).

[19] D.V. Widder, *The heat equation*, New York, Academic Press, 1975.

© 2025 World Scientific Publishing Company
https://doi.org/10.1142/9789819812202_0009

Chapter 9

On the Symmetric Laplace Integral and Its Application to Trigonometric Series

Sougata Mahanta

Department of Mathematics,
Visva-Bharati, 731235, India
Email: sougatamahanta1@gmail.com

A generalized symmetric integral, the symmetric Laplace integral, is defined using the concept of symmetric Laplace derivative which is an integral with continuous primitives and is more general than the Henstock integral. After presenting the basic properties of this integral, a useful application to the trigonometric series is provided.

1. Introduction

Isaac Newton (1642-1727) and Gottfried Leibniz (1646-1716) first discovered, independently, the concept of differential and integral calculus. They used to view integration as an antiderivative. Later in the eighteenth century, Augustin-Louis Cauchy (1789-1857) founded a definition of integration for a bounded function $f \colon [a, b] \to \mathbb{R}$. He first defined the following sum:

$$C(f) = \sum_{k=1}^{n} f(x_{k-1})(x_k - x_{k-1}), \tag{1}$$

where $a = x_0 < x_1 < \cdots < x_n = b$ is a partition of $[a, b]$, and then he defined the integral of f as follows:

$$\int_a^b f = \lim_{\|\Delta x\| \to 0} \sum_{k=1}^{n} f(x_{k-1})(x_k - x_{k-1}),$$

provided that the limit exists when the norm of the partition $\|\Delta x\|$ tends to zero. He argued that if $F \colon [a, b] \to \mathbb{R}$ has a continuous derivative F'

175

everywhere on $[a, b]$ then we have

$$F(x) = \int_a^x F' \qquad \text{for } a \leqslant x \leqslant b,$$

i.e. F can be recovered from its continuous derivative.

In 1822, Joseph Fourier (1786-1830) was looking for a solution of the boundary value problem

$$\begin{cases} \frac{\partial^2 U}{\partial x^2} + \frac{\partial^2 U}{\partial y^2} = 0, & \text{on } (0, \pi) \times (0, \infty), \\ U(x, y) = 0, & \text{on } \{0\} \times (0, \infty) \cup \{\pi\} \times (0, \infty), \\ U(x, 0) = \phi(x), & \text{for } x \in (0, \pi), \end{cases} \qquad (2)$$

and he arrived at the conclusion that $\phi(x) = \sum b_k \sin(kx)$, where

$$b_k = \frac{2}{\pi} \int_0^\pi \phi(x) \sin(kx) \, dx.$$

Apparently, the Cauchy integral was not sufficient to solve (2) with discontinuous ϕ, and this motivated Bernhard Riemann (1826-1866) to redefine the method of integration. He made obvious changes to the definition of the Cauchy Sum (1). He replaced $f(x_{k-1})$ by $f(c_k)$, where c_k is an arbitrary point of $[x_{k-1}, x_k]$ and kept everything else unchanged. Now the question is how discontinuous a Riemann integrable function can be? In 1881, Vito Volterra provided an example of a function $V \colon [0, 1] \to \mathbb{R}$ with bounded derivative $V' \colon [0, 1] \to \mathbb{R}$ such that V' is not continuous on a Cantor-like set $C \subseteq [0, 1]$ of Lebesgue measure $1/2$, and V' is not Riemann integrable. Thus, the Riemann integral cannot integrate all bounded derivatives. Later, in 1902, Lebesgue proved that a bounded function $f \colon [a, b] \to \mathbb{R}$ is Riemann integrable if and only if it is continuous a.e. on $[a, b]$. He further gave a completely new method of integration, the Lebesgue integration, by partitioning the range of a function instead of partitioning its domain. This revolutionized Analysis. The Lebesgue integration is also known as an absolute integral since a function f is Lebesgue integrable if and only if $|f|$ is so. But the function

$$f(x) = \begin{cases} x^2 \cos \frac{\pi}{x^2}, & 0 < x \leqslant 1, \\ 0, & x = 0, \end{cases}$$

is not Lebesgue integrable though it has a derivative $f'(x)$ everywhere on $[0, 1]$. Thus, the Lebesgue integral cannot include the Newton integral, and so this was the time when Mathematicians understood that to include the Newton integral and Lebesgue integral they need to introduce non-absolute integrals, i.e., f can be integrable even when $|f|$ is not.

In 1912, the French mathematician A. Denjoy invented a process of integration which is more general than that of the Lebesgue integral. A function $f\colon [a, b] \to \mathbb{R}$ is said to be Denjoy integrable if there is an ACG^* (see Chapter 6 of [5]) function $F\colon [a, b] \to \mathbb{R}$ such that $F' = f$ a.e. on $[a, b]$, and then we write

$$F(b) - F(a) = \int_a^b f.$$

Theorem 6.22 of [5, p. 103] indicates that if F has a derivative F' everywhere on $[a, b]$ then, we have

$$F(x) = \int_a^x F' \qquad \text{for } a \leqslant x \leqslant b.$$

So, the Denjoy integral recovered the paradise of the Newton integral, and by definition it also includes the Lebesgue integral.

In 1915, a few ideas of de la Vallée Poussin motivated a German mathematician O. Perron to invent a new process (see Chapter 8 of [5]) of integration. He invented this while he was working on the theory of differential equations. Later H. Hake, P.S. Aleksandrov, and H. Looman proved (see Chapter 11 of [5]) the equivalency of Denjoy and Perron integrals, and for this reason these integrals are now known as the Denjoy-Perron integral.

The definition of Denjoy integral was extended to higher dimensions and even on abstract spaces but they were not work-worthy and satisfactory (see [1, p. 6]). So, the same was done using Perron's method but it also could not satisfy the mathematicians. It is possible to construct a Perron integrable function of two variables which looses the integrability while the axes are rotated by $\pi/4$. Moreover, if a multidimensional integral satisfies the Fubini's theorem and if its one-dimensional restriction is nothing but the Denjoy integral then it is possible to construct a differentiable function [8] whose partial derivatives are not integrable; and this implies that Fubini's theorem enforces us to lose the purpose (recovering a function from its derivative) of no-absolute integral. Kurzweil (see [7]) and Henstock (see [6]) constructed an integral, independently, by modifying the Riemann sum, and this integral is now known as the Henstock integral or the Henstock-Kurzweil integral. Although the Henstock integral is equivalent to the Denjoy-Perron integral, its definition is much easier. However, even this easy characterisation of Denjoy-Perron integral is not able to provide a pleasant extension to $\mathbb{R}^n$.

There is another interesting reason (possibly the main reason) behind the generalisation of integration. Assume for $x \in [-\pi, \pi]$ the following trigonometric series

$$\frac{a_0}{2} + \sum_{k=0}^{\infty} (a_k \cos kx + b_k \sin kx) \tag{3}$$

converges to the function $f(x)$. Then we all know

$$\begin{cases} a_k = \frac{1}{\pi} \int_{-\pi}^{\pi} f(x) \cos kx \, dx, & \text{for } k = 0, 1, \cdots, \\ b_k = \frac{1}{\pi} \int_{-\pi}^{\pi} f(x) \sin kx \, dx, & \text{for } k = 1, 2, \cdots. \end{cases} \tag{4}$$

But for (4), f need to be integrable in some sense, and this phenomena gave birth of the following two questions which kept mathematicians busy for a few years:

(1) Does (4) hold if f is assumed to be integrable?
(2) Is the sum function f even integrable?

However, under the assumptions that f is Lebesgue, or Denjoy integrable, (I) has an affirmative answer (see [10, p. 326(vol I), 84(vol II)]). But what about (II)? It is well-known that if

$$f(x) = \frac{a_0}{2} + \sum_{k=0}^{\infty} (a_k \cos kx + b_k \sin kx)$$

for $x \in [-\pi, \pi]$, then $D^2 F(x) = f(x)$ for $x \in [-\pi, \pi]$, where

$$F(x) = \frac{a_0 x^2}{4} - \sum_{k=1}^{\infty} \left(\frac{a_k \cos kx + b_k \sin kx}{n^2} \right)$$

and $\qquad D^2 F(x) = \lim_{h \to 0} \dfrac{F(x + 2h) + F(x - 2h) - 2F(x)}{4h^2},$

provided the limit exists. The symmetric derivative $D^2 F$ influenced mathematicians to think about symmetric integrals, i.e., the integrals which are defined in terms of symmetric derivatives. There are lots of symmetric integrals but the important one is the Burkill's SCP-integral [2]. His integration completely solved the questions (I) and (II).

In the previous paragraph, we mentioned that Burkill's SCP-integral completely solved the coefficient problem for trigonometric series. So, theoretically it is meaningless to attack the same problem again. But the SCP-integral has a serious flaw, that for this integral, we can know the primitives only $a.e.$ on the domain, and not all over the domain which may create complexity in calculating the Fourier coefficients. Keeping this thought in mind, a symmetric type integral is defined in this chapter and instead of proving a general theorem, a simple and easy version of the coefficient problem is provided.

2. Preliminaries

Definition 1 (First Order Symmetric Laplace Derivative [3]). Let a function $f(x)$ be Laplace or Denjoy integrable in some neighbourhood of $x_0 \in \mathbb{R}$. For $\delta > 0$, define

$$\overline{SLD^1}f(x_0) = \limsup_{s \to \infty} s^2 \int_0^\delta e^{-st} \left[\frac{f(x_0 + t) - f(x_0 - t)}{2} \right] dt,$$

$$\underline{SLD^1}f(x_0) = \liminf_{s \to \infty} s^2 \int_0^\delta e^{-st} \left[\frac{f(x_0 + t) - f(x_0 - t)}{2} \right] dt.$$

If $\overline{SLD^1}f(x_0) = \underline{SLD^1}f(x_0)$, then this common value is called the first order symmetric Laplace derivative of $f(x)$ at x_0 and is denoted by $SLD^1 f(x_0)$.

Definition 2 (Second Order Symmetric Laplace Derivative [3]). Let a function $f(x)$ be Laplace or Denjoy integrable in some neighbourhood of $x_0 \in \mathbb{R}$. For $\delta > 0$, define

$$\overline{SLD^2}f(x_0) = \limsup_{s \to \infty} s^3 \int_0^\delta e^{-st} \left[\frac{f(x_0 + t) + f(x_0 - t) - 2f(x_0)}{2} \right] dt,$$

$$\underline{SLD^2}f(x_0) = \liminf_{s \to \infty} s^3 \int_0^\delta e^{-st} \left[\frac{f(x_0 + t) + f(x_0 - t) - 2f(x_0)}{2} \right] dt.$$

If $\overline{SLD^2}f(x_0) = \underline{SLD^2}f(x_0)$, then this common value is called the second order symmetric Laplace derivative of $f(x)$ at x_0 and is denoted by $SLD^2 f(x_0)$.

Definition 3 (Laplace Smoothness of Order Two [3]). Let a function $f(x)$ be Laplace or Denjoy integrable in some neighbourhood of $x_0 \in \mathbb{R}$. If for some $\delta > 0$, we have

$$\lim_{s \to \infty} s^2 \int_0^\delta e^{-st} \left[f(x_0 + t) + f(x_0 - t) - 2f(x_0) \right] dt = 0,$$

then we say f is Laplace smooth of order two at x_0.

Theorem 1 ([4]). *Let $f : [a, b] \mapsto \mathbb{R}$ be such that*

(1) f is continuous,
(2) $\overline{SLD^2}f(x) \geqslant 0$, $\forall \, x \in (a, b) \setminus N$, where N is possibly denumerable subset of (a, b),
(3) f is Laplace smooth of order two on N.

Then f is convex on (a, b).

Theorem 2. *Let $f\colon [a,b] \to \mathbb{R}$ be such that*

(1) f is continuous,
(2) $\overline{SLD^2}f \geqslant 0$ a.e.,
(3) $\overline{SLD^2}f(x) > -\infty \ \forall \ x \in (a,b) \setminus N$, where N is possibly denumerable subset of (a,b) and on which f is Laplace smooth of order two.

Then f is convex.

Proof. The proof is almost similar to that of Corollary 5.31 of [9] but for completeness we give detailed proof here.

For $\epsilon = 1/n(b-a)$ and $k = 1, 2, \cdots$, choose an open set $G_k \supset E$ such that the Lebesgue measure of G_k is $\epsilon 4^{-k}$, where $E = \{x \in (a,b) \mid \overline{SLD^2}f(x) < 0\}$. Define

$$h_k(x) = \begin{cases} 2^k, & \text{for } x \in G_k, \\ 0, & \text{otherwise.} \end{cases}$$

Let $H_k(x) = \int_a^x h_k(t)\,dt$. Then H_k is not only continuous but also increasing with $H_k(a) = 0$ and $H_k(b) < \epsilon 2^{-k}$. Now define

$$H(x) = \sum_{k=0}^{\infty} H_k(x) \quad \text{and} \quad g_n(x) = \int_a^x H(t)\,dt.$$

Then we have $SLD^2 g_n(x) = \infty$, everywhere in E, and $g_n(b) < \epsilon(b-a) = 1/n$. Thus by previous theorem, we have $f = \lim_{n\to\infty}(f + g_n)$ is convex on (a,b). $\qquad \square$

Corollary 1. *Let $f : [a,b] \mapsto \mathbb{R}$ be continuous such that $\overline{SLD^1}f \geqslant 0$ a.e. and $\overline{SLD^1}f(x) > -\infty \ \forall \ x \in (a,b) \setminus N$, where N is possibly denumerable on which f is symmetric Laplace continuous. Then f is increasing.*

3. Symmetric Laplace Integral

Definition 4. A continuous function $M : [a,b] \mapsto \mathbb{R}$ is called a SL-major function of f if

(1) $M(a) = 0$,
(2) $\underline{SLD^1}M \geqslant f$ a.e.,
(3) $\underline{SLD^1}M(x) > -\infty \ \forall \ x \in (a,b) \setminus N$, where N is possibly denumerable.

A continuous function $m : [a,b] \mapsto \mathbb{R}$ is called a SL-minor function of f if $-m$ is SL-major function of f. Moreover, if we have

$$\sup_m\{m(b)\} = \inf_M\{M(b)\}$$

then we say f is SL-integrable and write $\int_a^b f = \sup_m\{m(b)\} = \inf_M\{M(b)\}$.

Remark 1. One can redefine a SL-major function of f by rewriting the Definition 4 by relaxing the condition $M(a) = 0$. In that case, we say f is SL-integrable if we have

$$\sup_m\{m(b) - m(a)\} = \inf_M\{M(b) - M(a)\}$$

and we write $\int_a^b f = \sup_m\{m(b) - m(a)\} = \inf_M\{M(b) - M(a)\}$, where $-m$ is a SL-major function of f.

The symmetric Laplace integral will satisfy all the basic properties of integration, and their proofs are easy. So, instead of enlisting them all here we discuss only the two important ones – the fundamental theorem of calculus, and the integration by parts.

Theorem 3 (Fundamental Theorem of Calculus). *If* $F(x) = \int_a^x f(t)dt$ *then* $SLD^1 F = f$ *a.e. on* (a, b).

Proof. Assume $f - \underline{SLD^1}F > k$ for a positive constant k on a subset E of (a, b) of measure $\sigma(> 0)$. Take $\epsilon < \frac{k\sigma}{2}$. Let $R = M - F < \epsilon$ on (a, b) where M is a major function of f. Then R is increasing and

$$\int_a^b R' \leqslant R(b) - R(a) = R(b) < \epsilon.$$

Hence the set $\{x \mid R'(x) > k/2\}$ has measure less than σ. Therefore, E has a positive measure subset $E_1 = \{x \mid 0 \leqslant R'(x) \leqslant k/2\}$. Hence a.e. on E_1 we have

$$f(x) \leqslant \underline{SLD^1}M(x) = \underline{SLD^1}F(x) + R'(x) \leqslant \underline{SLD^1}F(x) + \frac{k}{2}.$$

Which is a contradiction to our assumption. Therefore $f \leqslant \underline{SLD^1}F$ a.e.. Proof of $f \geqslant \underline{SLD^1}F$ a.e. is similar. Therefore,

$$SLD^1 F = f \quad \text{a.e.} \qquad \square$$

Theorem 4 (Integration by Parts). *Let* $f : [a, b] \mapsto \mathbb{R}$ *be* SL-integrable *and* g *is of bounded variation. Then* fG *is* SL-integrable *and*

$$\int_a^b fG = FG|_a^b - (D)\int_a^b Fg,$$

where $F(x) = \int_a^x f$ *and* $G(x) = \int_a^x g$.

Proof. It's enough to consider g is positive; otherwise, we can replace g by
$$\tilde{g} := g - \inf_{x \in (a,b)} g(x).$$
Hence G is monotone increasing and positive. Let M be a SL-major function of f. We show that MG is a major function of $fG + Fg$.
$$M(x+t)G(x+t) - M(x-t)G(x-t)$$
$$= (M(x+t) - M(x-t))G(x) + M(x+t)(G(x+t) - G(x))$$
$$- M(x-t)(G(x-t) - G(x))$$
Since M is continuous so for $\epsilon > 0$ there is $\delta > 0$ such that
$$-\epsilon + M(x) < M(x+h) < M(x) + \epsilon, \quad \forall\, |h| < \delta.$$
Which implies
$$- \epsilon g(x) + M(x)g(x)$$
$$\leqslant \liminf_{s \to \infty} s^2 \int_0^\delta e^{-st} M(x+t)(G(x+t) - G(x))dt$$
$$\leqslant \limsup_{s \to \infty} s^2 \int_0^\delta e^{-st} M(x+t)(G(x+t) - G(x))dt$$
$$\leqslant M(x)g(x) + \epsilon g(x),$$
nearly everywhere on (a, b). As ϵ is arbitrary, we have
$$\lim_{s \to \infty} s^2 \int_0^\delta e^{-st} M(x+t)(G(x+t) - G(x))dt = M(x)g(x),$$
nearly everywhere on (a, b). Similarly,
$$- \lim_{s \to \infty} s^2 \int_0^\delta e^{-st} M(x-t)(G(x-t) - G(x))dt = M(x)g(x),$$
nearly everywhere on (a, b). Hence, we have
$$\liminf_{s \to \infty} s^2 \int_0^\delta e^{-st} \left(\frac{M(x+t)G(x+t) - M(x-t)G(x-t)}{2} \right) dt$$
$$\geqslant (SLD^1 M)G + Mg$$
$$\geqslant fG + Fg \quad \text{a.e.}$$
$$\geqslant -\infty \quad \text{nearly everywhere}$$
In similar fashion we can show that mG is a SL-minor function of $fG + Fg$. Now as $m(b)G(b) \leqslant F(b)G(b) \leqslant M(b)G(b)$, the integrability of f ensures that $fG + Fg$ is integrable and
$$\int_a^b fG = FG|_a^b - (D) \int_a^b Fg. \qquad \square$$

Theorem 5 (Trigonometric Series). *Let* $f\colon [-\pi, \pi] \to \mathbb{R}$ *be SL-integrable such that*

$$f(x) = \frac{a_0}{2} + \sum_{k=0}^{\infty} (a_k \cos kx + b_k \sin kx),$$

$$and \quad F(x) = \frac{a_0 x}{2} + \sum_{k=0}^{\infty} \left(\frac{a_k \sin kx - b_k \cos kx}{k} \right),$$

for all $x \in [-\pi, \pi]$. *Furthermore, assume* F *is continuous. Then, we have*

$$a_k = \frac{1}{\pi} \int_{-\pi}^{\pi} f(x) \cos kx \, dx \quad \text{for } k = 0, 1, \cdots,$$

$$b_k = \frac{1}{\pi} \int_{-\pi}^{\pi} f(x) \sin kx \, dx, \quad \text{for } k = 1, 2, \cdots.$$

Proof. For all $x \in [-\pi, \pi]$, let

$$\Phi(x) = \frac{a_0 x^2}{4} - \sum_{k=1}^{\infty} \left(\frac{a_k \cos kx + b_k \sin kx}{n^2} \right).$$

Then, we have $D^2\Phi = SLD^1 F = f$ on $[-\pi, \pi]$ (see Theorem 2.4 [10, p. 319(vol I)]). Let $G(x) = \int_{-\pi}^{x} f$, then we have $SLD^1(G - F) = 0$ and $G - F$ is continuous. So, by Corollary 2.11 of [3] and Theorem 5.2 of [2], we have

$$\frac{\pi a_n}{n} = \int_{-\pi}^{\pi} F(x) \sin nx \, dx$$

$$= \int_{-\pi}^{\pi} G(x) \sin nx \, dx = \int_{-\pi}^{\pi} f(x) \frac{\cos nx}{n} \, dx. \qquad \square$$

References

[1] P. S. Bullen. Nonabsolute integrals in the twentieth century. Mathematical review subject classification 26A39, 2000.

[2] J. C. Burkill. Integrals and trigonometric series. Proc. London Math. Soc., 3(1):46-57, 1951.

[3] A. Garai and S. Ray. On the symmetric Laplace derivative. Acta Math. Hungar., 133(1-2):166–184, 2011.

[4] A. Garai and S. Ray. On the symmetric Laplace derivative. Acta Mathematica Hungarica, 133, 2011.

[5] R. A. Gordon. The Integrals of Lebesgue, Denjoy, Perron, and Henstock. American Mathematical Soc., 1994. 9

[6] Ralph Henstock. The equivalence of generalized forms of the ward, variational, Denjoy-Stieltjes, and Perron-Stieltjes integrals. Proceedings of the London Mathematical Society, 3(1):281-303, 1960.

[7] Jaroslav Kurzweil. Generalized ordinary differential equations and continuous dependence on a parameter. Czechoslovak Mathematical Journal, 07(3):418-449, 1957.

[8] W. F. Pfeffer. The divergence theorem. Transactions of the American Mathematical Society, 295(2):665-685, 1986.

[9] Brian S. Thomson. Symmetric properties of real functions. CRC Press, 2020.

[10] Antoni Zygmund. Trigonometric series, volume I-II. Cambridge university press, 2002.

© 2025 World Scientific Publishing Company
https://doi.org/10.1142/9789819812202_0010

Chapter 10

Finite and Infinite Integral Formulae Associated with the Family of Incomplete I-Functions

Sanjay Bhatter

Department of Mathematics,
Malaviya National Institute of Technology Jaipur, India
Email: sbhatter.maths@mnit.ac.in

Nishant

Department of Mathematics,
Malaviya National Institute of Technology Jaipur, India
Email: nishantjangra1996@gmail.com

Sunil Dutt Purohit

Department of HEAS (Mathematics),
Rajasthan Technical University, Kota, India
and
Department of Computer Science and Mathematics,
Lebanese American University, Beirut, Lebanon
Email: sunil_a_purohit@yahoo.com

Hundreds of special functions have been employed in applied mathematics and computing sciences for many centuries due to their outstanding features and wide range of applications. It is important to illustrate image formulas involving one or more variable special functions under a variety of definite integrals, given the importance of their consequences in the assessment of generalized integrals, applied physics, and many engineering fields. In this paper, we have invented an integral representation of the incomplete I-function. Also, a variety of finite and infinite integrals involving a family of incomplete I-functions have been investigated in this study, mostly as a result of various applications of these findings. Due to the fact that our results are unified, a substantial number of new results can be constructed as special instances from our leading results.

185

1. Introduction and Preliminaries

In several areas of the mathematical sciences, generalized special functions (GSF) are related to many types of issues. Many scientists have been inspired to research integrals and associated GSF as a result of the relationships between GSF and other fields of study. The analysis of crystallographic minimum surface difficulties, the study of electromagnetic waves through elliptic disks, and elliptic fracture problems are just a few examples of physical and technical issues where a number of integral formulas using GSF play a significant role. The generalized special function with contrasting line of research has inspired the researchers to look into the area of integrals and linked GSF (see, e.g. [2, 4, 17–19]). The findings from this research are of a generic nature and are extremely beneficial in the fields of engineering, economics, and chemical sciences, digital signals, image processing, finance, and ship target recognition by sonar system and radar signals.

In the year 1997, Rathie [16] discovered the I-function which is defined as follows:

$$I_{p,\,q}^{m,\,n}(z) = I_{p,\,q}^{m,\,n} \left[z \,\middle|\, \begin{array}{l} (\mathfrak{f}_1, \alpha_1; \mathfrak{F}_1), \cdots, (\mathfrak{f}_p, \alpha_p; \mathfrak{F}_p) \\ (\mathfrak{g}_1, \beta_1; \mathfrak{G}_1), \cdots, (\mathfrak{g}_q, \beta_q; \mathfrak{G}_q) \end{array} \right]$$

$$= I_{p,\,q}^{m,\,n} \left[z \,\middle|\, \begin{array}{l} (\mathfrak{f}_j, \alpha_j; \mathfrak{F}_j)_{1,\,p} \\ (\mathfrak{g}_j, \beta_j; \mathfrak{G}_j)_{1,\,q} \end{array} \right] = \frac{1}{2\pi i} \int_{\mathcal{L}} \psi(\mathsf{h}) \, z^{\mathsf{h}} \, ds, \tag{1}$$

where,

$$\psi(\mathsf{h}) = \frac{\prod\limits_{j=1}^{m} \{\Gamma(\mathfrak{g}_j - \beta_j s)\}^{\mathfrak{G}_j} \prod\limits_{j=1}^{n} \{\Gamma(1 - \mathfrak{f}_j + \alpha_j s)\}^{\mathfrak{F}_j}}{\prod\limits_{j=n+1}^{p} \{\Gamma(\mathfrak{f}_j - \alpha_j s)\}^{\mathfrak{F}_j} \prod\limits_{j=m+1}^{q} \{\Gamma(1 - \mathfrak{g}_j + \beta_j s)\}^{\mathfrak{G}_j}}. \tag{2}$$

$\gamma(\mathsf{h}, \mathsf{t})$ and $\Gamma(\mathsf{h}, \mathsf{t})$ are the familiar lower and upper incomplete gamma function respectively and are represented by:

$$\gamma(\mathsf{h}, \mathsf{t}) = \int_0^{\mathsf{t}} \mathfrak{y}^{\mathsf{h}-1} \, e^{-\mathfrak{y}} \, d\mathfrak{y}, \qquad (\Re(\mathsf{h}) > 0; \ \mathsf{t} \geqq 0), \tag{3}$$

and

$$\Gamma(\mathsf{h}, \mathsf{t}) = \int_{\mathsf{t}}^{\infty} \mathfrak{y}^{\mathsf{h}-1} \, e^{-\mathfrak{y}} \, d\mathfrak{y}, \qquad (\mathsf{t} \geqq 0; \ \Re(\mathsf{h}) > 0 \quad \text{when} \quad \mathsf{t} = 0). \tag{4}$$

The following relation is established by the incomplete gamma functions (known as decomposition formula):

$$\gamma(\mathsf{h}, \mathsf{t}) + \Gamma(\mathsf{h}, \mathsf{t}) = \Gamma(\mathsf{h}), \qquad (\Re(\mathsf{h}) > 0). \tag{5}$$

Finite and Infinite Integral Formulae Involving Family of Incomplete I-Functions 187

A family of incomplete I-functions ${}^{\gamma}I_{p,q}^{m,n}(z)$ and ${}^{\Gamma}I_{p,q}^{m,n}(z)$, newly described by Jangid et al. [13], is an extension of the well-known Fox's H-function [9] and numerous other special functions. It is the general form of the I-function described by Rathie [16] and is stated below in the form of Mellin-Barnes type contour integrals.

$$
{}^{\gamma}I_{p,q}^{m,n}(z) = {}^{\gamma}I_{p,q}^{m,n}\left[z \left|\begin{array}{l} (\mathfrak{c}_1, \varsigma_1; \mathfrak{C}_1 : t), (\mathfrak{c}_2, \varsigma_2; \mathfrak{C}_2), \cdots, (\mathfrak{c}_p, \varsigma_p; \mathfrak{C}_p) \\ (\mathfrak{d}_1, \varrho_1; \mathfrak{D}_1), \cdots, (\mathfrak{d}_q, \varrho_q; \mathfrak{D}_q) \end{array}\right.\right]
$$

$$
= {}^{\gamma}I_{p,q}^{m,n}\left[z \left|\begin{array}{l} (\mathfrak{c}_1, \varsigma_1; \mathfrak{C}_1 : t), (\mathfrak{c}_j, \varsigma_j; \mathfrak{C}_j)_{2,p} \\ (\mathfrak{d}_j, \varrho_j; \mathfrak{D}_j)_{1,q} \end{array}\right.\right] = \frac{1}{2\pi i}\int_{\mathcal{L}} \phi(\mathsf{h}, \mathsf{t})\, z^{\mathsf{h}}\, d\mathsf{h}, \quad (6)
$$

and

$$
{}^{\Gamma}I_{p,q}^{m,n}(z) = {}^{\Gamma}I_{p,q}^{m,n}\left[z \left|\begin{array}{l} (\mathfrak{c}_1, \varsigma_1; \mathfrak{C}_1 : t), (\mathfrak{c}_2, \varsigma_2; \mathfrak{C}_2), \cdots, (\mathfrak{c}_p, \varsigma_p; \mathfrak{C}_p) \\ (\mathfrak{d}_1, \varrho_1; \mathfrak{D}_1), \cdots, (\mathfrak{d}_q, \varrho_q; \mathfrak{D}_q) \end{array}\right.\right]
$$

$$
= {}^{\Gamma}I_{p,q}^{m,n}\left[z \left|\begin{array}{l} (\mathfrak{c}_1, \varsigma_1; \mathfrak{C}_1 : t), (\mathfrak{c}_j, \varsigma_j; \mathfrak{C}_j)_{2,p} \\ (\mathfrak{d}_j, \varrho_j; \mathfrak{D}_j)_{1,q} \end{array}\right.\right] = \frac{1}{2\pi i}\int_{\mathcal{L}} \Phi(\mathsf{h}, \mathsf{t})\, z^{\mathsf{h}}\, d\mathsf{h}, \quad (7)
$$

for all $z \neq 0$, where

$$
\phi(\mathsf{h}, \mathsf{t}) = \frac{\{\gamma(1 - \mathfrak{c}_1 + \varsigma_1 \mathsf{h}, \mathsf{t})\}^{\mathfrak{C}_1} \prod\limits_{j=1}^{m} \{\Gamma(\mathfrak{d}_j - \varrho_j \mathsf{h})\}^{\mathfrak{D}_j} \prod\limits_{j=2}^{n} \{\Gamma(1 - \mathfrak{c}_j + \varsigma_j \mathsf{h})\}^{\mathfrak{C}_j}}{\prod\limits_{j=n+1}^{p} \{\Gamma(\mathfrak{c}_j - \varsigma_j \mathsf{h})\}^{\mathfrak{C}_j} \prod\limits_{j=m+1}^{q} \{\Gamma(1 - \mathfrak{d}_j + \varrho_j \mathsf{h})\}^{\mathfrak{D}_j}},
$$

$$(8)$$

and

$$
\Phi(\mathsf{h}, \mathsf{t}) = \frac{\{\Gamma(1 - \mathfrak{c}_1 + \varsigma_1 \mathsf{h}, \mathsf{t})\}^{\mathfrak{C}_1} \prod\limits_{j=1}^{m} \{\Gamma(\mathfrak{d}_j - \varrho_j \mathsf{h})\}^{\mathfrak{D}_j} \prod\limits_{j=2}^{n} \{\Gamma(1 - \mathfrak{c}_j + \varsigma_j \mathsf{h})\}^{\mathfrak{C}_j}}{\prod\limits_{j=n+1}^{p} \{\Gamma(\mathfrak{c}_j - \varsigma_j \mathsf{h})\}^{\mathfrak{C}_j} \prod\limits_{j=m+1}^{q} \{\Gamma(1 - \mathfrak{d}_j + \varrho_j \mathsf{h})\}^{\mathfrak{D}_j}},
$$

$$(9)$$

where $\gamma(., \mathsf{t})$ and $\Gamma(., \mathsf{t})$ are the lower and upper incomplete gamma functions described in (3) and (4). The incomplete I-functions ${}^{\gamma}I_{p,q}^{m,n}(z)$ and ${}^{\Gamma}I_{p,q}^{m,n}(z)$ exist for all $\mathsf{t} \geq 0$ under the same conditions and contour as mentioned in Rathie [16]. The incomplete I-functions comply with the relation below (known as the decomposition formula)

$$
{}^{\gamma}I_{p,q}^{m,n}(z) + {}^{\Gamma}I_{p,q}^{m,n}(z) = I_{p,q}^{m,n}(z), \tag{10}
$$

for the well known I-function given by Rathie [16]. Furthermore, the I-function [16] specified in (1) is obtained if we set $\mathsf{t} = 0$ in (7).

Next we recall the Mellin transform for the incomplete I-function that Kamlesh Jangid et al. [13] provided.

Mellin Transform

On p. 9, equation (36) in [13], Kamlesh Jangid et al. provided the following formula for the Mellin transform of an incomplete I-function:

$$\mathcal{M}\left\{{}^{\Gamma}I_{p,q}^{m,n}\left[cz^{\nu}\left|\begin{array}{c}(\mathfrak{c}_1,\varsigma_1;\mathfrak{C}_1:\mathfrak{t}),(\mathfrak{c}_2,\varsigma_2;\mathfrak{C}_2),\cdots,(\mathfrak{c}_p,\varsigma_p;\mathfrak{C}_p)\\(\mathfrak{d}_1,\varrho_1;\mathfrak{D}_1),\cdots,(\mathfrak{d}_q,\varrho_q;\mathfrak{D}_q)\end{array}\right.\right];p\right\}$$

$$=\frac{c^{-p}}{\nu}\Phi\left(\frac{-p}{\nu},\mathfrak{t}\right)\tag{11}$$

assuming the conditions are as stated in Kamlesh Jangid et al. [13] and $\Phi(\mathsf{h},\mathsf{t})$ is given in (9).

We established the family of incomplete I-function finite and infinite integral formulas, which have been influenced by Bansal's work [3]. For the incomplete I-function, we developed the integral representation in section 2. The incomplete I-function has been used to create a number of finite and infinite integral formulas in section 3. Additionally, we talked about several particular cases involving $\overline{I}$-function, incomplete $\overline{H}$-function and I-function for both the sections. The particular case of our primary findings is represented by the conclusions reached by Bansal et al. [3] and Jain et al. [12].

2. Integral Representation of the Incomplete I-Function

Recently, Bansal et al. [3] and Srivastava et al. [19] developed the integral representation of the incomplete H-functions and incomplete Gauss hypergeometric functions respectively. Inspired by their work, we develop the integral representation for the incomplete I-function.

Theorem 1. *If* $\Re(\mathfrak{f}_1) > 0$ *and* $\mathsf{t} \geq 0$, *then the integral representation formula is as follows:*

$$
{}^{\Gamma}I_{p,q}^{m,n}\left[z\left|\begin{array}{c}(1-\mathfrak{f}_1,\alpha_1;\mathfrak{F}_1:\mathsf{t}),(\mathfrak{f}_j,\alpha_j;\mathfrak{F}_j)_{2,p}\\(\mathfrak{g}_j,\beta_j;\mathfrak{G}_j)_{1,q}\end{array}\right.\right]
$$

$$
=\int_{\mathsf{t}}^{\infty}l^{\mathfrak{f}_1-1}\,\mathfrak{g}^{-l}\,I_{p-1,q}^{m,n-1}\left[l^{\alpha_1}z\left|\begin{array}{c}(\mathfrak{f}_j,\alpha_j;\mathfrak{F}_j)_{2,p}\\(\mathfrak{g}_j,\beta_j;\mathfrak{G}_j)_{1,q}\end{array}\right.\right]dl.\tag{12}
$$

Finite and Infinite Integral Formulae Involving Family of Incomplete I-Functions 189

Proof. The RHS of the equation (12) is:

$$G = \int_t^\infty l^{\mathfrak{f}_1-1}\, \mathfrak{g}^{-l} I_{p-1,\,q}^{m,\,n-1}\left[l^{\alpha_1} z \,\middle|\, \begin{matrix} (\mathfrak{f}_j,\alpha_j;\mathfrak{F}_j)_{2,\,p} \\ (\mathfrak{g}_j,\beta_j;\mathfrak{G}_j)_{1,\,q} \end{matrix}\right] dl. \tag{13}$$

Replace the *I*-function by (1), we get:

$$G = \int_t^\infty l^{\mathfrak{d}_1-1}\, \mathfrak{g}^{-l} \frac{1}{2\pi i}\int_{\mathcal{L}} \psi(\mathsf{h})\, z^{\mathsf{h}}\, d\mathsf{h}\, dl, \tag{14}$$

where $\psi(\mathsf{h})$ is given in (2).
Interchanging the integration order allowed within the given set of parameters and using the definition of incomplete gamma function (4), we get:

$$G - \frac{1}{2\pi i}\int_{\mathcal{L}} \frac{(\mathfrak{f}_1 - \alpha_1 \mathsf{h}, t)\prod\limits_{j=1}^{m}\{\Gamma(\mathfrak{g}_j - \beta_j \mathsf{h})\}^{\mathfrak{G}_j}\prod\limits_{j=1}^{n}\{\Gamma(1 - \mathfrak{f}_j + \alpha_j \mathsf{h})\}^{\mathfrak{F}_j}}{\prod\limits_{j=n+1}^{p}\{\Gamma(\mathfrak{f}_j - \alpha_j \mathsf{h})\}^{\mathfrak{F}_j}\prod\limits_{j=m+1}^{q}\{\Gamma(1 - \mathfrak{g}_j + \beta_j \mathsf{h})\}^{\mathfrak{G}_j}} z^{\mathsf{h}} d\mathsf{h}. \tag{15}$$

Now, with the help of (7), we arrive at the result of (12). □

Theorem 2. *If $\Re(\mathfrak{f}_1) > 0$ and $t \geq 0$, then the integral representation formula is as follows:*

$$\begin{aligned} {}^{\gamma}I_{p,\,q}^{m,\,n}&\left[z \,\middle|\, \begin{matrix} (1 - \mathfrak{f}_1,\alpha_1;\mathfrak{F}_1 : t),(\mathfrak{f}_j,\alpha_j;\mathfrak{F}_j)_{2,\,p} \\ (\mathfrak{g}_j,\beta_j;\mathfrak{G}_j)_{1,\,q} \end{matrix}\right] \\ &= \int_0^t l^{\mathfrak{f}_1-1}\, \mathfrak{g}^{-l} I_{p-1,\,q}^{m,\,n-1}\left[l^{\alpha_1} z \,\middle|\, \begin{matrix} (\mathfrak{f}_j,\alpha_j;\mathfrak{F}_j)_{2,\,p} \\ (\mathfrak{g}_j,\beta_j;\mathfrak{G}_j)_{1,\,q} \end{matrix}\right] dl. \end{aligned} \tag{16}$$

Theorem 2 can be proved in the similar way as that of Theorem 1.
Now we obtain the following corollaries from the above theorems.

Corollary 1. *If we put $\mathfrak{G}_j = 1, \forall j = 1, \cdots, m$ in Theorem 1, then our result can be expressed in the form of the incomplete $\bar{I}$-function $\left({}^{\Gamma}\bar{I}_{p,\,q}^{m,\,n}(z)\right)$, which is described as:*

$$\begin{aligned} {}^{\Gamma}\bar{I}_{p,\,q}^{m,\,n}&\left[z \,\middle|\, \begin{matrix} (1 - \mathfrak{f}_1,\alpha_1;\mathfrak{F}_1 : t),(\mathfrak{f}_j,\alpha_j;\mathfrak{F}_j)_{2,\,p} \\ (\mathfrak{g}_j,\beta_j;1)_{1,\,m},(\mathfrak{g}_j,\beta_j;\mathfrak{G}_j)_{m+1,\,q} \end{matrix}\right] \\ &= \int_t^\infty l^{\mathfrak{f}_1-1}\, \mathfrak{g}^{-l} I_{p-1,\,q}^{m,\,n-1}\left[l^{\alpha_1} z \,\middle|\, \begin{matrix} (\mathfrak{f}_j,\alpha_j;\mathfrak{F}_j)_{2,\,p} \\ (\mathfrak{g}_j,\beta_j;1)_{1,\,m},(\mathfrak{g}_j,\beta_j;\mathfrak{G}_j)_{m+1,\,q} \end{matrix}\right] dl. \end{aligned} \tag{17}$$

190   S. Bhatter, Nishant, S. D. Purohit

For detailed studies about the incomplete $\overline{I}$-function, the interested reader can refer to [5, 14].

Corollary 2. *If we put* $\mathfrak{G}_j = 1, \forall j = 1, \cdots, m,$ *and* $\mathfrak{F}_j = 1, \forall j = n + 1, \cdots, p$ *in Theorem 1, then our result can be expressed in the form of the incomplete* $\overline{H}$*-function* $\left(\overline{\Gamma}_{p,\,q}^{m,\,n}(z)\right),$ *which is described as:*

$$\overline{\Gamma}_{p,\,q}^{m,\,n}\left[z \,\middle|\, \begin{array}{c} (1 - \mathfrak{f}_1, \alpha_1; \mathfrak{F}_1 : \mathfrak{t}), (\mathfrak{f}_j, \alpha_j; \mathfrak{F}_j)_{2,\,n}, (\mathfrak{f}_j, \alpha_j; 1)_{n+1,\,p} \\ (\mathfrak{g}_j, \beta_j; 1)_{1,\,m}, (\mathfrak{g}_j, \beta_j; \mathfrak{G}_j)_{m+1,\,q} \end{array}\right]$$

$$= \int_t^\infty l^{\mathfrak{f}_1 - 1}\, \mathfrak{g}^{-l} I_{p-1,\,q}^{m,\,n-1}\left[l^{\alpha_1} z \,\middle|\, \begin{array}{c} (\mathfrak{f}_j, \alpha_j; \mathfrak{F}_j)_{2,\,n}, (\mathfrak{f}_j, \alpha_j; 1)_{n+1,\,p} \\ (\mathfrak{g}_j, \beta_j; 1)_{1,\,m}, (\mathfrak{g}_j, \beta_j; \mathfrak{G}_j)_{m+1,\,q} \end{array}\right] dl. \quad (18)$$

For detailed studies about the incomplete $\overline{H}$-function, the interested reader can refer to [20].

Remark:

(1) If we put $\mathfrak{G}_j = 1, \forall 1 \le j \le q,$ and $\mathfrak{F}_j = 1 \,\forall 1 \le j \le p$ in Theorem 1, then the result can be obtained in the form of incomplete H-function, which is established by Bansal et al. [3].

(2) If we put $\mathfrak{t} = 0, \mathfrak{G}_j = 1, \forall 1 \le j \le q,$ and $\mathfrak{F}_j = 1, \forall 1 \le j \le p$ in Theorem 1, then the result can be obtained in the form of Fox's H-function, which is mentioned as special case of [12].

(3) If we replace the upper gamma function $\Gamma(., \mathfrak{t})$ by the lower gamma function $\gamma(., \mathfrak{t})$ in the first two corollaries then we get special case for Theorem 2.

3. Finite and Infinite Integrals Related to the Incomplete I-Functions

Numerous finite and infinite integrals related to the incomplete I-function are evaluated in this section:

Theorem 3. *For* $\theta, \beta > 0,\ -\theta \min_{1 \le j \le m} \Re\left(\frac{\partial_j}{\varrho_j}\right) < \theta\Re\left(\frac{1-\mathfrak{c}_j}{\varsigma_j}\right),$ *then the improper integral is as follows holds for* $\mathfrak{t} \ge 0.$

$$\int_0^\infty l^{\alpha-1}(1 + \mathfrak{c})^{-\beta}\, \Gamma I_{p,\,q}^{m,\,n}\left[wl^\theta \,\middle|\, \begin{array}{c} (\mathfrak{c}_1, \varsigma_1; \mathfrak{C}_1 : \mathfrak{t}), (\mathfrak{c}_j, \varsigma_j; \mathfrak{C}_j)_{2,\,p} \\ (\partial_j, \varrho_j; \mathfrak{D}_j)_{1,\,q} \end{array}\right] dl$$

$$= \frac{\mathfrak{c}^{\alpha-\beta}}{\Gamma(\beta)}\, \Gamma I_{p+1,\,q+1}^{m+1,\,n+1}\left[w\mathfrak{c}^\theta \,\middle|\, \begin{array}{c} (\mathfrak{c}_1, \varsigma_1; \mathfrak{C}_1 : \mathfrak{t}), (1 - \alpha, \theta, 1), (\mathfrak{c}_j, \varsigma_j; \mathfrak{C}_j)_{2,\,p} \\ (\beta - \alpha, \theta, 1), (\partial_j, \varrho_j; \mathfrak{D}_j)_{1,\,q} \end{array}\right] \quad (19)$$

and

$$\int_0^\infty l^{\alpha-1}(l+\mathfrak{c})^{-\beta}\,{}^\gamma I_{p,\,q}^{m,\,n}\left[wl^\theta\;\middle|\;\begin{array}{c}(\mathfrak{c}_1,\varsigma_1;\mathfrak{C}_1:\mathsf{t}),(\mathfrak{c}_j,\varsigma_j;\mathfrak{C}_j)_{2,\,p}\\(\mathfrak{d}_j,\varrho_j;\mathfrak{D}_j)_{1,\,q}\end{array}\right]$$

$$=\frac{\mathfrak{c}^{\alpha-\beta}}{\Gamma(\beta)}\,{}^\gamma I_{p+1,\,q+1}^{m+1,\,n+1}\left[w\mathfrak{c}^\theta\;\middle|\;\begin{array}{c}(\mathfrak{c}_1,\varsigma_1;\mathfrak{C}_1:\mathsf{t}),(1-\alpha,\theta,1),(\mathfrak{c}_j,\varsigma_j;\mathfrak{C}_j)_{2,\,p}\\(\beta-\alpha,\theta,1),(\mathfrak{d}_j,\varrho_j;\mathfrak{D}_j)_{1,\,q}\end{array}\right].$$

$$(20)$$

Proof. The LHS of equation (19) is:

$$G'=\int_0^\infty l^{\alpha-1}(l+\mathfrak{c})^{-\beta}\,{}^\Gamma I_{p,\,q}^{m,\,n}\left[wl^\theta\;\middle|\;\begin{array}{c}(\mathfrak{c}_1,\varsigma_1;\mathfrak{C}_1:\mathsf{t}),(\mathfrak{c}_j,\varsigma_j;\mathfrak{C}_j)_{2,\,p}\\(\mathfrak{d}_j,\varrho_j;\mathfrak{D}_j)_{1,\,q}\end{array}\right]dl.$$

$$(21)$$

Replacing the incomplete I-function by (7), we get:

$$G'=\int_0^\infty l^{\alpha-1}(l+\mathfrak{c})^{-\beta}\,\frac{1}{2\pi i}\int_{\mathcal{L}}\Phi(\mathsf{h},\mathsf{t})\,(wl^\theta)^{\mathsf{h}}\,d\mathsf{h}dl,\qquad(22)$$

where $\Phi(\mathsf{h},\mathsf{t})$ is given by (9).

Changing the integration order and using the definition of beta function, we get:

$$G'=\frac{\mathfrak{c}^{\alpha-\beta}}{\Gamma(\beta)}\frac{1}{2\pi i}\int_{\mathcal{L}}\Phi(\mathsf{h},\mathsf{t})\,(w\mathfrak{c}^\theta)^{\mathsf{h}}\,\Gamma(\alpha+\theta\mathsf{h})\Gamma(\beta-\alpha-\theta\mathsf{h})d\mathsf{h}.\qquad(23)$$

Now convert equation (23) into an incomplete I-function to obtain the required outcome.

Equation (20) can be proved in the similar way as equation (19). $\qquad\square$

Theorem 4. *For* $\theta,\nu>0$, $\Re(\alpha)+\theta\min_{1\le j\le m}\Re\left(\frac{\mathfrak{d}_j}{\varrho_j}\right)>0$, $\Re(\beta)+\nu\min_{1\le j\le m}\Re\left(\frac{\mathfrak{c}_j}{\varsigma_j}\right)>0$, *then the improper integral is as follows holds for* $\mathsf{t}\ge0$.

$$\int_0^l x^{\alpha-1}(l-x)^{\beta-1}\,{}^\Gamma I_{p,\,q}^{m,\,n}\left[wx^\theta(l-x)^\nu\;\middle|\;\begin{array}{c}(\mathfrak{c}_1,\varsigma_1;\mathfrak{C}_1:\mathsf{t}),(\mathfrak{c}_j,\varsigma_j;\mathfrak{C}_j)_{2,\,p}\\(\mathfrak{d}_j,\varrho_j;\mathfrak{D}_j)_{1,\,q}\end{array}\right]dx$$

$$=l^{\alpha+\beta-1}\,{}^\Gamma I_{p+2,\,q+1}^{m,\,n+2}\left[wl^{\theta+\nu}\;\middle|\;\begin{array}{c}(\mathfrak{c}_1,\varsigma_1;\mathfrak{C}_1:\mathsf{t}),(1-\alpha,\theta,1),(1-\beta,\nu,1),\\(\mathfrak{d}_j,\varrho_j;\mathfrak{D}_j)_{1,\,q},\end{array}\right.$$

$$\left.\begin{array}{c}(\mathfrak{c}_j,\varsigma_j;\mathfrak{C}_j)_{2,\,p}\\(1-\alpha-\beta,\theta+\nu,1)\end{array}\right]$$

$$=l^{\alpha+\beta-1}\,{}^\Gamma I_{p+1,q+2}^{m+2,n}\left[wl^{\theta+\nu}\;\middle|\;\begin{array}{c}(\mathfrak{c}_1,\varsigma_1;\mathfrak{C}_1:\mathsf{t}),(\alpha+\beta,\theta+\nu,1),(\mathfrak{c}_j,\varsigma_j;\mathfrak{C}_j)_{2,\,p}\\(\alpha,\theta,1),(\beta,\nu,1),(\mathfrak{d}_j,\varrho_j;\mathfrak{D}_j)_{1,\,q},\end{array}\right]$$

$$(24)$$

and

$$\int_0^l x^{\alpha-1}(l-x)^{\beta-1}\,{}^\gamma I_{p,\,q}^{m,\,n}\left[wx^\theta(l-x)^\nu \,\middle|\, \begin{matrix} (\mathfrak{c}_1,\varsigma_1;\mathfrak{C}_1:t),(\mathfrak{c}_j,\varsigma_j;\mathfrak{C}_j)_{2,p} \\ (\mathfrak{d}_j,\varrho_j;\mathfrak{D}_j)_{1,q} \end{matrix}\right] dx$$

$$= l^{\alpha+\beta-1}\,{}^\gamma I_{p+2,\,q+1}^{m,\,n+2}\left[wl^{\theta+\nu} \,\middle|\, \begin{matrix} (\mathfrak{c}_1,\varsigma_1;\mathfrak{C}_1:t),(1-\alpha,\theta,1),(1-\beta,\nu,1), \\ (\mathfrak{d}_j,\varrho_j;\mathfrak{D}_j)_{1,q}, \end{matrix}\right.$$

$$\left.\begin{matrix} (\mathfrak{c}_j,\varsigma_j;\mathfrak{C}_j)_{2,p} \\ (1-\alpha-\beta,\theta+\nu,1) \end{matrix}\right]$$

$$= l^{\alpha+\beta-1}\,{}^\gamma I_{p+1,\,q+2}^{m+2,\,n}\left[wl^{\theta+\nu} \,\middle|\, \begin{matrix} (\mathfrak{c}_1,\varsigma_1;\mathfrak{C}_1:t),(\alpha+\beta,\theta+\nu,1),(\mathfrak{c}_j,\varsigma_j;\mathfrak{C}_j)_{2,p} \\ (\alpha,\theta,1),(\beta,\nu,1),(\mathfrak{d}_j,\varrho_j;\mathfrak{D}_j)_{1,q}, \end{matrix}\right].$$

$$(25)$$

Proof. The LHS of the equation is:

$$G'' = \int_0^l x^{\alpha-1}(l-x)^{\beta-1}$$

$$\times\,{}^\Gamma I_{p,\,q}^{m,\,n}\left[wx^\theta(l-x)^\nu \,\middle|\, \begin{matrix} (\mathfrak{c}_1,\varsigma_1;\mathfrak{C}_1:t),(\mathfrak{c}_j,\varsigma_j;\mathfrak{C}_j)_{2,p} \\ (\mathfrak{d}_j,\varrho_j;\mathfrak{D}_j)_{1,q} \end{matrix}\right] dx. \qquad (26)$$

Replacing the incomplete I-function by (7), we get:

$$G'' = \int_0^t x^{\alpha-1}(l-x)^{\beta-1}\frac{1}{2\pi i}\int_{\mathcal{L}}\Phi(\mathsf{h},\mathsf{t})\,(wx^\theta(l-x)^\nu)^{\mathsf{h}}\,d\mathsf{h}\,dx. \qquad (27)$$

where $\Phi(\mathsf{h},\mathsf{t})$ is given by (9).

Changing the integration order and using the definition of beta function, we get:

$$G'' = l^{\alpha+\beta-1}\frac{1}{2\pi i}\int_0^l \Phi(\mathsf{h},\mathsf{t})\,w^{\mathsf{h}}\,l^{\theta\mathsf{h}+\nu\mathsf{h}}\frac{\Gamma(\alpha+\theta\mathsf{h})\Gamma(\beta+\nu\mathsf{h})}{\Gamma(\alpha+\beta+\theta\mathsf{h}+\nu\mathsf{h})}\,d\mathsf{h}. \qquad (28)$$

Now, converting equation (28) into an incomplete I-function, we get the requied outcome after a simple modification.

Equation (25) can be proved in a similar manner as equation (24). $\qquad\square$

Theorem 5. *If* $-\varsigma\,\max\limits_{1\leq j\leq n}\left(\frac{1-\mathfrak{c}_j}{\varsigma_j}\right) - \min\limits_{1\leq j\leq M}\Re\left(\frac{\mathfrak{g}_j}{\beta_j}\right) < \Re(\alpha) < \varsigma\,\min\limits_{1\leq j\leq m}\left(\frac{\mathfrak{d}_j}{\varrho_j}\right) + \max\limits_{1\leq j\leq N}\Re\left(\frac{1-\mathfrak{d}_j}{\alpha_j}\right)$ *and* $\varsigma > 0$, *then the improper integral is as follows holds*

Finite and Infinite Integral Formulae Involving Family of Incomplete I-Functions 193

for $t \geq 0$.

$$\int_0^\infty l^{\alpha-1} {}^\Gamma I_{p,\,q}^{m,\,n} \left[k\, l \, \middle| \, \begin{array}{c} (\mathfrak{c}_1, \varsigma_1; \mathfrak{C}_1 : t), (\mathfrak{c}_j, \varsigma_j; \mathfrak{C}_j)_{2,\,p} \\ (\mathfrak{d}_j, \varrho_j; \mathfrak{D}_j)_{1,\,q} \end{array} \right]$$

$$\times I_{P,\,Q}^{M,\,N} \left[s\, l^\varsigma \, \middle| \, \begin{array}{c} (\mathfrak{f}_j, \alpha_j; \mathfrak{F}_j)_{1,\,P} \\ (\mathfrak{g}_j, \beta_j; \mathfrak{G}_j)_{1,\,Q} \end{array} \right] = k^{-\alpha\Gamma} I_{p+P,\,q+Q}^{m+M,\,n+N} \left[\frac{k}{s^\varsigma} \, \middle| \, \begin{array}{c} (\mathfrak{f}_j, \alpha_j; \mathfrak{F}_j)_{1,\,P}, \\ (\mathfrak{g}_j, \beta_j; \mathfrak{G}_j)_{1,\,Q}, \end{array} \right.$$

$$\left. \begin{array}{c} (\mathfrak{c}_1 + \varsigma_1\alpha, -\varsigma_1\varsigma, \mathfrak{C}_1 : t), (\mathfrak{c}_j + \varsigma_j\alpha, -\varsigma_j\varsigma, \mathfrak{C}_j)_{2,\,p} \\ (\mathfrak{d}_j + \alpha\varrho_j, -\varsigma\varrho_j, \mathfrak{D}_j)_{1,\,q} \end{array} \right] \tag{29}$$

and

$$\int_0^\infty l^{\alpha-1}\,{}^\gamma I_{p,\,q}^{m,\,n} \left[k\, l \, \middle| \, \begin{array}{c} (\mathfrak{c}_1, \varsigma_1; \mathfrak{C}_1 : t), (\mathfrak{c}_j, \varsigma_j; \mathfrak{C}_j)_{2,\,p} \\ (\mathfrak{d}_j, \varrho_j; \mathfrak{D}_j)_{1,\,q} \end{array} \right]$$

$$\times I_{P,\,Q}^{M,\,N} \left[s\, l^\varsigma \, \middle| \, \begin{array}{c} (\mathfrak{f}_j, \alpha_j; \mathfrak{F}_j)_{1,\,P} \\ (\mathfrak{g}_j, \beta_j; \mathfrak{G}_j)_{1,\,Q} \end{array} \right] = k^{-\alpha\gamma} I_{p+P,\,q+Q}^{m+M,\,n+N} \left[\frac{k}{s^\varsigma} \, \middle| \, \begin{array}{c} (\mathfrak{f}_j, \alpha_j; \mathfrak{F}_j)_{1,\,P}, \\ (\mathfrak{g}_j, \beta_j; \mathfrak{G}_j)_{1,\,Q}, \end{array} \right.$$

$$\left. \begin{array}{c} (\mathfrak{c}_1 + \varsigma_1\alpha, -\varsigma_1\varsigma, \mathfrak{C}_1 : t), (\mathfrak{c}_j + \varsigma_j\alpha, -\varsigma_j\varsigma, \mathfrak{C}_j)_{2,\,p}, \\ (\mathfrak{d}_j + \alpha\varrho_j, -\varsigma\varrho_j, \mathfrak{D}_j)_{1,\,q} \end{array} \right]. \tag{30}$$

Proof. The LHS of the equation (29) is:

$$H = \int_0^\infty l^{\alpha-1}\,{}^\gamma I_{p,\,q}^{m,\,n} \left[k\, l \, \middle| \, \begin{array}{c} (\mathfrak{c}_1, \varsigma_1; \mathfrak{C}_1 : t), (\mathfrak{c}_j, \varsigma_j; \mathfrak{C}_j)_{2,\,p} \\ (\mathfrak{d}_j, \varrho_j; \mathfrak{D}_j)_{1,\,q} \end{array} \right]$$

$$\times I_{P,\,Q}^{M,\,N} \left[s\, l^\varsigma \, \middle| \, \begin{array}{c} (\mathfrak{f}_j, \alpha_j; \mathfrak{F}_j)_{1,\,P} \\ (\mathfrak{g}_j, \beta_j; \mathfrak{G}_j)_{1,\,Q} \end{array} \right] dl. \tag{31}$$

Replacing the *I*-function with (1), we get:

$$H = \int_0^\infty l^{\alpha-1} \frac{1}{2\pi i} \int_{\mathcal{L}} \psi(\xi)\, (s\, l^\varsigma)^\xi$$

$$\times {}^\Gamma I_{p,\,q}^{m,\,n} \left[k\, l \, \middle| \, \begin{array}{c} (\mathfrak{c}_1, \varsigma_1; \mathfrak{C}_1 : t), (\mathfrak{c}_j, \varsigma_j; \mathfrak{C}_j)_{2,\,p} \\ (\mathfrak{d}_j, \varrho_j; \mathfrak{D}_j)_{1,\,q} \end{array} \right] d\xi\, dl, \tag{32}$$

where $\psi(\xi)$ is given by equation (2).
Interchange the integration order to get:

$$\frac{1}{2\pi i} \int_{\mathcal{L}} \psi(\xi) s^{\xi} \int_0^{\infty} l^{\alpha + \varsigma \xi - 1} \, {}^{\Gamma}I_{p,q}^{m,n} \left[kl \, \middle| \, \begin{matrix} (\mathfrak{c}_1, \varsigma_1; \mathfrak{C}_1 : \mathfrak{t}), (\mathfrak{c}_j, \varsigma_j; \mathfrak{C}_j)_{2,p} \\ (\mathfrak{d}_j, \varrho_j; \mathfrak{D}_j)_{1,q} \end{matrix} \right] dl \, d\xi.$$

$$(33)$$

Use Mellin Transformation to get:

$$H = k^{-\alpha} \frac{1}{2\pi i} \int_{\mathcal{L}} \left(\frac{s}{k^{\varsigma}} \right) \times \left(\frac{\prod_{j=1}^{m} \{\Gamma(\mathfrak{g}_j - \beta_j \xi)\}^{\mathfrak{G}_j} \prod_{j=1}^{n} \{\Gamma(1 - \mathfrak{f}_j + \alpha_j \xi)\}^{\mathfrak{F}_j}}{\prod_{j=n+1}^{p} \{\Gamma(\mathfrak{f}_j - \alpha_j \xi)\}^{\mathfrak{F}_j} \prod_{j=m+1}^{q} \{\Gamma(1 - \mathfrak{g}_j + \beta_j \xi)\}^{\mathfrak{G}_j}} \right)$$

$$\times \frac{\{\Gamma(1 - \mathfrak{c}_1 + \varsigma_1(-\xi\varsigma - \alpha), \mathfrak{t})\}^{\mathfrak{C}_1} \prod_{j=1}^{m} \{\Gamma(\mathfrak{d}_j - \varrho_j(-\xi\varsigma - \alpha))\}^{\mathfrak{D}_j}}{\prod_{j=n+1}^{p} \{\Gamma(\mathfrak{c}_j - \varsigma_j(-\xi\varsigma - \alpha))\}^{\mathfrak{C}_j}}$$

$$\times \frac{\prod_{j=2}^{n} \{\Gamma(1 - \mathfrak{c}_j + \varsigma_j(-\xi\varsigma - \alpha))\}^{\mathfrak{C}_j}}{\prod_{j=m+1}^{q} \{\Gamma(1 - \mathfrak{d}_j + \varrho_j(-\xi\varsigma - \alpha))\}^{\mathfrak{D}_j}} \, d\xi. \tag{34}$$

Now, converting equation (34) into an incomplete I-function, we get the required outcome after a simple modification.

Equation (30) can be proved in a similar manner as equation (29). □

Theorem 6. *If* $\Re(\alpha) + \theta \min\limits_{1 \le j \le m} \Re\left(\frac{\mathfrak{d}_j}{\varrho_j}\right) > 0$, $\Re(\beta) + \nu \min\limits_{1 \le j \le m} \Re\left(\frac{\mathfrak{c}_j}{\varsigma_j}\right) > 0$, $\varrho, \varsigma > 0$ *and* $\theta + \nu \ge 1$, *then the improper integral for* $u, \mathfrak{t} \ge 0$ *is as follows.*

$$\int_0^l x^{\alpha-1}(l-x)^{\beta-1} \, {}_P\Gamma_Q \left[\begin{matrix} (g_1, u), g_2, g_3, ..., g_P \\ h_1, h_2, ..., h_Q \end{matrix} \, a\, x^{\theta}(l-x)^{\nu} \right]$$

$$\times {}^{\Gamma}I_{p,q}^{m,n} \left[w\, x^{\varrho}(l-x)^{\varsigma} \, \middle| \, \begin{matrix} (\mathfrak{c}_1, \varsigma_1; \mathfrak{C}_1 : \mathfrak{t}), (\mathfrak{c}_j, \varsigma_j; \mathfrak{C}_j)_{2,p} \\ (\mathfrak{d}_j, \varrho_j; \mathfrak{D}_j)_{1,q} \end{matrix} \right] = l^{\alpha+\beta-1}$$

$$\times \frac{\prod_{j=1}^{Q} \Gamma(h_j)}{\prod_{j=1}^{P} \Gamma(g_j)} \sum_{r=0}^{\infty} h(r) \, t^{(\theta+\nu)r} \, {}^{\Gamma}I_{p+2, q+1}^{m, n+2} \left[wl^{\theta+\nu} \, \middle| \, \begin{matrix} (\mathfrak{c}_1, \varsigma_1; \mathfrak{C}_1 : \mathfrak{t}), \\ (\mathfrak{d}_j, \varrho_j; \mathfrak{D}_j)_{1,q}, \end{matrix} \right.$$

$$\left. \begin{matrix} (1 - \alpha - \theta r, \varrho, 1), (1 - \beta - \nu r, \varsigma, 1), (\mathfrak{c}_j, \varsigma_j; \mathfrak{C}_j)_{2,p} \\ (1 - \alpha - \theta r - \beta - \nu r, \varrho + \varsigma, 1) \end{matrix} \right] \tag{35}$$

and

Finite and Infinite Integral Formulae Involving Family of Incomplete I-Functions   195

$$
\int_0^l x^{\alpha-1}(l-x)^{\beta-1} \, {}_P\Gamma_Q \left[\begin{array}{c} (g_1, u), g_2, g_3, \ldots, g_P \\ h_1, h_2, \ldots, h_Q \end{array} a\, x^\theta (l-x)^\nu \right]
$$

$$
\times \, {}^\gamma I_{p,\,q}^{m,\,n} \left[w\, x^\varrho (l-x)^\varsigma \left| \begin{array}{c} (\mathfrak{c}_1, \varsigma_1; \mathfrak{C}_1 : \mathfrak{t}), (\mathfrak{c}_j, \varsigma_j; \mathfrak{C}_j)_{2,\,p} \\ (\mathfrak{d}_j, \varrho_j; \mathfrak{D}_j)_{1,\,q} \end{array} \right. \right] = l^{\alpha+\beta-1}
$$

$$
\times \, \frac{\prod_{j=1}^{Q} \Gamma(h_j)}{\prod_{j=1}^{P} \Gamma(g_j)} \sum_{r=0}^{\infty} h(r)\, t^{(\theta+\nu)r} \, {}^\gamma I_{p+2,\,q+1}^{m,\,n+2} \left[w l^{\theta+\nu} \left| \begin{array}{c} (\mathfrak{c}_1, \varsigma_1; \mathfrak{C}_1 : \mathfrak{t}), \\ (\mathfrak{d}_j, \varrho_j; \mathfrak{D}_j)_{1,\,q}, \end{array} \right. \right.
$$

$$
\left. \begin{array}{c} (1-\alpha-\theta r, \varrho, 1), (1-\beta-\nu r, \varsigma, 1), (\mathfrak{c}_j, \varsigma_j; \mathfrak{C}_j)_{2,\,p} \\ (1-\alpha-\theta r - \beta - \nu r, \varrho + \varsigma, 1) \end{array} \right] \tag{36}
$$

where,

$$
h(r) = \frac{\Gamma(g_1 + r, u) \prod_{j=2}^{P} \Gamma(g_j + r)}{\prod_{j=1}^{Q} \Gamma(h_j + r)} \frac{\mathfrak{c}^r}{r!}. \tag{37}
$$

Proof. The LHS of the equation is:

$$
H' = \int_0^l x^{\alpha-1}(l-x)^{\beta-1} \, {}_P\Gamma_Q \left[\begin{array}{c} (g_1, u), g_2, g_3, \ldots, g_P \\ h_1, h_2, \ldots, h_Q \end{array} \mathfrak{c}\, x^\theta (l-x)^\nu \right]
$$

$$
\times \, {}^\Gamma I_{p,\,q}^{m,\,n} \left[w\, x^\varrho (l-x)^\varsigma \left| \begin{array}{c} (\mathfrak{c}_1, \varsigma_1; \mathfrak{C}_1 : \mathfrak{t}), (\mathfrak{c}_j, \varsigma_j; \mathfrak{C}_j)_{2,\,p} \\ (\mathfrak{d}_j, \varrho_j; \mathfrak{D}_j)_{1,\,q} \end{array} \right. \right]. \tag{38}
$$

Representing the incomplete hypergeometric function in the form of a series, we obtain:

$$
H' = \int_0^l x^{\alpha-1}(l-x)^{\beta-1} \left[\frac{\prod_{j=1}^{Q} \Gamma(h_j)}{\prod_{j=1}^{P} \Gamma(g_j)} \sum_{r=0}^{\infty} \frac{\Gamma(g_1 + r, u) \prod_{j=2}^{P} \Gamma(g_j + r)}{\prod_{j=1}^{Q} \Gamma(h_j + r)} \right.
$$

$$
\left. \frac{(a\mathfrak{c}x^\theta (l-x)^\nu)^r}{r!} \right] \times \, {}^\Gamma I_{p,\,q}^{m,\,n} \left[w\, x^\varrho (l-x)^\varsigma \left| \begin{array}{c} (\mathfrak{c}_1, \varsigma_1; \mathfrak{C}_1 : \mathfrak{t}), (\mathfrak{c}_j, \varsigma_j; \mathfrak{C}_j)_{2,\,p} \\ (\mathfrak{d}_j, \varrho_j; \mathfrak{D}_j)_{1,\,q} \end{array} \right. \right].
$$

$$
\tag{39}
$$

Further changing the order of integration and summation:

$$
H' = \left[\frac{\prod_{j=1}^{Q} \Gamma(h_j)}{\prod_{j=1}^{P} \Gamma(g_j)} \sum_{r=0}^{\infty} \frac{\Gamma(g_1 + r, u) \prod_{j=2}^{P} \Gamma(g_j + r)}{\prod_{j=1}^{Q} \Gamma(h_j + r)} \frac{\mathfrak{c}^r}{r!} \right]
$$

$$
\times \int_0^l x^{\alpha+\theta r - 1}(l-x)^{\beta+\nu r - 1} \times \, {}^\Gamma I_{p,\,q}^{m,\,n} \left[w\, x^\varrho (l-x)^\varsigma \right|
$$

$$
\left. \begin{array}{c} (\mathfrak{c}_1, \varsigma_1; \mathfrak{C}_1 : \mathfrak{t}), (\mathfrak{c}_j, \varsigma_j; \mathfrak{C}_j)_{2,\,p} \\ (\mathfrak{D}_j, \varrho_j; \mathfrak{D}_j)_{1,\,q} \end{array} \right]. \tag{40}
$$

Now, after using equation (24) and making some modifications, we get the required outcome. $\qquad\square$

Equation (36) can be proved similar manner as Equation (35).

Now we obtain certain special cases for the above theorems.

Corollary 3. *If we set $\mathfrak{D}_j = 1, \forall 1 \le j \le m$ in Theorem 3, then the outcome can be depicted as an incomplete $\overline{I}$-function $\left({}^{\Gamma}\overline{I}^{m,\,n}_{p,\,q}(z) \right)$, which is described as:*

$$\int_0^\infty l^{\alpha-1}(1+\mathfrak{c})^{-\beta}\, {}^{\Gamma}\overline{I}^{m,\,n}_{p,\,q} \left[wl^\theta \,\middle|\, \begin{array}{l} (\mathfrak{c}_1,\varsigma_1;\mathfrak{C}_1:\mathfrak{t}),(\mathfrak{c}_j,\varsigma_j;\mathfrak{C}_j)_{2,\,p} \\ (\mathfrak{d}_j,\varrho_j;1)_{1,\,m},(\mathfrak{d}_j,\varrho_j;\mathfrak{D}_j)_{m+1,\,q} \end{array} \right] dl$$

$$= \frac{\mathfrak{c}^{\alpha-\beta}}{\Gamma(\beta)}\, {}^{\Gamma}\overline{I}^{m+1,\,n+1}_{p+1,\,q+1} \left[w\mathfrak{c}^\theta \,\middle|\, \begin{array}{l} (\mathfrak{c}_1,\varsigma_1;\mathfrak{C}_1:\mathfrak{t}),(1-\alpha,\theta,1),(\mathfrak{c}_j,\varsigma_j;\mathfrak{C}_j)_{2,\,p} \\ (\beta-\alpha,\theta,1),(\mathfrak{d}_j,\varrho_j;1)_{1,\,m},(\mathfrak{d}_j,\varrho_j;\mathfrak{D}_j)_{m+1,\,q} \end{array} \right]$$

$$(41)$$

and

$$\int_0^\infty l^{\alpha-1}(1+\mathfrak{c})^{-\beta}\, {}^{\gamma}\overline{I}^{m,\,n}_{p,\,q} \left[wl^\theta \,\middle|\, \begin{array}{l} (\mathfrak{c}_1,\varsigma_1;\mathfrak{C}_1:\mathfrak{t}),(\mathfrak{c}_j,\varsigma_j;\mathfrak{C}_j)_{2,\,p} \\ (\mathfrak{d}_j,\varrho_j;1)_{1,\,m},(\mathfrak{d}_j,\varrho_j;\mathfrak{D}_j)_{m+1,\,q} \end{array} \right] dl$$

$$= \frac{\mathfrak{c}^{\alpha-\beta}}{\Gamma(\beta)}\, {}^{\gamma}\overline{I}^{m+1,\,n+1}_{p+1,\,q+1} \left[w\mathfrak{c}^\theta \,\middle|\, \begin{array}{l} (\mathfrak{c}_1,\varsigma_1;\mathfrak{C}_1:\mathfrak{t}),(1-\alpha,\theta,1),(\mathfrak{c}_j,\varsigma_j;\mathfrak{C}_j)_{2,\,p} \\ (\beta-\alpha,\theta,1),(\mathfrak{d}_j,\varrho_j;1)_{1,\,m},(\mathfrak{d}_j,\varrho_j;\mathfrak{D}_j)_{m+1,\,q} \end{array} \right].$$

$$(42)$$

Corollary 4. *If we set $\mathfrak{D}_j = 1, \forall 1 \le j \le m$ and $\mathfrak{C}_j = 1, \forall n+1 \le j \le p$ in Theorem 3, then the outcome can be depicted as an incomplete $\overline{H}$-function $\left({}^{\Gamma}\overline{H}^{m,\,n}_{p,\,q}(z) \right)$, which is described as:*

$$\int_0^\infty l^{\alpha-1}(1+\mathfrak{c})^{-\beta}\, {}^{\Gamma}\overline{H}^{m,\,n}_{p,\,q} \left[wl^\theta \,\middle|\, \begin{array}{l} (\mathfrak{c}_1,\varsigma_1;\mathfrak{C}_1:\mathfrak{t}),(\mathfrak{c}_j,\varsigma_j;\mathfrak{C}_j)_{2,\,n}, \\ (\mathfrak{d}_j,\varrho_j;1)_{1,\,m}, \end{array} \right.$$

$$\left. \begin{array}{l} (\mathfrak{c}_j,\varsigma_j;1)_{n+1,\,p} \\ (\mathfrak{d}_j,\varrho_j;\mathfrak{D}_j)_{m+1,\,q} \end{array} \right] dl = \frac{\mathfrak{c}^{\alpha-\beta}}{\Gamma(\beta)}\, {}^{\Gamma}\overline{H}^{m+1,\,n+1}_{p+1,\,q+1} \left[w\mathfrak{c}^\theta \,\middle|\, \begin{array}{l} (\mathfrak{c}_1,\varsigma_1;\mathfrak{C}_1:\mathfrak{t}),(1-\alpha,\theta,1), \\ (\beta-\alpha,\theta,1), \end{array} \right.$$

$$\left. \begin{array}{l} (\mathfrak{c}_j,\varsigma_j;\mathfrak{C}_j)_{2,\,n},(\mathfrak{c}_j,\varsigma_j;1)_{n+1,\,p} \\ (\mathfrak{d}_j,\varrho_j;1)_{1,\,m},(\mathfrak{d}_j,\varrho_j;\mathfrak{D}_j)_{m+1,\,q} \end{array} \right] \qquad (43)$$

and

$$\int_0^\infty l^{\alpha-1}(l+\mathfrak{c})^{-\beta}\,\overline{\gamma}_{p,\,q}^{m,\,n}\left[wl^\theta\;\middle|\;\begin{array}{l}(\mathfrak{c}_1,\varsigma_1;\mathfrak{C}_1:\mathfrak{t}),(\mathfrak{c}_j,\varsigma_j;\mathfrak{C}_j)_{2,\,n},\\ (\eth_j,\varrho_j;1)_{1,\,m},\end{array}\right.$$

$$\left.\begin{array}{l}(\mathfrak{c}_j,\varsigma_j;1)_{n+1,\,p}\\ (\eth_j,\varrho_j;\mathfrak{D}_j)_{m+1,\,q}\end{array}dl=\frac{\mathfrak{c}^{\alpha-\beta}}{\Gamma(\beta)}\,\overline{\gamma}_{p+1,\,q+1}^{m+1,\,n+1}\left[w\mathfrak{c}^\theta\;\middle|\;\begin{array}{l}(\mathfrak{c}_1,\varsigma_1;\mathfrak{C}_1:\mathfrak{t}),(1-\alpha,\theta,1),\\ (\beta-\alpha,\theta,1),\end{array}\right.\right.$$

$$\left.\begin{array}{l}(\mathfrak{c}_j,\varsigma_j;\mathfrak{C}_j)_{2,\,n},(\mathfrak{c}_j,\varsigma_j;1)_{n+1,\,p}\\ (\eth_j,\varrho_j;1)_{1,\,m},(\eth_j,\varrho_j;\mathfrak{D}_j)_{m+1,\,q}\end{array}\right] \tag{44}$$

Corollary 5. *If we set* $\mathfrak{D}_j=1,\forall\,1\le j\le m$ *in Theorem 4, then the outcome can be depicted as an incomplete $\overline{I}$-function* $\left(^\Gamma\overline{I}_{p,\,q}^{m,\,n}(z)\right)$, *which is described as:*

$$\int_0^l x^{\alpha-1}(l-x)^{\beta-1}\,{}^\Gamma\overline{I}_{p,\,q}^{m,\,n}\left[wx^\theta(l-x)^\nu\;\middle|\;\begin{array}{l}(\mathfrak{c}_1,\varsigma_1;\mathfrak{C}_1:\mathfrak{t}),(\mathfrak{c}_j,\varsigma_j;\mathfrak{C}_j)_{2,\,p}\\ (\eth_j,\varrho_j;1)_{1,\,m},(\eth_j,\varrho_j;\mathfrak{D}_j)_{m+1,\,q}\end{array}\right]dx$$

$$=l^{\alpha+\beta-1}\,{}^\Gamma\overline{I}_{p+2,\,q+1}^{m,\,n+2}\left[wl^{\theta+\nu}\;\middle|\;\begin{array}{l}(\mathfrak{c}_1,\varsigma_1;\mathfrak{C}_1:\mathfrak{t}),(1-\alpha,\theta,1),(1-\beta,\nu,1),\\ (\eth_j,\varrho_j;1)_{1,\,m},(\eth_j,\varrho_j;\mathfrak{D}_j)_{m+1,\,q},\end{array}\right.$$

$$\left.\begin{array}{c}(\mathfrak{c}_j,\varsigma_j;\mathfrak{C}_j)_{2,\,p}\\ (1-\alpha-\beta,\theta+\nu,1)\end{array}\right]$$

$$=l^{\alpha+\beta-1}\,{}^\Gamma\overline{I}_{p+1,\,q+2}^{m+2,\,n}\left[wl^{\theta+\nu}\;\middle|\;\begin{array}{l}(\mathfrak{c}_1,\varsigma_1;\mathfrak{C}_1:\mathfrak{t}),(\alpha+\beta,\theta+\nu,1),\\ (\alpha,\theta,1),(\beta,\nu,1),(\eth_j,\varrho_j;1)_{1,\,m},\end{array}\right.$$

$$\left.\begin{array}{c}(\mathfrak{c}_j,\varsigma_j;\mathfrak{C}_j)_{2,\,p}\\ (\eth_j,\varrho_j;\mathfrak{D}_j)_{m+1,\,q}\end{array}\right] \tag{45}$$

and

$$\int_0^l x^{\alpha-1}(l-x)^{\beta-1}\,{}^\gamma\overline{I}_{p,\,q}^{m,\,n}\left[wx^\theta(l-x)^\nu\;\middle|\;\begin{array}{l}(\mathfrak{c}_1,\varsigma_1;\mathfrak{C}_1:\mathfrak{t}),(\mathfrak{c}_j,\varsigma_j;\mathfrak{C}_j)_{2,\,p}\\ (\eth_j,\varrho_j;1)_{1,\,m},(\eth_j,\varrho_j;\mathfrak{D}_j)_{m+1,\,q}\end{array}\right]dx$$

$$=l^{\alpha+\beta-1}\,{}^\gamma\overline{I}_{p+2,\,q+1}^{m,\,n+2}\left[wl^{\theta+\nu}\;\middle|\;\begin{array}{l}(\mathfrak{c}_1,\varsigma_1;\mathfrak{C}_1:\mathfrak{t}),(1-\alpha,\theta,1),(1-\beta,\nu,1),\\ (\eth_j,\varrho_j;1)_{1,\,m},(\eth_j,\varrho_j;\mathfrak{D}_j)_{m+1,\,q},\end{array}\right.$$

$$\left.\begin{array}{c}(\mathfrak{c}_j,\varsigma_j;\mathfrak{C}_j)_{2,\,p}\\ (1-\alpha-\beta,\theta+\nu,1)\end{array}\right]$$

$$=l^{\alpha+\beta-1}\,{}^\gamma\overline{I}_{p+1,\,q+2}^{m+2,\,n}\left[wl^{\theta+\nu}\;\middle|\;\begin{array}{l}(\mathfrak{c}_1,\varsigma_1;\mathfrak{C}_1:\mathfrak{t}),(\alpha+\beta,\theta+\nu,1),\\ (\alpha,\theta,1),(\beta,\nu,1),(\eth_j,\varrho_j;1)_{1,\,m},\end{array}\right.$$

$$\left.\begin{array}{c}(\mathfrak{c}_j,\varsigma_j;\mathfrak{C}_j)_{2,\,p}\\ (\eth_j,\varrho_j;\mathfrak{D}_j)_{m+1,\,q}\end{array}\right] \tag{46}$$

Corollary 6. *If we set $\mathfrak{D}_j = 1, \forall 1 \leq j \leq m$ and $\mathfrak{C}_j = 1, \forall n+1 \leq j \leq p$ in Theorem 4, then the outcome can be depicted as an incomplete $\overline{H}$-function $\left(\overline{\Gamma}_{p,\,q}^{m,\,n}(z)\right)$, which is described as:*

$$\int_0^l x^{\alpha-1}(l-x)^{\beta-1}\,\overline{\Gamma}_{p,\,q}^{m,\,n}\left[wx^\theta(l-x)^\nu \;\middle|\; \begin{array}{c} (\mathfrak{c}_1,\varsigma_1;\mathfrak{C}_1:t),(\mathfrak{c}_j,\varsigma_j;\mathfrak{C}_j)_{2,\,n}, \\ (\mathfrak{d}_j,\varrho_j;1)_{1,\,m}, \end{array}\right.$$

$$\left.\begin{array}{c} (\mathfrak{c}_j,\varsigma_j;1)_{n+1,\,p} \\ (\mathfrak{d}_j,\varrho_j;\mathfrak{D}_j)_{m+1,\,q} \end{array}\right]dx$$

$$= l^{\alpha+\beta-1}\,\overline{\Gamma}_{p+2,\,q+1}^{m,\,n+2}\left[wl^{\theta+\nu} \;\middle|\; \begin{array}{c} (1-\alpha,\theta,1),(1-\beta,\nu,1),(\mathfrak{c}_1,\varsigma_1;\mathfrak{C}_1:t), \\ (\mathfrak{d}_j,\varrho_j;1)_{1,\,m},(\mathfrak{d}_j,\varrho_j;\mathfrak{D}_j)_{m+1,\,q}, \end{array}\right.$$

$$\left.\begin{array}{c} (\mathfrak{c}_j,\varsigma_j;\mathfrak{C}_j)_{2,\,n},(\mathfrak{c}_j,\varsigma_j;1)_{n+1,\,p} \\ (1-\alpha-\beta,\theta+\nu,1) \end{array}\right]$$

$$= l^{\alpha+\beta-1}\,\overline{\Gamma}_{p+1,\,q+2}^{m+2,\,n}\left[wl^{\theta+\nu} \;\middle|\; \begin{array}{c} (\alpha+\beta,\theta+\nu,1),(\mathfrak{c}_1,\varsigma_1;\mathfrak{C}_1:t),(\mathfrak{c}_j,\varsigma_j;\mathfrak{C}_j)_{2,\,n}, \\ (\alpha,\theta,1),(\beta,\nu,1),(\mathfrak{d}_j,\varrho_j;1)_{1,\,m}, \end{array}\right.$$

$$\left.\begin{array}{c} (\mathfrak{c}_j,\varsigma_j;1)_{n+1,\,p} \\ (\mathfrak{d}_j,\rho_j;\mathfrak{D}_j)_{m+1,\,q} \end{array}\right] \tag{47}$$

and

$$\int_0^l x^{\alpha-1}(l-x)^{\beta-1}\,\overline{\gamma}_{p,\,q}^{m,\,n}\left[wx^\theta(l-x)^\nu \;\middle|\; \begin{array}{c} (\mathfrak{c}_1,\varsigma_1;\mathfrak{C}_1:t),(\mathfrak{c}_j,\varsigma_j;\mathfrak{C}_j)_{2,\,n}, \\ (\mathfrak{d}_j,\varrho_j;1)_{1,\,m}, \end{array}\right.$$

$$\left.\begin{array}{c} (\mathfrak{c}_j,\varsigma_j;1)_{n+1,\,p} \\ (\mathfrak{d}_j,\varrho_j;\mathfrak{D}_j)_{m+1,\,q} \end{array}\right]dx$$

$$= l^{\alpha+\beta-1}\,\overline{\gamma}_{p+2,\,q+1}^{m,\,n+2}\left[wl^{\theta+\nu} \;\middle|\; \begin{array}{c} (1-\alpha,\theta,1),(1-\beta,\nu,1),(\mathfrak{c}_1,\varsigma_1;\mathfrak{C}_1:t), \\ (\mathfrak{d}_j,\varrho_j;1)_{1,\,m},(\mathfrak{d}_j,\varrho_j;\mathfrak{D}_j)_{m+1,\,q}, \end{array}\right.$$

$$\left.\begin{array}{c} (\mathfrak{c}_j,\varsigma_j;\mathfrak{C}_j)_{2,\,n},(\mathfrak{c}_j,\varsigma_j;1)_{n+1,\,p} \\ (1-\alpha-\beta,\theta+\nu,1) \end{array}\right]$$

$$= l^{\alpha+\beta-1}\,\overline{\gamma}_{p+1,\,q+2}^{m+2,\,n}\left[wl^{\theta+\nu} \;\middle|\; \begin{array}{c} (\alpha+\beta,\theta+\nu,1),(\mathfrak{c}_1,\varsigma_1;\mathfrak{C}_1:t),(\mathfrak{c}_j,\varsigma_j;\mathfrak{C}_j)_{2,\,n}, \\ (\alpha,\theta,1),(\beta,\nu,1),(\mathfrak{d}_j,\varrho_j;1)_{1,\,m}, \end{array}\right.$$

$$\left.\begin{array}{c} (\mathfrak{c}_j,\varsigma_j;1)_{n+1,\,p} \\ (\mathfrak{d}_j,\rho_j;\mathfrak{D}_j)_{m+1,\,q} \end{array}\right]. \tag{48}$$

Corollary 7. *If we set $\mathfrak{D}_j = 1, \forall 1 \leq j \leq m$ in Theorem 5, then the outcome can be depicted as an incomplete $\overline{I}$-function $\left({}^\Gamma\overline{I}_{p,\,q}^{m,\,n}(z)\right)$, which*

Finite and Infinite Integral Formulae Involving Family of Incomplete I-Functions 199

is described as:

$$\int_0^\infty l^{\alpha-1}\, {}^\Gamma I_{p,\,q}^{m,\,n}\left[k\,l\;\middle|\;\begin{array}{l}(\mathfrak{c}_1,\varsigma_1;\mathfrak{C}_1:\mathfrak{t}),(\mathfrak{c}_j,\varsigma_j;\mathfrak{C}_j)_{2,\,p}\\(\mathfrak{d}_j,\varrho_j;1)_{1,\,m},(\mathfrak{d}_j,\varrho_j;\mathfrak{D}_j)_{m+1,\,q}\end{array}\right]$$

$$\times I_{P,\,Q}^{M,\,N}\left[s\,l^\varsigma\;\middle|\;\begin{array}{l}(\mathfrak{f}_j,\alpha_j;\mathfrak{F}_j)_{1,\,P}\\(\mathfrak{g}_j,\beta_j;\mathfrak{G}_j)_{1,\,Q}\end{array}\right]=k^{-\alpha}$$

$$\times {}^\Gamma I_{p+P,\,q+Q}^{m+M,\,n+N}\left[\frac{k}{s^\varsigma}\;\middle|\;\begin{array}{l}(\mathfrak{c}_1+\varsigma_1\alpha,-\varsigma_1\varsigma,\mathfrak{C}_1:\mathfrak{t}),(\mathfrak{c}_j+\varsigma_j\alpha,-\varsigma_j\varsigma,\mathfrak{C}_j)_{2,p},\\(\mathfrak{d}_j+\alpha\varrho_j,-\varsigma\varrho_j,1)_{1,m},(\mathfrak{d}_j+\alpha\varrho_j,-\varsigma\varrho_j,\mathfrak{D}_j)_{m+1,q},\end{array}\right.$$

$$\left.\begin{array}{l}(\mathfrak{f}_j,\alpha_j;\mathfrak{F}_j)_{1,\,P}\\(\mathfrak{g}_j,\beta_j;\mathfrak{G}_j)_{1,\,Q}\end{array}\right] \tag{49}$$

and

$$\int_0^\infty l^{\alpha-1}\, {}^\gamma I_{p,\,q}^{m,\,n}\left[k\,l\;\middle|\;\begin{array}{l}(\mathfrak{c}_1,\varsigma_1;\mathfrak{C}_1:\mathfrak{t}),(\mathfrak{c}_j,\varsigma_j;\mathfrak{C}_j)_{2,\,p}\\(\mathfrak{d}_j,\varrho_j;1)_{1,\,m},(\mathfrak{d}_j,\varrho_j;\mathfrak{D}_j)_{m+1,\,q}\end{array}\right]$$

$$\times I_{P,\,Q}^{M,\,N}\left[s\,l^\varsigma\;\middle|\;\begin{array}{l}(\mathfrak{f}_j,\alpha_j;\mathfrak{F}_j)_{1,\,P}\\(\mathfrak{g}_j,\beta_j;\mathfrak{G}_j)_{1,\,Q}\end{array}\right]=k^{-\alpha}$$

$$\times {}^\gamma I_{p+P,\,q+Q}^{m+M,\,n+N}\left[\frac{k}{s^\varsigma}\;\middle|\;\begin{array}{l}(\mathfrak{c}_1+\varsigma_1\alpha,-\varsigma_1\varsigma,\mathfrak{C}_1:\mathfrak{t}),(\mathfrak{c}_j+\varsigma_j\alpha,-\varsigma_j\varsigma,\mathfrak{C}_j)_{2,p},\\(\mathfrak{d}_j+\alpha\varrho_j,-\varsigma\varrho_j,1)_{1,m},(\mathfrak{d}_j+\alpha\varrho_j,-\varsigma\varrho_j,\mathfrak{D}_j)_{m+1,q},\end{array}\right.$$

$$\left.\begin{array}{l}(\mathfrak{f}_j,\alpha_j;\mathfrak{F}_j)_{1,\,P}\\(\mathfrak{g}_j,\beta_j;\mathfrak{G}_j)_{1,\,Q}\end{array}\right] \tag{50}$$

Corollary 8. *If we set $\mathfrak{D}_j = 1, \forall\, 1 \le j \le m$ and $\mathfrak{C}_j = 1, \forall\, n+1 \le j \le p$ in Theorem 5, then the outcome can be depicted as an incomplete $\overline{H}$-function $\left(\overline{\Gamma}_{p,\,q}^{m,\,n}(z)\right)$, which is described as:*

$$\int_0^\infty l^{\alpha-1}\, {}^\Gamma I_{p,\,q}^{m,\,n}\left[k\,l\;\middle|\;\begin{array}{l}(\mathfrak{c}_1,\varsigma_1;\mathfrak{C}_1:\mathfrak{t}),(\mathfrak{c}_j,\varsigma_j;\mathfrak{C}_j)_{2,\,n},(\mathfrak{c}_j,\varsigma_j;1)_{n+1,\,p}\\(\mathfrak{d}_j,\varrho_j;1)_{1,\,m},(\mathfrak{d}_j,\varrho_j;\mathfrak{D}_j)_{m+1,\,q}\end{array}\right]$$

$$\times I_{P,\,Q}^{M,\,N}\left[s\,l^\varsigma\;\middle|\;\begin{array}{l}(\mathfrak{f}_j,\alpha_j;\mathfrak{F}_j)_{1,\,P}\\(\mathfrak{g}_j,\beta_j;\mathfrak{G}_j)_{1,\,Q}\end{array}\right]=k^{-\alpha}$$

$$\times {}^\Gamma I_{p+P,\,q+Q}^{m+M,\,n+N}\left[\frac{k}{s^\varsigma}\;\middle|\;\begin{array}{l}(\mathfrak{c}_1+\varsigma_1\alpha,-\varsigma_1\varsigma,\mathfrak{C}_1:\mathfrak{t}),(\mathfrak{c}_j+\varsigma_j\alpha,-\varsigma_j\varsigma,\mathfrak{C}_j)_{2,n},\\(\mathfrak{d}_j+\alpha\varrho_j,-\varsigma\varrho_j,1)_{1,m},(\mathfrak{d}_j+\alpha\varrho_j,-\varsigma\varrho_j,\mathfrak{D}_j)_{m+1,q},\end{array}\right.$$

$$\left.\begin{array}{l}(\mathfrak{c}_j+\varsigma_j\alpha,-\varsigma_j\varsigma,1)_{n+1,p},(\mathfrak{f}_j,\alpha_j;\mathfrak{F}_j)_{1,\,P}\\(\mathfrak{g}_j,\beta_j;\mathfrak{G}_j)_{1,\,Q}\end{array}\right] \tag{51}$$

and

$$\int_0^\infty l^{\alpha-1}\,{}^\gamma I_{p,\,q}^{m,\,n}\left[kl\;\middle|\;\begin{matrix}(\mathfrak{c}_1,\varsigma_1;\mathfrak{C}_1:\mathfrak{t}),(\mathfrak{c}_j,\varsigma_j;\mathfrak{C}_j)_{2,\,n},(\mathfrak{c}_j,\varsigma_j;1)_{n+1,\,p}\\(\mathfrak{d}_j,\varrho_j;1)_{1,\,m},(\mathfrak{d}_j,\varrho_j;\mathfrak{D}_j)_{m+1,\,q}\end{matrix}\right]$$

$$\times\,I_{P,\,Q}^{M,\,N}\left[s\,l^\varsigma\;\middle|\;\begin{matrix}(\mathfrak{f}_j,\alpha_j;\mathfrak{F}_j)_{1,\,P}\\(\mathfrak{g}_j,\beta_j;\mathfrak{G}_j)_{1,\,Q}\end{matrix}\right]=k^{-\alpha}$$

$$\times\,{}^\gamma I_{p+P,\,q+Q}^{m+M,\,n+N}\left[\begin{matrix}k\\\overline{s^\varsigma}\end{matrix}\;\middle|\;\begin{matrix}(\mathfrak{c}_1+\varsigma_1\alpha,-\varsigma_1\varsigma,\mathfrak{C}_1:\mathfrak{t}),(\mathfrak{c}_j+\varsigma_j\alpha,-\varsigma_j\varsigma,\mathfrak{C}_j)_{2,n},\\(\mathfrak{d}_j+\alpha\varrho_j,-\varsigma\varrho_j,1)_{1,m},(\mathfrak{d}_j+\alpha\varrho_j,-\varsigma\varrho_j,\mathfrak{D}_j)_{m+1,q},\end{matrix}\right.$$

$$\left.\begin{matrix}(\mathfrak{c}_j+\varsigma_j\alpha,-\varsigma_j\varsigma,1)_{n+1,p},(\mathfrak{f}_j,\alpha_j;\mathfrak{F}_j)_{1,\,P}\\(\mathfrak{g}_j,\beta_j;\mathfrak{G}_j)_{1,\,Q}\end{matrix}\right]. \tag{52}$$

Corollary 9. *If we set $\mathfrak{D}_j=1,\forall j=1,\cdots,m$ in Theorem 6, then the outcome can be depicted as an incomplete $\overline{I}$-function $\left({}^{\Gamma}\overline{I}_{p,\,q}^{m,\,n}(z)\right)$, which is described as:*

$$\int_0^l x^{\alpha-1}(l-x)^{\beta-1}\,{}_P\Gamma_Q\left[\begin{matrix}(g_1,u),g_2,g_3,\cdots,g_P\\h_1,h_2,\cdots,h_Q\end{matrix}\;\mathfrak{c}\,x^\theta(l-x)^\nu\right]$$

$$\times\,{}^{\Gamma}I_{p,\,q}^{m,\,n}\left[w\,x^\rho(l-x)^\sigma\;\middle|\;\begin{matrix}(\mathfrak{c}_1,\varsigma_1;\mathfrak{C}_1:\mathfrak{t}),(\mathfrak{c}_j,\varsigma_j;\mathfrak{C}_j)_{2,\,p}\\(\mathfrak{d}_j,\varrho_j;1)_{1,\,m},(\mathfrak{d}_j,\varrho_j;\mathfrak{D}_j)_{m+1,\,q}\end{matrix}\right]=l^{\alpha+\beta-1}$$

$$\frac{\prod_{j=1}^Q\Gamma(h_j)}{\prod_{j=1}^P\Gamma(g_j)}\sum_{r=0}^\infty h(r)\,t^{(\theta+\nu)r}\,{}^{\Gamma}I_{p+2,\,q+1}^{m,\,n+2}\left[wl^{\theta+\nu}\;\middle|\;\begin{matrix}(\mathfrak{c}_1,\varsigma_1;\mathfrak{C}_1:\mathfrak{t}),\\(\mathfrak{d}_j,\varrho_j;1)_{1,\,m},\end{matrix}\right.$$

$$\left.\begin{matrix}(1-\alpha-\theta r,\varrho,1),(1-\beta-\nu r,\varsigma,1),(\mathfrak{c}_j,\varsigma_j;\mathfrak{C}_j)_{2,\,p}\\(\mathfrak{d}_j,\varrho_j;\mathfrak{D}_j)_{m+1,\,q},(1-\alpha-\theta r-\beta-\nu r,\varrho+\varsigma,1)\end{matrix}\right] \tag{53}$$

and

$$\int_0^l x^{\alpha-1}(l-x)^{\beta-1}\,{}_P\Gamma_Q\left[\begin{matrix}(g_1,u),g_2,g_3,\cdots,g_P\\h_1,h_2,\cdots,h_Q\end{matrix}\;\mathfrak{c}\,x^\theta(l-x)^\nu\right]$$

$$\times\,{}^\gamma I_{p,\,q}^{m,\,n}\left[w\,x^\rho(l-x)^\sigma\;\middle|\;\begin{matrix}(\mathfrak{c}_1,\varsigma_1;\mathfrak{C}_1:\mathfrak{t}),(\mathfrak{c}_j,\varsigma_j;\mathfrak{C}_j)_{2,\,p}\\(\mathfrak{d}_j,\varrho_j;1)_{1,\,m},(\mathfrak{d}_j,\varrho_j;\mathfrak{D}_j)_{m+1,\,q}\end{matrix}\right]=l^{\alpha+\beta-1}$$

$$\frac{\prod_{j=1}^Q\Gamma(h_j)}{\prod_{j=1}^P\Gamma(g_j)}\sum_{r=0}^\infty h(r)\,t^{(\theta+\nu)r}\,{}^\gamma I_{p+2,\,q+1}^{m,\,n+2}\left[wl^{\theta+\nu}\;\middle|\;\begin{matrix}(\mathfrak{c}_1,\varsigma_1;\mathfrak{C}_1:\mathfrak{t}),\\(\mathfrak{d}_j,\varrho_j;1)_{1,\,m},\end{matrix}\right.$$

$$\left.\begin{matrix}(1-\alpha-\theta r,\varrho,1),(1-\beta-\nu r,\varsigma,1),(\mathfrak{c}_j,\varsigma_j;\mathfrak{C}_j)_{2,\,p}\\(\mathfrak{d}_j,\varrho_j;\mathfrak{D}_j)_{m+1,\,q},(1-\alpha-\theta r-\beta-\nu r,\varrho+\varsigma,1)\end{matrix}\right] \tag{54}$$

Finite and Infinite Integral Formulae Involving Family of Incomplete I-Functions 201

$h(r)$ is as defined in equation (37).

Corollary 10. *If we set* $\mathfrak{D}_j = 1, \forall 1 \le j \le m$ *and* $\mathfrak{C}_j = 1, \forall n + 1 \le j \le p$ *in Theorem 5, then the outcome can be depicted as an incomplete* $\overline{H}$*-function* $\left(\overline{\Gamma}_{p,\,q}^{m,\,n}(z)\right)$*, which is described as:*

$$\int_0^l x^{\alpha-1}(1-x)^{\beta-1}\,{}_P\Gamma_Q\left[\begin{array}{c}(g_1, u), g_2, g_3, \cdots, g_P \\ h_1, h_2, \cdots, h_Q\end{array}\,\mathfrak{c}\,x^\theta(1-x)^\nu\right]$$

$$\times\,{}^\Gamma I_{p,\,q}^{m,\,n}\left[w\,x^\rho(1-x)^\sigma\,\left|\,\begin{array}{c}(\mathfrak{c}_1, \varsigma_1; \mathfrak{C}_1 : \mathfrak{t}), (\mathfrak{c}_j, \varsigma_j; \mathfrak{C}_j)_{2,\,n}, (\mathfrak{c}_j, \varsigma_j; 1)_{n+1,\,p} \\ (\mathfrak{d}_j, \varrho_j; 1)_{1,\,m}, (\mathfrak{d}_j, \varrho_j; \mathfrak{D}_j)_{m+1,\,q}\end{array}\right.\right]$$

$$= l^{\alpha+\beta-1}\frac{\prod_{j=1}^Q\Gamma(h_j)}{\prod_{j=1}^P\Gamma(g_j)}\sum_{r=0}^\infty h(r)\times t^{(\theta+\nu)r}$$

$$\times\,{}^\Gamma I_{p+2,\,q+1}^{m,\,n+2}\left[wl^{\theta+\nu}\,\left|\,\begin{array}{c}(\mathfrak{c}_1, \varsigma_1; \mathfrak{C}_1 : \mathfrak{t}), (\mathfrak{c}_j, \varsigma_j; \mathfrak{C}_j)_{2,\,n}, \\ (\mathfrak{d}_j, \varrho_j; 1)_{1,\,m}, (\mathfrak{d}_j, \varrho_j; \mathfrak{D}_j)_{m+1,\,q},\end{array}\right.\right.$$

$$\left.\left.\begin{array}{c}(1-\alpha-\theta r, \varrho, 1), (1-\beta-\nu r, \varsigma, 1), (\mathfrak{c}_j, \varsigma_j; 1)_{n+1,\,p} \\ (1-\alpha-\theta r-\beta-\nu r, \varrho+\varsigma, 1)\end{array}\right]\right.$$

$$\tag{55}$$

and

$$\int_0^l x^{\alpha-1}(1-x)^{\beta-1}\,{}_P\Gamma_Q\left[\begin{array}{c}(g_1, u), g_2, g_3, \cdots, g_P \\ h_1, h_2, \cdots, h_Q\end{array}\,\mathfrak{c}\,x^\theta(1-x)^\nu\right]$$

$$\times\,{}^\gamma I_{p,\,q}^{m,\,n}\left[w\,x^\rho(1-x)^\sigma\,\left|\,\begin{array}{c}(\mathfrak{c}_1, \varsigma_1; \mathfrak{C}_1 : \mathfrak{t}), (\mathfrak{c}_j, \varsigma_j; \mathfrak{C}_j)_{2,\,n}, (\mathfrak{c}_j, \varsigma_j; 1)_{n+1,\,p} \\ (\mathfrak{d}_j, \varrho_j; 1)_{1,\,m}, (\mathfrak{d}_j, \varrho_j; \mathfrak{D}_j)_{m+1,\,q}\end{array}\right.\right]$$

$$= l^{\alpha+\beta-1}\frac{\prod_{j=1}^Q\Gamma(h_j)}{\prod_{j=1}^P\Gamma(g_j)}\sum_{r=0}^\infty h(r)\times t^{(\theta+\nu)r}$$

$$\times\,{}^\gamma I_{p+2,\,q+1}^{m,\,n+2}\left[wl^{\theta+\nu}\,\left|\,\begin{array}{c}(\mathfrak{c}_1, \varsigma_1; \mathfrak{C}_1 : \mathfrak{t}), (\mathfrak{c}_j, \varsigma_j; \mathfrak{C}_j)_{2,\,n}, \\ (\mathfrak{d}_j, \varrho_j; 1)_{1,\,m}, (\mathfrak{d}_j, \varrho_j; \mathfrak{D}_j)_{m+1,\,q},\end{array}\right.\right.$$

$$\left.\left.\begin{array}{c}(1-\alpha-\theta r, \varrho, 1), (1-\beta-\nu r, \varsigma, 1), (\mathfrak{c}_j, \varsigma_j; 1)_{n+1,\,p} \\ (1-\alpha-\theta r-\beta-\nu r, \varrho+\varsigma, 1)\end{array}\right]\right. .$$

$$\tag{56}$$

$h(r)$ *is same as defined in equation (37).*

Remark

(1) If we set $\mathfrak{D}_j = 1, \forall 1 \le j \le q$ and $\mathfrak{C}_j = 1, \forall 1 \le j \le p$ in Theorem 3, then the same conclusion is arrived as of Theorem 3.2. in Bansal et al. [3].

(2) If we set $\mathfrak{D}_j = 1, \forall 1 \le j \le q$ and $\mathfrak{C}_j = 1, \forall 1 \le j \le p$ in Theorem 4, then the same conclusion is arrived as of Theorem 3.3. in Bansal et al. [3].

(3) If we set $\mathfrak{D}_j = 1, \forall 1 \le j \le q$ and $\mathfrak{C}_j = 1, \forall 1 \le j \le p$ in Theorem 6, then the same conclusion is arrived as of Theorem 3.4. in Bansal et al. [3].

4. Concluding Remarks

Nowadays, research is focused on the integral presentation of different types of special functions due to their use in numerous disciplines. Numerous authors have written various research articles on integral, including different types of incomplete special functions. In this contest, with the family of incomplete I-functions, we investigate a variety of finite and infinite integrals that can be used to obtain many other integrals with real-world applications. Also, we have derived various interesting integrals involving the incomplete $\overline{I}$-function and incomplete $\overline{H}$-function as the special instance of our primary findings. Furthermore, we highlight a number of existing and novel special cases of these integrals. Additionally, in terms of the I-function, we developed the integral representation of incomplete I-functions. If we take $t = 0$ in Theorem 1, then we have the result in the form of Rathie's I-function [16], which is described as:

$$
\begin{aligned}
I_{p,q}^{m,n} & \left[z \left| \begin{array}{l} (1 - \mathfrak{f}_1, \alpha_1; \mathfrak{F}_1), (\mathfrak{f}_j, \alpha_j; \mathfrak{F}_j)_{2,p} \\ (\mathfrak{g}_j, \beta_j; \mathfrak{G}_j)_{1,q} \end{array} \right. \right] \\
& = \int_t^\infty l^{\mathfrak{f}_1 - 1} \mathfrak{g}^{-l} I_{p-1,q}^{m,n-1} \left[l^{\alpha_1} z \left| \begin{array}{l} (\mathfrak{f}_j, \alpha_j; \mathfrak{F}_j)_{2,p} \\ (\mathfrak{g}_j, \beta_j; \mathfrak{G}_j)_{1,q} \end{array} \right. \right] dl.
\end{aligned} \tag{57}
$$

The integral involving generalized special function plays an important role in the diverse fields of geometric function theory, applied mathematics, physics, statistics, and engineering. Hence, the results presented in this paper can be useful in the same direction of physics and engineering sciences.

References

[1] A. Atangana & D. Baleanu, *New fractional derivatives with nonlocal and non-singular kernel: theory and application to heat transfer model*, Therm. Sci., 20(2), 763-769, (2016).

[2] M. K. Bansal, D. Kumar, *On the integral operators pertaining to a family of incomplete I-functions*, AIMS Math., 5(2), 1247-1259, (2020).

[3] M.K. Bansal, D. Kumar, J. Singh, & K. S. Nisar, . *Finite and infinite integral formulas involving the family of incomplete H-Functions*, Appl. Appl. Math., 15(3), 2, (2020).

[4] S. Bhatter, Nishant, D. L. Suthar, & S. D. Purohit, *Boros integral involving the product of family of polynomials and the incomplete I-Function*, J. Comput. Anal. Appl., 31(1), 400-412, (2023).

[5] S. Bhatter, K. Jangid, S. Kumawat, D. Baleanu, D. L. Suthar, & S. D. Purohit, *Analysis of the family of integral equation involving incomplete types of I and $\overline{I}$-functions*, Applied Mathematics in Science and Engineering, 31(1), 2165280, (2023).

[6] C. Cattani, *A review on harmonic wavelets and their fractional extension*, J. Adv. Engg. Comput. 2(4), 224-238,(2018).

[7] A. K. Dizicheh, S. Salahshour, A. Ahmadian, & D. Baleanu, *A novel algorithm based on the Legendre wavelets spectral technique for solving the Lane-Emden equations*, Appl. Numer. Math., 153, 443-456, (2020).

[8] J. Edward, *A treatise on the integral calculus*, Vol. II, Chelsea Publication Company, New York, (1922).

[9] C. Fox, *The G and H-functions as symmetrical Fourier kernels*, Trans. Amer. Math. Soc. 98(3), 395-429, (1961).

[10] W. Gao, P. Veeresha, D. G. Prakasha, H. M. Baskonus, & G. Yel, *A powerful approach for fractional Drinfeld-Sokolov-Wilson equation with Mittag-Leffler law*. Alexandria Engg. J. 58(4), 1301-1311, (2019).

[11] G. Gumah, S. Al-Omari, & D. Baleanu, *Soft computing technique for a system of fuzzy Volterra integro-differential equations in a Hilbert space*, Appl. Numer. Math., 152, 310-322.

[12] P. Jain, & V. Jat, *Finite and infinite integral formulas associated with a family of incomplete I-functions*, Indian Journal of Advanced Physics (IJAP), 1(1), (2021).

[13] K. Jangid, S. Bhatter, S. Meena, & S. D. Purohit, *Certain classes of the incomplete I-functions and their properties*, Discontin. Nonlinearity Complex., 12(2), 437-454, (2023).

[14] K. Jangid, S. Bhatter, S. Meena, D. Baleanu, M. Al Qurashi, & S. D. Purohit, *Some fractional calculus findings associated with the incomplete I-functions*, Adv. Difference Equ., 2020, 265, (2020).

[15] Y. X. Jun, D. Baleanu, M. P. Lazarevic, & M. Cajic, *Fractal boundary value problems for integral and differential equations with local fractional operators*, Thermal Sci. 19(3), 959-966, (2015).

[16] A. K. Rathie, *A new generalization of generalized hypergeometric functions*, Le Math.;LII: 297-310, (1997).

[17] Srivastava, R., Cho, N. E., *Generating functions for a certain class of incomplete hypergeometric polynomials*, Appl. Math. Comput., 219(6), 3219-3225, (2012).

[18] R. Srivastava, *Some properties of a family of incomplete hypergeometric functions*. Russ. J. Math. Phys., 20(1), 121-128, (2013).

[19] H. M. Srivastava, M. A. Chaudhry, & R. P. Agarwal, *The incomplete Pochhammer symbols and their applications to hypergeometric and related functions*, Integral Transforms Spec. Funct., 23(9), 659-683, (2012).

[20] H. M. Srivastava, R. K. Saxena, & R. K. Parmar, *Some families of the incomplete H-functions and the incomplete $\overline{H}$-function and associated integral transforms and operators of fractional calculus with applications*, Russ. J. Math. Phys., 25(1), 116-138, (2018).

© 2025 World Scientific Publishing Company
https://doi.org/10.1142/9789819812202_0011

Chapter 11

Homotopy Analysis Method for Solving Nonlinear Fredholm Integral Equations of Second Kind

Subhendu Paul

Department of Mathematics,
Birla Institute of Technology and Science-Pilani,
Hyderabad Campus, Hyderabad-500078, India
Email: p20220028@hyderabad.bits-pilani.ac.in

Santanu Koley

Department of Mathematics,
Birla Institute of Technology and Science-Pilani,
Hyderabad Campus, Hyderabad-500078, India
Email: santanu@hyderabad.bits-pilani.ac.in

1. Introduction

Most of the scientific and engineering problems demonstrated in mathematical form or converted into mathematical models are defined by some set of governing equations and initial/boundary conditions associated with the governing equations. For a number of problems, these governing equations turn out to be mostly integral equations (such as problems arising in the theory of elasticity, electrostatics, potential flow theory, radiative heat transfer, etc). Most of the time, the problems arising from real life situations are highly nonlinear in nature, making them often very difficult to handle and solve. There are a few analytical techniques available to solve mostly linear and a very few particular types of nonlinear integral equations. We also have many numerical methods to solve these nonlinear integral equations like the Trapezoidal Method, the Nystrom Method, the Galerkin Method, the iterative boundary element method, etc [2, 5, 7, 26]. Further, we also have polynomial approximation methods [6, 7] using different functions, Chebyshev Polynomials [28], Bernstein Polynomials [22, 29], numerical methods of integration [9, 13, 25]. In recent years, numerical methods are frequently used in dealing with problems related to water waves interaction with marine structures [27]. A Fredholm Integral Equation technique [14, 15, 17] and an Integro-differential Equation technique [16] have been used by Koley et. al. to study the scattering of water waves by plates and to perform hydroelastic analysis. An Integral Equation technique [18] has been used for wave scattering by vertical barriers. But these methods require a lot of codes that run on some very powerful processors which make them costly and time consuming. It is also hard to have a whole and essential understanding of the nonlinear problem. Even more difficulties may arise if the nonlinear problem contains singularities or has multiple solutions. There are also some analytical and semi-analytical approaches like the Linear Superposition Technique [22], the Laplace Decomposition Method [12], the Adomian Decomposition Method [19], the Variational Iterations Method [1], etc. for solving nonlinear integral equations [24]. But these methods only work well for some particular types of functions and have many limitations like: having no freedom to choose base functions and no convenient way to adjust the convergence region and rate of approximation series. Therefore, we need an adaptable analytical method which can solve a wider spectrum of nonlinear integral equations.

The Homotopy Analysis Method was formulated by Liao Shijun of Shanghai Jiaotong University in his PhD dissertation in 1992. It was further modified and strengthened in 1997 [20]. This is an analytical method based on homotopy in Topology, which involves continuous deformation of one mathematical object to another with some similar properties. In our case, we will continuously deform a solution of a known system to the unknown solution of the nonlinear integral equation. The solutions to the nonlinear integral equations are obtained in the form of series expansions. This is a relatively new approach and holds the potential to solve a very large number of complex and difficult problems. The advantages of using the Homotopy Analysis Method over the other solution techniques are as follows:

i. We can use the present method directly without any assumptions or transformations.

ii. This method is highly generalized as it is a union of Lyapunov's Small Artificial Parameter method [21], the Delta Expansion Method [11], the Adomian Decomposition Method [19] and the Homotopy Perturbation Method [8].

iii. The present method is independent of any small/large physical parameters.

iv. This method allows for the flexibility of constructing a variety of equations for continuous variation as there is a large freedom to choose auxiliary linear operators, initial guesses and other auxiliary functions.

v. The present method guarantees of convergence up to a great extent as we can control the convergence of the approximation series solution using the convergence control parameter.

In this section, we will study the q-Homotopy Analysis Method, a generalised and more powerful variation of Homotopy Analysis Method to solve the Nonlinear Fredholm Integral Equations of the Second type (NFIES) and test the efficiency and accuracy of the method with some examples. The q-Homotopy Analysis Method is a stronger technique as it increases the interval of convergence that exists in the Homotopy Analysis Method and the approximation series solutions are more likely to converge [10].

2. Concept of Homotopy

The concept of homotopy has its origin in Topology and Differential Geometry. In Topology, homotopy means continuous deformation from one mathematical object to another. It is the connection between different mathematical things with similarities in some aspects of their properties. For example, we can deform a circle into an ellipse or a square, while maintaining that they are continuous simple closed curve. In general, we can continuously deform a continuous function into another continuous function without disrupting the continuity of the functions, or an equation to another equation.

2.1. *Examples*

(1) ***Homotopy of functions:*** We can connect two different real functions, say $f(x) = \sin(\pi x)$ and $g(x) = 8x(x-1)$, in the interval $x \in [0,1]$ by constructing a family of functions:

$$\mathcal{H}(x;q) : (1-q)\sin(\pi x) + q[8x(x-1)], \tag{1}$$

where $q \in [0,1]$ is the embedding parameter [20].

When $q = 0 : \mathcal{H}(x;0) = \sin(\pi x), x \in [0,1]$ and at $q = 1 : \mathcal{H}(x;1) = 8x(x-1), x \in [0,1]$. So, as the embedding parameter $q \in [0,1]$ increases from 0 to 1, the real function $\mathcal{H}(x;q)$ varies continuously from a trigonometric function $\sin(\pi x)$ to a polynomial $8x(x-1)$.

In Topology, $\mathcal{H}(x;q)$ is called a homotopy; $\sin(\pi x)$ and $8x(x-1)$ are called homotopic, and are denoted by [20]:

$$\mathcal{H} : \sin(\pi x) \sim 8x(x-1). \tag{2}$$

(2) **Homotopy of Equations:** Since curves can be defined by algebraic, differential or integral equations, the concept of homotopy can be easily extended to equations.

We consider a family of algebraic equations given by [20],

$$\widetilde{\mathcal{E}}(q) : (1+3q)x^2 + \frac{y^2}{(1+3q)} = 1, \tag{3}$$

where $q \in [0,1]$ is the embedding parameter.

When $q = 0$, $\mathcal{E}_0 : x^2 + y^2 = 1$, which is the equation for a circle with solution $y = \pm\sqrt{1-x^2}$ and when $q = 1$, $\mathcal{E}_1 : 4x^2 + \frac{y^2}{4} = 1$ is the equation for an ellipse with solution $y = \pm 2\sqrt{1-4x^2}$.

So, as $\widetilde{\mathcal{E}}(q)$ continuously deforms from a circle $\mathcal{E}_0$ to an ellipse $\mathcal{E}_1$, we see that the solutions also vary continuously from $y = \pm\sqrt{1-x^2}$ to $y = \pm 2\sqrt{1-4x^2}$. Hence we obtain two homotopies; one is the homotopy of equation [20]:

$$\widetilde{\mathcal{E}}(q) : \mathcal{E}_0 \sim \mathcal{E}_1 \tag{4}$$

and the other is homotopy of the solution functions [20]:

$$\widetilde{y}(x;q) : \pm\sqrt{1-x^2} \sim \pm 2\sqrt{1-4x^2}. \tag{5}$$

2.2. Definitions

(1) **Homotopy:** Homotopy between two continuous functions $f(x)$ and $g(x)$ from a topological space X to a topological space Y is formally defined to be a continuous function: $\mathcal{H}: X \times [0,1] \longrightarrow Y$ i.e., from the product of X with the unit interval to Y such that whenever $x \in X$, $\mathcal{H}(x;0) = f(x)$ and $\mathcal{H}(x;1) = g(x)$.

It is denoted as $\mathcal{H} : f(x) \sim g(x)$ and the functions $f(x)$ and $g(x)$ are said to be *homotopic* [20].

(2) **Homotopy Parameter:** The embedding parameter in a homotopy of equations or functions is called the homotopy parameter and is generally denoted by q, where $q \in [0,1]$ [20].

(3) **Zeroth Order Deformation Equation:** If there exists a homotopy equation $\widetilde{\mathcal{E}}(q) : \mathcal{E}_0 \sim \mathcal{E}_1$, with the homotopy parameter $q \in [0,1]$ and $\mathcal{E}_1$ be a given equation with at least one solution u, and $\mathcal{E}_0$ be a proper simpler equation (called the *Initial Equation*) with known solution u_0 (called the *Base Function*) such that, when q varies from 0 to 1, $\widetilde{\mathcal{E}}(q)$ varies/deforms continuously from the initial equation $\mathcal{E}_0$ to the original equation $\mathcal{E}_1$ while the solution also continuously varies from the known solution u_0 of $\mathcal{E}_0$ to the unknown solution u of $\mathcal{E}_1$, then this type of homotopy of equations is called zeroth order deformation equation [20].

(4) **Nth Order Deformation Equation:** Given a nonlinear equation $\mathcal{E}_1$ with at least one solution $u(z,t)$, let $q \in [0,1]$ denote a homotopy parameter, and $\widetilde{\mathcal{E}}(q)$ denote the zeroth order deformation equation, which connects the original equation $\mathcal{E}_1$ and an initial equation $\mathcal{E}_0$ with known initial approximation $u_0(z,t)$.

Assuming $\widetilde{\mathcal{E}}(q)$ is properly constructed, i.e., its solution $\phi(z,t;q)$ exists and is analytic at $q = 0$, then we can obtain a *Homotopy Maclaurin Series:*

$$\phi(z,t;q) \sim u_0(z,t) + \sum_{n=1}^{\infty} u_n(z,t)q^n \tag{6}$$

and a *Homotopy Series:*

$$\phi(z,t;1) \sim u_0(z,t) + \sum_{n=1}^{\infty} u_n(z,t). \tag{7}$$

The equations related to the unknown $u_n(z,t)$ are called the nth order deformation equation [20].

(5) **Homotopy Series Solution:** The Homotopy Maclaurin Series

$$\phi(z,t;1) \sim u_0(z,t) + \sum_{n=1}^{\infty} u_n(z,t), \tag{8}$$

if exists for the zeroth order deformation equation $\widetilde{\mathcal{E}}(q) : \mathcal{E}_0 \sim \mathcal{E}_1$, then the homotopy series solution for the original equation $\mathcal{E}_1$ is [20]:

$$u(z,t) = u_0(z,t) + \sum_{n=1}^{\infty} u_n(z,t) \tag{9}$$

and the Mth-order Homotopy –approximation is [20]:

$$u(z,t) \approx u_0(z,t) + \sum_{n=1}^{M} u_n(z,t) = U_M(z,t). \tag{10}$$

2.3. Theory of HAM

Let us take a Nonlinear Fredholm Integral Equation of the Second kind (NFIES), say:

$$u(x) = f(x) + \int_a^b k(x,t)F(u(t))\,dt\ , \tag{11}$$

where x is an independent variable, $u(x)$ is the unknown function, $f(x)$ is some given function of $x \in [a,b]$, $k(x,t)$ is the kernel of the integral equation defined over the square G: $[a,b] \times [a,b]$ and $F(u)$ is the nonlinearity of the unknown function u. Now consider the equation:

$$N\,[u(x)] = 0, \tag{12}$$

where N is a nonlinear operator such that

$$N\,[u(x;q)] = [u(x;q) - f(x) - \int_a^b k(x,t)F(u(t))\,dt]\ . \tag{13}$$

We have the zeroth order deformation equation for the standard Homotopy Analysis Method [20]:

$$\mathcal{H}(x;q)\ : (1-q)L\,[\theta(x;q) - u_0(x;q)] - qhN\,[\theta(x;q)] = 0, \tag{14}$$

where $\mathcal{H}$ is the Homotopy, q is the embedding parameter or homotopy parameter, L is an auxiliary linear operator (usually $L[0] = 0$), $\theta(x;q) - u_0(x;q) = 0$ is the initial equation with the initial approximation $\theta(x;0) = u_0(x)$, h is the convergence control parameter.

But in our q-Homotopy Analysis Method, we will modify the standard zeroth order deformation equation to obtain a more generalized deformation equation which is comparatively stronger and works with a vast range of nonlinear equations [10],

$$q\mathcal{H}(x;q)\ : (1-rq)L\,[\theta(x;q) - u_0(x;q)] - qhH(x)N\,[\theta(x;q)] = 0, \tag{15}$$

where r is another parameter and here $q \in [0, 1/r]$, $H(x)$ is an non-zero auxiliary function. When $r = 1$ and $H(x) = 1$ the qHAM approaches the standard HAM. After imposing: $q\mathcal{H}(x;q) = 0$, we see that as q varies from 0 to $1/r$: the q-homotopy varies from $\mathcal{H}(x;0) : L[\theta(x;0) - u_0(x;0)] = 0$ at $q = 0$. (where $\theta(x;0) = u_0(x;0) = u_0(x)$, since $L[0] = 0$) to $\mathcal{H}\left(x;\frac{1}{r}\right) :$ $N\left[\theta\left(x;\frac{1}{r}\right)\right] = 0$ at $q = \frac{1}{r}$ (where $\theta\left(x;\frac{1}{r}\right) = u(x)$) and accordingly, the solution $\theta(x;q)$ varies from the initial guess $u_0(x)$ to the qHAM solution $u(x)$. Upon expanding $\theta(x;q)$ as a Taylor Series in terms of q, we yield Series Solution [3, 10, 20]:

$$\theta(x;q) = u_0(x) + \sum_{i=1}^{\infty} u_i(x)q^i, \tag{16}$$

where

$$u_i(x) = \frac{1}{i!} \left.\frac{\partial^i \theta(x;q)}{\partial q^i}\right|_{q=0}. \tag{17}$$

Assuming that h, $H(x)$, L, and $u_0(x)$ are so properly chosen that the above series converges at $q = \frac{1}{r}$ and [10]:

$$u(x) = \theta\left(x;\frac{1}{r}\right) = u_0(x) + \sum_{i=1}^{\infty} u_i(x)\left(\frac{1}{r}\right)^i. \tag{18}$$

Defining the vector $u_n(x) = \{u_0(x), u_1(x), u_2(x), \ldots, u_n(x)\}$ [20], to obtain the ith order deformation equation, we differentiate the zeroth order deformation equation i times with respect to q, evaluate the result at $q = 0$ and finally divide them by $i!$.
The ith order deformation equation [10, 20]:

$$L[u_i(x) - \chi_i u_{i-1}(x)] = hH(x)\mathcal{R}_i\left(\overline{u_{i-1}}(x)\right), \tag{19}$$

where

$$\mathcal{R}_i\left(\overline{u_{i-1}}(x)\right) = \frac{1}{(i-1)!} \left.\frac{\partial^{i-1} N[\theta(x;q)]}{\partial q^{i-1}}\right|_{q=0}, \tag{20}$$

$$\chi_i = \begin{cases} 0, & i \leq 1, \\ r, & otherwise. \end{cases}$$

It is to be specifically noted that $u_i(x)$ for $i \geq 1$ is governed by the linear equation with other conditions that come from the original problem. Due to the existence of the factor $\left(\frac{1}{r}\right)^i$, more chances for convergence may occur or even much faster convergence can be obtained better than the standard HAM.

3. Procedure of qHAM

We start with a Nonlinear Fredholm Integral Equation of Second kind [10]:

$$u(x) = f(x) + \int_a^b k(x,t)F\left(u(t)\right) dt. \tag{21}$$

We construct the zeroth order deformation equation:

$$(1-rq)L[u(x;q)-f(x)] = qh[u(x;q)-f(x)-\int_a^b k(x,t)F\left(u(t;q)\right) dt], \tag{22}$$

where we set the auxiliary linear operator $L\left[u(x)\right] = u(x)$, the initial guess $u_0(x) = f(x)$, the auxiliary function $H(x) = 1$, the nonlinear operator $N\left[u(x;q)\right] = [u(x;q) - f(x) - \int_a^b k(x,t)F\left(u(t;q)\right) dt]$ [10]. Now, setting $q = 0$ and $q = \frac{1}{r}$:

$$u(x;0) = f(x),$$
$$u\left(x;\tfrac{1}{r}\right) = u(x).$$

We have no restriction on the choice of $u_0(x)$; as a result, we can choose $u_0(x)$ as $f(x)$ or some portions of $f(x)$. Now, the Taylor Series for $u_0(x;q)$ in terms of q [3, 10]:

$$u(x;q) = u(x;0) + \sum_{i=1}^{\infty} \frac{u_i(x;h)}{i!}q^i, \tag{23}$$

where

$$u_i(x;h) = \frac{1}{i!} \left.\frac{\partial^i u(x;q;h)}{\partial q^i}\right|_{q=0}. \tag{24}$$

The ith order deformation equation [3, 10, 20]:

$$u_i(x) = \chi_i u_{i-1}(x) + h[\mathcal{R}_i(u_{i-1}(x))] \tag{25}$$

with

$$\chi_i = \begin{cases} 0, & i \le 1, \\ r, & otherwise, \end{cases}$$

and [10]

$$\mathcal{R}_i\left(u_{i-1}(x)\right) = u_{i-1}(x) - \int_a^b (k(x,t)F\left(u_{i-1}(t)\right))dt - (1-\chi_i)f(x). \tag{26}$$

Hence, we obtain [10]:

$$u_1(x;h) = hu_0(x) - hf(x) - \left[h\int_a^b (k(x,t)F(u_0(t)))\,dt\right], \qquad (27)$$

where the initial guess $u_0(x)$ is obtained from $f(x)$. Therefore, the solution series is given by [3, 10]:

$$u(x;h) = u_0(x) + \sum_{i=1}^{\infty} u_i(x;h) \left(\frac{1}{r}\right)^i. \qquad (28)$$

Plugging in the right value of h at which the series converges, we obtain the q-homotopy solution series [10]:

$$u(x) = \lim_{n\to\infty} \sum_{i=0}^{n} u_i(x). \qquad (29)$$

We get the right value of h by first finding the valid region of h from h–curves given by the ith order deformation equation and qHAM approximation at different values of x and r.

4. Problem Illustration

Let us consider the Nonlinear Fredholm Integral Equation of the Second kind (NFIES) Problem [10]:

$$u(x) = \cos x - \frac{\pi^2}{48} + \frac{1}{12}\int_0^{\pi} tu^2(t)dt, \qquad (30)$$

where we have: $f(x) = \cos x - \frac{\pi^2}{48}$, $\lambda = \frac{1}{12}$, $a = 0$ and $b = \pi$, $k(x,t) = t$, (kernel of the integral equation) and $F(u(t)) = u^2(t)$ is the nonlinearity of the integral equation; when compared with the standard NFIES:

$$u(x) = f(x) + \lambda \int_a^b k(x,t)F(u(t))\,dt. \qquad (31)$$

Now, with the given NFIES problem we create the zeroth order deformation equation:

$$(1 - rq)L\,[u(x;q) - u_0(x)] = qhH(x)N\,[u(x;q)].$$

With our vast freedom to choose the Auxiliary Linear operator (L), Base Function $(u_0(x))$, Auxiliary Function $(H(x))$ and Auxiliary Nonlinear Operator (N), we select: $L\left(u(x)\right) = u(x)$, $u_0(x) = f(x) = \cos x - \frac{\pi^2}{48}$, $H(x) = 1$ and $N\left[u(x)\right] = u(x) - f(x) - \lambda \int_a^b k(x,t)F\left(u(t)\right) dt = u(x) - \cos x + \frac{\pi^2}{48} - \frac{1}{12}\int_0^\pi tu^2(t)dt$.

Therefore our working zeroth order deformation equation becomes:

$$(1 - rq)\left[u(x;q) - \left(\cos x - \frac{\pi^2}{48}\right)\right] =$$
$$qh\left[u(x;q) - \cos x + \frac{\pi^2}{48} - \frac{1}{12}\int_0^\pi tu^2(t;q)dt\right].$$

Hence: at $q = 0$: $u(x;0) = u_0(x) = \cos x - \frac{\pi^2}{48}$ and at $q = \frac{1}{r}$: $u\left(x;\frac{1}{r}\right) = u(x)$; we obtain the required solution to our nonlinear problem as $N\left[u\left(x;\frac{1}{r}\right)\right] = 0$, since $q \neq 0$ and $h \neq 0$.

We start with the initial guess

$$u_0(x) = \cos x - \frac{\pi^2}{48}. \tag{32}$$

Now, to obtain $u_i(x)$, we differentiate the zeroth order deformation equation i times with respect to q, then plug $q = 0$ and finally divide the resulting equation by $i!$. Therefore, for $u_1(x)$, we differentiate our zeroth order deformation equation (31) with respect to q once and then plug in $q = 0$ and finally divide by 1!.

$$\frac{\partial}{\partial q}\left((1 - rq)\left[u(x;q) - \left(\cos x - \frac{\pi^2}{48}\right)\right]\right) =$$
$$qh[u(x;q) - \cos x + \frac{\pi^2}{48} - \frac{1}{12}\int_0^\pi tu^2(t;q)dt\right). \tag{33}$$

$$\Rightarrow (1 - rq)\left[\frac{\partial}{\partial q}u(x;q)\right] - r\left[u(x;q) - \left(\cos x - \frac{\pi^2}{48}\right)\right]$$
$$= qh\left[\frac{\partial}{\partial q}u(x;q) - \frac{1}{12}\int_0^\pi t\frac{\partial}{\partial q}\left(u^2(t;q)\right)dt\right]$$
$$+ h\left[u(x;q) - \cos x + \frac{\pi^2}{48} - \frac{1}{12}\int_0^\pi tu^2(t;q)dt\right].$$

This equation above is the 1st order deformation equation. Now, plugging in $q = 0$ and dividing by 1!, we get

$$u_1(x) = h \left[-\frac{1}{12} \int_0^\pi t \left(\cos t - \frac{\pi^2}{48} \right)^2 dt \right], \tag{34}$$

where: $u(x; 0) = u_0(x) = \cos x - \frac{\pi^2}{48}$ and $\frac{\partial}{\partial q} u(x; q) \Big|_{q=0} = u_1(x)$.

On evaluating the integral we get

$$u_1(x) = \frac{-h}{12} \left(\frac{\pi^2}{3} + \frac{\pi^6}{2.48^2} \right). \tag{35}$$

To obtain $u_2(x)$, we differentiate the zeroth order deformation equation (31) twice with respect to q, plug in $q = 0$ and divide by 2!

$$\frac{\partial^2}{\partial q^2} \left((1 - rq) \left[u(x; q) - \left(\cos x - \frac{\pi^2}{48} \right) \right] \right) =$$

$$qh[u(x; q) - \cos x + \frac{\pi^2}{48} - \frac{1}{12} \int_0^\pi t u^2(t; q) dt \Big). \tag{36}$$

$$\Rightarrow \frac{\partial}{\partial q} \left((1 - rq) \left[\frac{\partial}{\partial q} u(x; q) \right] - r \left[u(x; q) - \left(\cos x - \frac{\pi^2}{48} \right) \right] \right.$$

$$= qh \left[\frac{\partial}{\partial q} u(x; q) - \frac{1}{12} \int_0^\pi t \frac{\partial}{\partial q} \left(u^2(t; q) \right) dt \right]$$

$$+ h \left[u(x; q) - \cos x + \frac{\pi^2}{48} - \frac{1}{12} \int_0^\pi t u^2(t; q) dt \right] \Big).$$

$$\Rightarrow (1 - rq) \left[\frac{\partial^2}{\partial q^2} u(x; q) \right] - r \left[\frac{\partial}{\partial q} u(x; q) \right] - r \left[\frac{\partial}{\partial q} u(x; q) \right]$$

$$= qh \left[\frac{\partial^2}{\partial q^2} u(x; q) - \frac{1}{12} \int_0^\pi t \frac{\partial^2}{\partial q^2} \left(u^2(t; q) \right) dt \right] +$$

$$h \left[\frac{\partial}{\partial q} u(x; q) - \frac{1}{12} \int_0^\pi t \frac{\partial}{\partial q} \left(u^2(t; q) \right) dt \right] +$$

$$h \left[\frac{\partial}{\partial q} u(x; q) - \frac{1}{12} \int_0^\pi t \frac{\partial}{\partial q} \left(u^2(t; q) \right) dt \right].$$

This equation above is the 2nd order deformation equation. Now, plugging in $q = 0$ and dividing by 2!, we get

$$u_2(x) - ru_1(x) = h \left[u_1(x) - \frac{1}{12} \int_0^\pi t \left(\frac{\partial}{\partial q} \left(u^2(t; q) \right) \Big|_{q=0} \right) dt \right]$$

$$u_2(x) = (r+h)u_1(x) + h \left[-\frac{1}{12} \int_0^\pi t \left(2u(t;q)\frac{\partial}{\partial q}\left(u(t;q)\right)\Big|_{q=0} \right) dt \right]$$

$$u_2(x) = (r+h)u_1(x) - \frac{h}{12} \int_0^\pi t\left(2u_0(t)u_1(t)\right) dt, \tag{37}$$

where $u(x;0) = u_0(x) = \cos x - \frac{\pi^2}{48}$, $u_1(x) = \frac{\partial}{\partial q}u(x;q)\Big|_{q=0}$ and $u_2(x) = \frac{1}{2!}\frac{\partial^2}{\partial q^2}u(x;q)\Big|_{q=0}$.

On evaluating the integral we get

$$u_2(x) = (r+h)u_1(x) + \frac{h}{12}\left(4 + \frac{\pi^4}{48}\right)u_1(x). \tag{38}$$

Similarly, we obtain the 3rd order deformation equation

$$(1 - rq)\left[\frac{\partial^3}{\partial q^3}u(x;q)\right] - r\left[\frac{\partial^2}{\partial q^2}u(x;q)\right] - 2r\left[\frac{\partial^2}{\partial q^2}u(x;q)\right]$$
$$= qh\left[\frac{\partial^3}{\partial q^3}u(x;q) - \frac{1}{12}\int_0^\pi t\frac{\partial^3}{\partial q^3}\left(u^2(t;q)\right)dt\right] +$$
$$h\left[\frac{\partial^2}{\partial q^2}u(x;q) - \frac{1}{12}\int_0^\pi t\frac{\partial^2}{\partial q^2}\left(u^2(t;q)\right)dt\right] +$$
$$2h\left[\frac{\partial^2}{\partial q^2}u(x;q) - \frac{1}{12}\int_0^\pi t\frac{\partial^2}{\partial q^2}\left(u^2(t;q)\right)dt\right].$$

Now, plugging in $q = 0$ and dividing by 3!, we get:

$$u_3(x) - ru_2(x) = h\left[u_2(x) - \frac{1}{2!}\cdot\frac{1}{12}\int_0^\pi t\left(\frac{\partial^2}{\partial q^2}\left(u^2(t;q)\right)\Big|_{q=0}\right)dt\right]$$

$$\Rightarrow u_3(x) = (r+h)u_2(x) + \frac{h}{2!}\left[-\frac{1}{12}\int_0^\pi t\left(\frac{\partial^2}{\partial q^2}\left(u^2(t;q)\right)\Big|_{q=0}\right)dt\right]$$

$$\Rightarrow u_3(x) = (r+h)u_2(x) +$$
$$\frac{h}{2!}\left[-\frac{1}{12}\int_0^\pi t\left(\frac{\partial}{\partial q}\left(2u(t;q)\frac{\partial}{\partial q}\left(u(t;q)\right)\right)\Big|_{q=0}\right)dt\right]$$

$$\Rightarrow u_3(x) = (r+h)u_2(x) +$$

$$\frac{h}{2!}\left[-\frac{1}{12}\int_0^\pi t\left(\left(2\frac{\partial}{\partial q}(u(t;q))\frac{\partial}{\partial q}(u(t;q)) + 2u(t;q)\frac{\partial^2}{\partial q^2}u(t;q)\right)\Big|_{q=0}\right)dt\right]$$

$$u_3(x) = (r+h)u_2(x) - \frac{h}{12.2!}\int_0^\pi t\left(2u_1^2(t) + 4u_0(t)u_2(t)\right)dt. \tag{39}$$

On evaluating which we get

$$u_3(x) = (r+h)u_2(x) + \frac{h}{12}\left(4 + \frac{\pi^4}{48}\right)u_2(x) - \frac{h}{12.2!}\left(\pi^2\right)u_1^2(x). \tag{40}$$

So, as we proceed we can observe the pattern to generate $u_n(x)$ for $n \geq 2$:

$$u_n(x) = (r+h)u_{n-1}(x) + \frac{h}{(n-1)!}\left[-\frac{1}{12}\int_0^\pi t\left(\frac{\partial^{n-1}}{\partial q^{n-1}}\left(u^2(t;q)\right)\Big|_{q=0}\right)dt\right]. \tag{41}$$

So using this formula (41) we obtain

$$u_4(x) = (r+h)u_3(x) - \frac{h}{12.3!}\int_0^\pi t\left(12u_1(t)u_2(t) + 12u_0(t)u_3(t)\right)dt \tag{42}$$

$$\Rightarrow u_4(x) = (r+h)u_3(x) + \frac{h}{12}\left(4 + \frac{\pi^4}{48}\right)u_3(x) - \frac{h}{12.3!}\left(\pi^2\right)6u_1(x)u_2(x). \tag{43}$$

Hereafter we find the approximation series solution $u_m = \sum_{i=0}^m u_i(x)\left(\frac{1}{r}\right)^i$ in terms of the convergence parameter h and r. The $\left(\frac{1}{r}\right)^i$ factor highly increases the chances of convergence.

For this approximation we will put the value of $r = 1$ and $h = -1$ and we obtain the following values from calculating (32), (35), (38), (40), and (43):

$$u_0(x) = \cos x - 0.205616,$$
$$u_1(x) = 0.2915419,$$
$$u_2(x) = -0.146484,$$
$$u_3(x) = 0.108553,$$
$$u_4(x) = -0.089666$$

and so on.

Therefore we have:

The 1st order homotopy-approximation as: $U_1(x) = u_0(x) + u_1(x) = \cos x + 0.0859259$.

The 2nd order homotopy-approximation as: $U_2(x) = u_0(x) + u_1(x) + u_2(x) = \cos x - 0.0605581$.

The 3rd order homotopy-approximation as: $U_3(x) = u_0(x) + u_1(x) + u_2(x) + u_3(x) = \cos x + 0.0479949$.

The 4th order homotopy-approximation as: $U_4(x) = u_0(x) + u_1(x) + u_2(x) + u_3(x) + u_4(x) = \cos x - 0.0416711$.

So, we see as we increase the order M of homotopy-approximation, our Series solution approaches closer to the Exact Solution $u(x) = \cos x$ as the error $|U_M(x) - u(x)|$ decreases with increasing M.

Therefore to obtain a better approximation we must increase the order M of the approximation series.

5. Conclusions

The Homotopy Analysis Method and q-HAM is a strong semi analytical technique that has the potential to provide very accurate solutions for highly nonlinear problems. As we have seen in the above example that even with smaller orders of the approximation series solution, we approach the exact solution quite fast. We have not touched on the other controlling factors of the homotopy equation in this discussion, like the linear and nonlinear operators and the auxiliary function. These factors provide even more flexibility in obtaining exact solutions in closed forms or at the least highly accurate numerical approximations.

Acknowledgements:

SK acknowledge the financial support received through the DST Project: DST/INSPIRE/04/2017/002460 to pursue this research work. Further, SK acknowledges the financial support received through the SERB Project: CRG/2021/001550. Moreover, SK acknowledge the financial support received through the CDRF Project: C1/23/112 funded by BITS-Pilani.

References

[1] Chen, Xumei, and Linjun Wang. "The variational iteration method for solving a neutral functional-differential equation with proportional delays." *Computers & Mathematics with Applications* 59, no. 8 (2010): 2696-2702.

[2] Delves, Leonard Michael, and Joan Walsh. "Numerical solution of integral equations." (1974).

[3] El-Tawil, M. A., and S. N. Huseen. "The q-homotopy analysis method (q-HAM)." *Int. J. Appl. Math. Mech* 8, no. 15 (2012): 51-75.

[4] El-Tawil, Magdy A., and Shaheed N. Huseen. "On convergence of the q-homotopy analysis method." *Int. J. Contemp. Math. Sci* 8, no. 10 (2013): 481-497.

[5] Elliott, D., C. T. H. Baker, and G. F. Miller. "Treatment of Integral Equations by Numerical Methods." (1982).

[6] Han, Weimin, and Kendall E. Atkinson. *Theoretical numerical analysis: A functional analysis framework.* Springer New York, 2009.

[7] Ikebe, Yasuhiko. "The Galerkin method for the numerical solution of Fredholm integral equations of the second kind." *Siam Review* 14, no. 3 (1972): 465-491.

[8] Jafari, Hossein, M. Zabihi, and M. Saidy. "Application of homotopy perturbation method for solving gas dynamics equation." *Applied Mathematical Sciences* 2, no. 48 (2008): 2393-2396.

[9] Kaligatla, R. B., S. Koley, and T. Sahoo. "Trapping of surface gravity waves by a vertical flexible porous plate near a wall." Zeitschrift für angewandte Mathematik und Physik 66 (2015): 2677-2702.

[10] Karim Jbr, Rajaa, and Adil AL-Rammahi. "q-homotopy analysis method for solving nonlinear Fredholm integral equation of the second kind." *International Journal of Nonlinear Analysis and Applications* 12, no. 2 (2021): 2145-2152.

[11] Karmishin, A. V., A. I. Zhukov, and V. G. Kolosov. "Methods of dynamics calculation and testing for thin-walled structures." *Mashinostroyenie, Moscow* (1990): 137-149.

[12] Khan, Yasir, and Naeem Faraz. "Application of modified Laplace decomposition method for solving boundary layer equation." *Journal of King Saud University-Science* 23, no. 1 (2011): 115-119.

[13] Koley, S., R. B. Kaligatla, and T. Sahoo. "Oblique wave scattering by a vertical flexible porous plate." Studies in Applied Mathematics 135, no. 1 (2015): 1-34.

[14] Koley, S., Behera, H. and Sahoo, T. "Oblique wave trapping by porous structures near a wall." *J. Eng. Mech.* 141, no. 3 (2015): 04014122.

[15] Koley, S. and Sahoo, T. "Wave interaction with a submerged semicircular porous breakwater placed on a porous seabed." *J. Eng. Mech.* 80, (2017): 18-37.

[16] Koley, Santanu, and Trilochan Sahoo. "An integro-differential equation approach to study the scattering of water waves by a floating flexible porous plate." Geophysical & Astrophysical Fluid Dynamics 112, no. 5 (2018): 345-356.

[17] Koley, S., R. Mondal, and T. Sahoo. "Fredholm integral equation technique for hydroelastic analysis of a floating flexible porous plate." European Journal of Mechanics-B/Fluids 67 (2018): 291-305.

[18] Koley, Santanu, and Trilochan Sahoo. "Integral equation technique for water wave interaction by an array of vertical flexible porous wave barriers." ZAMM-Journal of Applied Mathematics and Mechanics/Zeitschrift für Angewandte Mathematik und Mechanik 101.5 (2021): e201900274.

[19] Li, Jian-Lin. "Adomian's decomposition method and homotopy perturbation method in solving nonlinear equations." *Journal of Computational and Applied Mathematics* 228, no. 1 (2009): 168-173.

[20] Liao, Shijun. *Homotopy analysis method in nonlinear differential equations.* Beijing: Higher education press, 2012.

[21] Lyapunov, A. M. "General Problem of Stability of Motion and Other Works in the Theory of Stability and the Theory of Ordinary Differential Equations." *Academy of Sciences of the US SR, Moscow* (1956).

[22] Ma, Wen-Xiu, and Engui Fan. "Linear superposition principle applying to Hirota bilinear equations." *Computers & Mathematics with Applications* 61, no. 4 (2011): 950-959.

[23] Mandal, Birendra Nath, and Subhra Bhattacharya. "Numerical solution of some classes of integral equations using Bernstein polynomials." *Applied Mathematics and computation* 190, no. 2 (2007): 1707-1716.

[24] Matinfar, M., M. Saeidy, and J. Vahidi. "Application of homotopy analysis method for solving systems of Volterra integral equations." *Advances in Applied Mathematics and Mechanics* 4, no. 1 (2012): 36-45.

[25] Maturi, Dalal Adnan. "The successive approximation method for solving nonlinear Fredholm integral equation of the second kind using maple." *Advances in Pure Mathematics* 9, no. 10 (2019): 832.

[26] Panduranga, K., Koley, S. and Sahoo, T. "Surface gravity wave scattering by multiple slatted screens placed near a caisson porous breakwater in the presence of seabed undulations." *Appl. Ocean Res.* 111, (2021): 102675.

[27] Panduranga, Kottala, and Santanu Koley. "Hydroelastic analysis of very large rectangular plate floating on shallow water." Zeitschrift für angewandte Mathematik und Physik 73 (2022): 1-22.

[28] Zarnan, Jumah Aswad. "On the numerical solution of Urysohn integral equation using Chebyshev polynomial." *International Journal of Basic & Applied Sciences IJBAS-IJENS* 16, no. 06 (2016): 23-27.

[29] Zarnan, Jumah Aswad. "A novel approach for the solution of a class of Urysohn integral equations using Bernstein polynomials." *Int. J. Adv. Res* 5, no. 1 (2017): 2156-2162.

© 2025 World Scientific Publishing Company
https://doi.org/10.1142/9789819812202_0012

Chapter 12

L^1-space of Vector Measures with Density Defined on δ-rings

Celia Avalos-Ramos

C. U. C. E. I., Universidad de Guadalajara,
Blvd. Gral. Marcelino García Barragán 1421,
Olímpica Guadalajara, 44430 México
Email: celia.avalos@academicos.udg.mx

Much of the theory of integration of scalar functions with respect to a vector measure defined on a σ-algebra has been extended to the case of measures defined δ-rings by several authors, among them D. R. Lewis, P. R. Masani and H. Niemi, O. Delgado and M. A. Juan. In these notes, certain properties that have not been analyzed are considered. Let us consider a vector measure ν defined on the δ-ring $\mathcal{R}$ of subsets of Ω taking values in a Banach space X. As usual, the variation and the semivariation of ν are denoted by $|\nu|$ and $||\nu||$, respectively. We study the relationship of the spaces $L^1(\nu)$ and $L^1(|\nu|)$. We show that the integral operator I_ν can be compact even though $|\nu|$ is unbounded. We also prove that if $|\nu|$ is bounded and X is separable space, then ν is σ-finite vector measure. On the other hand, given a locally ν-integrable function f we associate to it the vector measure ν_f defined in the same δ-ring R. We will study some lattice properties of the spaces of the integrable functions with respect to ν_f.

1. Introduction

The integration theory with respect to a vector measure ν defined in a δ-ring, which extends the integration theory with respect to classical vector measures, begins to be treated in 1972 by D. R. Lewis at [15] and later, in 1989 by P. R. Masani and H. Niemi at [17] and [18]. In these works, the spaces $L_w^1(\nu)$ and $L^1(\nu)$ of the weakly ν-integrable functions and the ν-integrable functions are defined. In 2005 O. Delgado analyzed some differences between these spaces and those related to vector measures in σ-algebras, she also studied how they affect certain properties of ν in the space $L^1(\nu)$ in [8]. On the other hand E. Jiménez Fernández, M. A.

Juan and E. A. Sánchez-Pérez provided, in 201, a description of the dual space of $L^1(\nu)$ in [13]. At the beginning of 2014 [5] by J. M. Calabuig, O. Delgado, M. A. Juan and E. A. Sánchez-Pérez studied lattice properties of the spaces $L^1_w(\nu)$ and $L^1(\nu)$.

In the present work we will consider vector measures defined on a δ-ring and we will work with scalar functions. As we have seen, much of the integration theory regarding classical vector measures has been extended by considering vector measures on δ-rings; however, there are still properties of the space $L^1(\nu)$ and of the operator I_ν to be studied. Some of the ones that we will consider are, for example, under what conditions the equality $L^1(|\nu|) = L^1(\nu)$ holds, where $|\nu|$ is the variation of the vector measure ν defined on the δ-ring $\mathcal{R}$, when $L^1(\nu)$ is weakly compactly generated and the compactness of the integral operator, among others.

In Section 2 we will do a quick tour of the theory that we will make use of throughout the work. Thus, we will recall the main concepts and state the fundamental results about Banach lattices, Banach functional spaces, measures defined in δ-rings and their spaces of corresponding integrable functions.

It is known that D. R. Lewis associates to each ν-integrable function f a vector measure $\tilde{\nu}_f$ defined on a σ-algebra. In Section 3 we will return to this idea but now we will consider the *locally ν-integrable* functions concept similar to that studied by P. R. Masani and H. Niemi with scalar measures, these functions are those that are integrable in the elements of $\mathcal{R}$. Then, in the same way as in the case of a ν-integrable function, we will associate a locally ν-integrable function with a vector measure ν_f defined in the same δ-ring $\mathcal{R}$. We will see how the property of being σ-finite is a property that ν inherits to the measure ν_f, The same does not happen with the property of being strongly additive. We will also analyze some lattice properties of the spaces of integrable functions with respect to ν_f and their relationship with the spaces $L^1_w(\nu)$ and $L^1(\nu)$, as well as some properties that the latter inherit from the spaces initially mentioned. As a consequence of this analysis we obtain a characterization of the ν-integrable functions in terms of the δ-ring $\mathcal{R}$.

Finally the generalizations concerning the results on spaces of integrable functions with respect to classical vector measures is what will occupy us in Section 4. We will especially keep in mind the study made by Okada, W. J. Ricker and E. Sánchez-Pérez in the first part of Chapter 3 in [20]. Thus we will define an equivalent norm in the space $L^1(\nu)$ that is not necessarily lattice. We will also work with positive vector measures. On the other hand D. R. Lewis established that $L^1(|\nu|) \subset L^1(\nu)$ and provided a condition under which an ν-integrable function is integrable with respect to the variation of ν. We will see that the conditions for equality to be fulfilled considered in [20, Chap. 3] are still valid in our case. We also present some examples where certain properties that hold when considering vector measures defined in σ-algebras fail to hold when passing vector measures defined in δ-rings.

2. Preliminary results

In this section we present the notation, concepts and results on Banach lattices, Banach function spaces, vector measures defined in δ-rings and integration with respect to these measures that we will need for our exposition.

2.1. *Banach lattices*

All the vector spaces that we will consider in this work will be over $\mathbb{K}$, where $\mathbb{K} = \mathbb{C}$, the field formed by the complex numbers, or $\mathbb{K} = \mathbb{R}$, the field of the real numbers.

Let X be a normed space. The open ball and the closed ball centered at $x \in X$, with radius $r > 0$ will be denoted by $V_r(x)$ and $B_r(x)$, respectively. By B_X we will indicate the unit ball on X and by X^* the dual space of X. We will represent by $\langle \cdot, \cdot \rangle$ the duality pairing, i.e.

$$\langle x, x^* \rangle := x^*(x), \forall \, x \in X, \; x^* \in X^*.$$

If Y is other normed space, to express that $X = Y$ as sets and with equal norms, we will write $X \equiv Y$.

Definition 1. A *real vector lattice* X is a vector space on $\mathbb{R}$ with a partial order $\leq$ satisfying:

i) If $f, g, h \in X$ and $f \leq g$, then $f + h \leq g + h$.
ii) If $f, g \in X$, $f \leq g$ and $\alpha \geq 0$, then $\alpha f \leq \alpha g$.
iii) If $f, g \in X$, then there exist the supremum and the infimum of f and g with respect to order, which will be denoted by $\sup\{f, g\}$ and $\inf\{f, g\}$, respectively.

A vector subspace of X which is closed under supremum and infimum with the order inherited by X will be called *sublattice*.

Let $A \subset X$. We will denote by A^+ the subset of X consisting of all $f \in A$ such that $f \geq 0$. Given $f \in X$, the *positive part* of f, the *negative part* of f and the *modulus* of f are defined by $f^+ := \sup\{f, 0\}$, $f^- := \sup\{-f, 0\}$ and $|f| := \sup\{f, -f\}$, respectively. Let us note that $f^+, f^-, |f| \in X^+$. Moreover $f = f^+ - f^-$, $|f| = f^+ + f^-$ and $|f + g| \leq |f| + |g|$.

Let $\{f_n\} \subset X$ be a sequence. To indicate that $f_{n+1} \leq f_n$, $\forall \, n \in \mathbb{N}$, we will use $f_n \downarrow$. If, in addition, there exists $\inf_n f_n = f \in X$ then we write $f_n \downarrow f$.

Similarly, if the sequence $\{f_n\}$ is such that $f_n \leq f_{n+1}$, $\forall \, n \in \mathbb{N}$, we will write $f_n \uparrow$ and if there exists $\sup_n f_n = f \in X$, we will use the notation $f_n \uparrow f$.

Let J be a directed set. Then a net $\{f_\tau\}_{\tau \in J} \subset X$ is an *upwards directed system* if for τ_1 and τ_2 in J, there exists $\tau_3 \in J$ such that $f_{\tau_1} \leq f_{\tau_3}$ and $f_{\tau_2} \leq f_{\tau_3}$. In this case we will use the notation $f_\tau \uparrow$. If additionally there exists $f = \sup_\tau f_\tau \in X$ we write $f_\tau \uparrow f$. Analogously, we define a *downward directed system* $\{f_\tau\}_{\tau \in J} \subset X$ and the notations $f_\tau \downarrow$ and $f_\tau \downarrow f$.

Let X a vector lattice. Then the following holds:

a) An element $e \in X^+$ is a *weak unit*, if $\inf\{e, f\} = 0$ implies that $f = 0$.
b) An *ideal* Y on X is a sublattice of X satisfying that if $f \in X$ and $|f| \leq |g|$ for some $g \in Y$, then $f \in Y$.

Definition 2. A *real normed lattice* Y is a normed space on $\mathbb{R}$ which is a vector lattice and whose norm $|\cdot|_Y$ is a *lattice* norm, i.e.,

$$\text{if } f, g \in Y \text{ are such that } |f| \leq |g|, \text{ then } |f|_Y \leq |g|_Y. \tag{1}$$

If additionally the normed space is completed, we will say that Y is a *real Banach lattice*.

In a normed lattice X, the subset X^+ is closed [21, Prop. 5.2 ii)].

Let X be a real normed vector lattice. Then X has the *Fatou property* if for each upwards directed system $\{f_\tau\} \subset X^+$ with $\sup_\tau |f_\tau|_X < \infty$, there exists $f = \sup_\tau f_\tau \in X$ y $|f|_X = \sup_n |f_n|_X$. Similarly, X has the σ-*Fatou property*, if given $\{f_n\} \subset X^+$ such that $f_n \uparrow$ and $\sup_n |f_n|_X < \infty$, then there exists $f = \sup_n f_n \in X$ and $|f|_X = \sup_n |f_n|_X$.

We say that X is *order continuous* if for any system $\{f_\tau\} \subset X$ satisfying $f_\tau \downarrow 0$, we have $|f_\tau|_X \downarrow 0$. Analogously, X is σ-*order continuous* if for any sequence $\{f_n\} \subset X$ satisfying $f_n \downarrow 0$ we have that $|f_n|_X \downarrow 0$.

Take a real Banach lattice X. Then in $Z := X+iX$, the complexification of X, the modulus is defined by $|h| := \sup\{|(\cos\theta)f+(\sin\theta)g| : 0 \le \theta < 2\pi\}$, $\forall\, h = f + ig \in Z$ [23], Ch. 14; Thm. 91.2, the norm by $|h|_Z = |\,|h|\,|_X$, $\forall\, h \in Z$, and the order is given by $f \le g$, if $f, g \in X$ and $f \le g$. In this case, Z is called a complex Banach lattice, X is its real part and we write $X = Z_{\mathbb{R}}$. Observe that $Z^+ = X^+$. We will say that a complex Banach lattice has one of the properties we introduced above if its real part has it. Henceforth we will say only Banach lattice (normed vector lattice) to refer to a complex or real Banach lattice (normed vector lattice).

Definition 3. Let X and Y be Banach lattices and $T : X \to Y$ a linear operator.

   a) T is a *lattice homomorphism*, if $T(X_{\mathbb{R}}) \subset Y_{\mathbb{R}}$ and $T(\sup\{f,g\}) = \sup\{Tf, Tg\}, \forall\, f, g \in X_{\mathbb{R}}$.

   b) T is *positive*, if $T(f) \in Y^+, \forall\, f \in X^+$.

Notice that in the case where X and Y are real Banach lattices, the first condition in the definiton of lattice homomorphism is redundant. Besides that

$$|Tg| = T|g|,\ \forall\, g \in X, \tag{2}$$

is satisfied is equivalent to the second condition [21, Ch. II Prop. 2.5].

The following result is fundamental in the theory of operators between Banach lattices [1, Lemma 3.22, Cor. 3.23].

Theorem 1. *Let X and Y be Banach lattices and $T : X \to Y$ a linear operator. If T is positive, then:*

   i) $T(X_{\mathbb{R}}) \subset Y_{\mathbb{R}}$.

   ii) T is bounded.

   iii) T preserves the order, i.e., $f \le g$ in X implies that $Tf \le Tg$ in Y.

   iv) For each $f \in X$

$$|Tf| \le T|f|. \tag{3}$$

If $T : X \to Y$ is an injective lattice homomorphism and $R(T)$ is a closed set, we will say that T is a *lattice isomorphism* and the spaces X and $R(T)$ are *lattice isomorphic*. In this case, by the previous theorem we have that T and $T^{-1}\colon R(T) \to X$ are bounded.

We will say that $T : X \to Y$ is a *lattice isometry*, if T is a lattice homomorphism and it is also an isometry. In this case, X and $T(X)$ will be *lattice isometric* spaces.

2.2. μ-Banach function spaces

Given a measurable space (Ω, Σ), we will denote by $L^0(\Sigma)$ the space formed by the Σ-measurable functions $f : \Omega \to \mathbb{K}$. If additionaly we have a complete positive measure μ defined on Σ, we indicate by $\mathcal{N}_0(\mu)$ the family of μ-null subsets, i.e. the sets $A \in \Sigma$ such that $\mu(A) = 0$. As usual a property holds μ-almost everywhere (briefly μ-a.e.), if it holds except on a μ-null set. We indicate by $L^0(\mu)$ the space of equivalence classes of functions in $L^0(\Sigma)$, where two functions are identified when they are equal μ-a.e.

Note that, when $\mathbb{K} = \mathbb{C}$, the space L^0 is the complexification of the real space

$$L^0_{\mathbb{R}}(\mu) := \{f \in \Lambda^0 : f \text{ take its values in } \mathbb{R} \ \mu\text{-a.e.}\}.$$

In $L^0_{\mathbb{R}}(\mu)$ we will always consider the μ-a.e. pointwise order. Hence, $L^0(\mu)$ is a complex vector lattice. Let $f \in L^0(\mu)$. So $\mathrm{Re}f, \mathrm{Im}f \in L^0_{\mathbb{R}}$ and $f = \mathrm{Re}f + i\mathrm{Im}f$. Moreover

$$|f| := \sup_{0 \le \theta < 2\pi} |(\cos \theta)\, \mathrm{Re}f + (\sin \theta)\mathrm{Im}f| = \sqrt{(\mathrm{Re}f)^2 + (\mathrm{Im}f)^2}.$$

Let us also observe that f is a weak unit if and only if $f > 0$ μ-a.e.

Definition 4. Let (Ω, Σ, μ) be a measure space. A normed space $X \subset L^0(\mu)$ is a *normed function space* related to μ (briefly μ-n.f.s.), if X is an ideal on $L^0(\mu)$ and its norm is lattice norm. If additionally X is complete we will call it a *Banach function space* related to μ (briefly μ-B.f.s.).

Remark 1. We must note that in the literature there appear other definitions with the same name such as in [16, Def. 1.b.17] and in [2, Def. I.1.3]; see also [9, p. 2]. The definition we work with does not impose conditions on the measure μ.

Let X be a μ-n.f.s. If X has the σ-Fatou property, then X is complete [22, Ch. 15 §65 Thm. 1]. Hence, X is a μ-B.f.s. with the σ-Fatou property.

Let X be a complex μ-B.f.s. Note that $X_{\mathbb{R}} = X \cap L^0_{\mathbb{R}}(\mu)$, with the μ-a.e. pointwise order, is a real Banach lattice and $X = X_{\mathbb{R}} + iX_{\mathbb{R}}$. Hence X is the complexification of $X_{\mathbb{R}}$. Moreover, $f \in X$ if and only if $(\mathrm{Re}f)^+, (\mathrm{Re}f)^-, (\mathrm{Im}f)^+, (\mathrm{Im}f)^- \in X^+$.

From the lattice property of the norm we obtain the following result.

Lemma 1. *Let X be a μ-B.f.s. and $g \in L^\infty(\mu)$. Then $fg \in X$, $\forall \ f \in X$ and*

$$|fg|_X \le |f|_X |g|_\infty. \tag{4}$$

Hence, the linear operator $M_g : X \to X$, given by $M_g(f) = fg$, is bounded and satisfies $|M_g| \leq |g|_\infty$. In particular, given $A \in \Sigma$, $\chi_A \in L^\infty(\mu)$. It follows that the linear operator $M_A := M_{\chi_A} : X \to X$, is bounded and satisfies that $|M_A| \leq 1$.

The following is an important result in the theory of Banach function spaces (see [23, Thm. 100.6], [9, p. 46], [18, Lemma 3.13]).

Proposition 1. *Let X be a μ-B.f.s., $\{f_n\} \subset X$ a sequence and $f \in X$. If $f_n \to f$ in X, then there exists a subsequence $\{f_{n_k}\}$ of $\{f_n\}$ such that $f_{n_k} \to f$ μ-a.e.*

2.3. Measures defined on δ-rings

2.3.1. *Scalar measures*

Part of the theory presented in this section was developed by N. Dinculeanu in [11] and another part by P. R. Masani and H. Niemi in [17].

Definition 5. Let Ω be a non empty set. A collection $\mathcal{R}$ of subsets of Ω is a δ-*ring*, if $\mathcal{R}$ is a ring and $\mathcal{R}$ "closed" under countable intersections.

Hereinafter $\mathcal{R}$ will be a δ-ring of subsets of a non-empty set Ω. We will now define a σ-algebra on Ω containing $\mathcal{R}$. This σ-algebra will serve as a starting point.

Proposition 2. *The collection*

$$\mathcal{R}^{loc} := \{A \subset \Omega : A \cap B \in \mathcal{R}, \ \forall \ B \in \mathcal{R}\}$$

is a σ-algebra on Ω and $\mathcal{R} \subset \mathcal{R}^{loc}$.

Given $A \in \mathcal{R}^{loc}$, we will use the notation,

$$\mathcal{R}_A := \{B \in \mathcal{R} : B \subset A\}.$$

If $A \in \mathcal{R}$, it turns out that $\mathcal{R}_A$ is a σ-algebra on A.

Remark 2. The following properties are equivalent:

i) $\mathcal{R}$ is a σ-algebra on Ω.
ii) $\mathcal{R} = \mathcal{R}^{loc}$.
iii) $\Omega \in \mathcal{R}$.

Definition 6. A *scalar measure (σ-additive) on $\mathcal{R}$* is a function $\lambda : \mathcal{R} \to \mathbb{K}$ such that if $\{B_n\} \subset \mathcal{R}$, where $B_n \cap B_m = \emptyset$ when $m \neq n$ and $\bigcup_{n=1}^{\infty} B_n \in \mathcal{R}$, then

$$\sum_{n=1}^{\infty} \lambda(B_n) = \lambda \left(\bigcup_{n=1}^{\infty} B_n \right). \tag{5}$$

Let us note that the restriction $\lambda_B := \lambda|_{\mathcal{R}_B} : \mathcal{R}_B \to \mathbb{K}$ is a scalar measure, $\forall\, B \in \mathcal{R}$.

To each scalar measure defined on a δ-ring $\mathcal{R}$ we will associate a positive measure defined on the σ-algebra $\mathcal{R}^{loc}$, which will allow us to use the theory of integration with respect to positive measures.

Notation. Given $A \in \mathcal{R}^{loc}$, we will denote by π_A the collection of finite families of pairwise disjoint sets in $\mathcal{R}_A$.

Definition 7. Let $\lambda : \mathcal{R} \to \mathbb{K}$ be a scalar measure. The *variation of λ* is the function $|\lambda| : \mathcal{R}^{loc} \to [0, \infty]$ defined by

$$|\lambda|(A) := \sup \left\{ \sum_{j=1}^{n} |\lambda(A_j)| : \{A_j\} \in \pi_A \right\}.$$

Proposition 3. *Let $\lambda : \mathcal{R} \to \mathbb{K}$ be a scalar measure.*

i) *The variation of λ is the smallest positive measure $|\lambda| : \mathcal{R}^{loc} \to [0, \infty]$ such that*
$$|\lambda(B)| \leq |\lambda|(B) = |\lambda_B|(B) < \infty \;\forall\, B \in \mathcal{R}.$$

ii) *If $A \in \mathcal{R}^{loc}$, then $|\lambda|(A) = \sup_{B \in \mathcal{R}_A} |\lambda|(B)$. Hence there exists an increasing sequence $\{B_n\} \subset \mathcal{R}_A$ such that $|\lambda|(A) = \lim_{n \to \infty} |\lambda|(B_n)$.*

iii) *Let $A \in \mathcal{R}^{loc}$. If $|\lambda|(A) < \infty$, then $A = \bigcup_{n=1}^{\infty} B_n \cup N$ where $\{B_n\} \subset \mathcal{R}$ and $N \in L^0(|\lambda|)$.*

Definition 8. A function $f \in \mathcal{M}$ is *λ-integrable* if $f \in L^1(|\lambda|)$.
We will denote by $L^1(\lambda)$ the subspace of $L^0(\lambda)$ formed by the λ-integrable functions.

Remark 3. Let $\Omega_{\mathcal{R}} = \bigcup_{B \in \mathcal{R}} B$ and $B \in \mathcal{R}$. Since $B \subset \Omega_{\mathcal{R}}$, $\Omega_{\mathcal{R}} \cap B = B \in \mathcal{R}$. Hence $\Omega_{\mathcal{R}} \in \mathcal{R}^{loc}$. Moreover $(\Omega \setminus \Omega_{\mathcal{R}}) \cap B = \emptyset$, $\forall\, B \in \mathcal{R}$ and so $\Omega \setminus \Omega_{\mathcal{R}} \in \mathcal{N}_0(|\lambda|)$. Then, for integration purposes, the subsets of interest are those of $\Omega_{\mathcal{R}}$. Therefore, if necessary, we can assume that $\Omega = \bigcup_{B \in \mathcal{R}} B$.

We will call the simple functions in $L^0(\mathcal{R}^{loc})$ with support in $\mathcal{R}$ *functions $\mathcal{R}$-simple* and the set of these functions will be denoted by $S(\mathcal{R})$. Note that $S(\mathcal{R})$ is a vector space.

Theorem 2. *The space $L^1(\lambda)$ with the norm given by $|\cdot|_{1,\lambda} := \int_\Omega |f| d|\lambda|$ is a $|\lambda|$-B.f.s., where $S(\mathcal{R}) \subset L^1(\lambda)$ is a dense subset. Moreover, $L^1(\lambda)$ is σ order-continuous and has the σ-Fatou property.*

Let $s \in S(\mathcal{R})$. If $s = \sum_{j=1}^{n} \chi_{B_j} x_j$, where $x_j \in \mathbb{K}$, $B_j \in \mathcal{R}$, $j = 1, \dots, n$, then

$$\int_\Omega s d\lambda = \sum_{j=1}^{n} \lambda(B_j) x_j.$$

Proposition 4. *The integral operator defines a linear bounded functional on $S(\mathcal{R}) \subset L^1(\lambda)$, with norm less than or equal to 1.*

The density of $S(\mathcal{R})$ in $L^1(\lambda)$ and the continuity of the integral operator on $S(\mathcal{R})$ allow us to define

$$\int_\Omega f d\lambda := \lim_{n \to \infty} \int_\Omega s_n d\lambda, \tag{6}$$

where $\{s_n\}$ is a sequence of functions $\mathcal{R}$-simple such that $s_n \to f$ in $L^1(\lambda)$.

The following result is the Dominated Convergence Theorem to the integral with respect to λ.

Theorem 3. *Let $\{f_n\} \subset L^0(\mathcal{R}^{loc})$, $g \in L^1(\lambda)$ and $f \in L^0(\mathcal{R}^{loc})$ be such that $f_n \to f$ and $|f_n| \le |g| \ |\lambda|$-a.e., $\forall \ n \in \mathbb{N}$. Then $f \in L^1(\lambda)$, $f_n \to f$ in $L^1(\lambda)$ and*

$$\int_\Omega f d\lambda = \lim_{n \to \infty} \int_\Omega f_n d\lambda. \tag{7}$$

The results that follow play a basic role in the theory. P. R. Masani and H. Niemi proved them in [17, Lemma 2.30, Thm. 2.32].

Proposition 5. *If $f \in L^0(\mathcal{R}^{loc})$, then*

$$\int_A |f| d|\lambda| = \sup_{B \in \mathcal{R}_A} \int_B |f| d|\lambda|, \ \forall \ A \in \mathcal{R}^{loc}. \tag{8}$$

Hence, $f \in L^1(\lambda)$ if and only if $\sup_{B \in \mathcal{R}} \int_B |f| d|\lambda| < \infty$.

The following definition was considered by P. R. Masani and H. Niemi in [17, Def. 2.14 c)].

Definition 9. A function $f \in L^0(\mathcal{R}^{loc})$ is *locally λ-integrable* if $f\chi_B$ is integrable with respect to $|\lambda|$, $\forall\, B \in \mathcal{R}$.

We will denote by $L^1_{loc}(\lambda)$ the subspace of $L^0(\lambda)$ formed by the locally λ-integrable functions.

Theorem 4. *Let $f \in L^1_{loc}(\lambda)$. The function $\lambda_f : \mathcal{R} \to \mathbb{K}$ defined by*

$$\lambda_f(B) := \int_B f d\lambda = \int_\Omega f\chi_B d\lambda, \forall\, B \in \mathcal{R},$$

is a scalar measure such that

$$|\lambda_f|(A) = \int_A |f|d|\lambda|, \ \forall\, A \in \mathcal{R}^{loc}. \tag{9}$$

We will say that a function $\lambda : \mathcal{R} \to [0, \infty]$ is a *positive measure in $\mathcal{R}$* if it satisfies the equality (5) for any disjoint collection $\{B_n\}$ of elements in $\mathcal{R}$ whose union remains on $\mathcal{R}$. In the same way that in the case of a scalar measure we define $|\lambda| : \mathcal{R}^{loc} \to [0, \infty]$, the variation of λ. Since $|\lambda|$ is a positive measure defined in a σ-algebra, the space $L^1(|\lambda|)$ is constructed. Proceeding similarly, the following results are obtained.

Proposition 6. *If $f \in L^0(\mathcal{R}^{loc})$, then*

$$\int_A |f|d|\lambda| = \sup_{B \in \mathcal{R}_A} \int_B |f|d|\lambda|, \ \forall\, A \in \mathcal{R}^{loc}. \tag{10}$$

Hence, $f \in L^1(|\lambda|)$ if and only if $\displaystyle\sup_{B \in \mathcal{R}} \int_B |f|d|\lambda| < \infty$.

Theorem 5. *Let $f \in L^1_{loc}(\lambda)$. The function $\lambda_f : \mathcal{R} \to \mathbb{K}$ defined by*

$$\lambda_f(B) := \int_B f d|\lambda| = \int_\Omega f\chi_B d|\lambda|, \forall\, B \in \mathcal{R},$$

is a scalar measure such that

$$|\lambda_f|(A) = \int_A |f|d|\lambda|, \ \forall\, A \in \mathcal{R}^{loc}. \tag{11}$$

2.3.2. *Vector measures*

In this section we present the basic theory with respect to vector measures defined on δ-rings. This is analogous to that known for vector measures defined on σ-algebras, which we will sometimes call classical vector measures. As in the previous section, most of these results were developed by N. Dinculeanu in [11] and by P. R. Masani and H. Niemi in [17] and [18].

Definition 10. Let X be an Banach space. A set function $\nu : \mathcal{R} \to X$ is a *vector measure* if

$$\sum_{n=1}^{\infty} \nu(B_n) = \nu\left(\bigcup_{n=1}^{\infty} B_n\right),$$

for any collection $\{B_n\} \subset \mathcal{R}$ of pairwise disjoint sets such that $\bigcup_{n=1}^{\infty} B_n \in \mathcal{R}$.

Lemma 2. *Let $\nu : \mathcal{R} \to X$ an additive function. The following sentences are equivalent:*

i) *The function ν is a vector measure.*

ii) *If $\{B_n\} \subset \mathcal{R}$ is an increasing sequence of sets such that $\bigcup_{n=1}^{\infty} B_n \in \mathcal{R}$, then*

$$\lim_{n \to \infty} \nu(B_n) = \nu\left(\bigcup_{n=1}^{\infty} B_n\right).$$

iii) *If $\{B_n\} \subset \mathcal{R}$ is a decreasing sequence of sets, then*

$$\lim_{n \to \infty} \nu(B_n) = \nu\left(\bigcap_{n=1}^{\infty} B_n\right).$$

Definition 11. Let $\nu : \mathcal{R} \to X$ be a vector measure.

a) The *variation* of ν is the function $|\nu| : \mathcal{R}^{loc} \to [0, \infty]$ defined by

$$|\nu|(A) := \sup \left\{ \sum_j |\nu(A_j)|_X : \{A_j\} \in \pi_A \right\}.$$

b) The *semi-variation* of ν is the function $||\nu|| : \mathcal{R}^{loc} \to [0, \infty]$ defined by

$$||\nu||(A) := \sup\{|\langle \nu, x^* \rangle|(A) : x^* \in B_{X^*}\},$$

where $|\langle \nu, x^* \rangle|$ is the variation of the scalar measure $\langle \nu, x^* \rangle : \mathcal{R} \to \mathbb{K}$, defined by

$$\langle \nu, x^* \rangle(B) = \langle \nu(B), x^* \rangle, \ \forall \, B \in \mathcal{R}.$$

232   *C. Avalos-Ramos*

c) The *quasi-variation* of ν is the function $|||\nu||| : \mathcal{R}^{loc} \to [0, \infty]$ defined by

$$|||\nu|||(A) := \sup\{|\nu(B)| : B \in \mathcal{R}_A\}.$$

Proposition 7. *Let $\nu : \mathcal{R} \to X$ a vector measure.*

i) *The variation de ν is the smallest positive measure $|\nu| : \mathcal{R}^{loc} \to [0, \infty]$ such that*

$$|\nu(B)|_X \leq |\nu|(B), \ \forall \ B \in \mathcal{R}. \tag{12}$$

ii) $|\nu|(A) = \sup\limits_{B \in \mathcal{R}_A} |\nu|(B)$ *holds for each $A \in \mathcal{R}^{loc}$. Hence there exists an increasing sequence $\{B_n\} \subset \mathcal{R}_A$ such that $|\nu|(A) = \lim\limits_{n \to \infty} |\nu|(B_n)$.*

iii) *Let $A \in \mathcal{R}^{loc}$. If $|\nu|(A) < \infty$, then $A = \bigcup\limits_{n=1}^{\infty} B_n \cup N$, where $\{B_n\} \subset \mathcal{R}$ y $N \in \mathcal{N}_0(|\nu|)$.*

Proposition 8. *[18, Lemma 3.4 c)] If $A \in \mathcal{R}^{loc}$, then*

$$||\nu||(A) = \sup\left\{ \left| \sum_{j=1}^{n} a_j \nu(B_j) \right|_X : \{a_j\}_1^n \subset B_{\mathbb{K}}, \ \{B_j\}_1^n \in \pi_A \right\}. \tag{13}$$

Proposition 9. *[18, Lemma 3.4, Cor. 3.5] The following sentences hold:*

i) *The semi-variation of ν is a monotone increasing function, σ-subadditive and finite on $\mathcal{R}$.*

ii) *For each $A \in \mathcal{R}^{loc}$, $|||\nu|||(A) \leq ||\nu||(A) \leq 4|||\nu|||(A)$ in the complex case and in the real case $|||\nu|||(A) \leq ||\nu||(A) \leq 2|||\nu|||(A)$.*

Remark 4. Let us note that from ii) in the previous proposition, it follows that ν is bounded if and only if $||\nu||(\Omega) < \infty$. Moreover

$$|\nu(B)|_X \leq ||\nu||(B), \ \forall \ B \in \mathcal{R}.$$

Notation. We will us $A_n \uparrow A$ to refer to an increasing sequence of sets $\{A_n\}$ such that $\bigcup\limits_{n} A_n = A$. Likewise, we will use $A_n \downarrow A$ to indicate a decreasing sequence of sets $\{A_n\}$ such that $\bigcap\limits_{n} A_n = A$.

Proposition 10. *[18, Lemma 3.4, Cor. 3.5] Let $A \in \mathcal{R}^{loc}$. Then:*

i) $||\nu||(A) \le |\nu|(A)$.

ii) $||\nu||(A) = \sup\limits_{B \in \mathcal{R}_A} ||\nu||(B)$. *Hence there exists an increasing sequence* $\{B_n\} \subset \mathcal{R}_A$ *such that*

$$||\nu||(A) = \lim_{n \to \infty} ||\nu||(B_n).$$

Definition 12. A set $A \in \mathcal{R}^{loc}$ is ν-*null* if $|\nu|(A) = 0$. We will denote by $\mathcal{N}_0(\nu)$ to the collection of ν-null sets.

From σ-subadditivity and the monotony of the semi-variation we obtain the following result.

Lemma 3. *Let* ν *be a vector measure defined on* $\mathcal{R}$.

i) *If* $\{A_n\} \subset \mathcal{N}_0(\nu)$, *then* $\bigcup\limits_n A_n \in \mathcal{N}_0(\nu)$.

ii) *If* $A \in \mathcal{N}_0(\nu)$ *and* $C \in (\mathcal{R}^{loc})_A$, *then* $C \in \mathcal{N}_0(\nu)$.

Proposition 11. *For* $A \in \mathcal{R}^{loc}$, *the following conditions are equivalent:*

i) A *is* ν-*null*.

ii) A *is* $|\nu|$-*null*.

iii) $\nu(B) = 0, \forall B \in \mathcal{R}_A$.

A positive measure $\lambda : \mathcal{R} \to [0, \infty]$ is a *local control measure* for ν, if $\mathcal{N}_0(|\lambda|) = \mathcal{N}_0(\nu)$ [8, p. 437].

From the previous proposition we have that $|\nu|$ is a local control measure for ν.

Finally we define $L^0(\nu)$ as the space of equivalence classes of functions in $\mathcal{M}$, where two functions are identified when they are equal ν-a.e. So, $L^0(\nu) = L^0(|\lambda|)$, where λ is any local control measure for ν.

2.4. Integration with respect to vector measures defined on δ-rings

2.4.1. Weakly integrable functions

In this section we will recall the concept of weak integrability with respect to a vector measure $\nu : \mathcal{R} \to X$, where $\mathcal{R}$ is a δ-ring of subsets of Ω and X is a Banach space.

Definition 13. A function $f \in L^0(\mathcal{R}^{loc})$ is *weakly integrable with respect to* ν, if $f \in L^1(\langle \nu, x^* \rangle)$ for each $x^* \in X^*$. We will denote by $L_w^1(\nu)$ the subspace of $L^0(\nu)$ of all weaky ν-integrable functions.

On $L^1_w(\nu)$ it defines the function

$$||f||_\nu := \sup\left\{\int_\Omega |f|d|\langle \nu, x^*\rangle| : x^* \in B_{X^*}\right\}, \tag{14}$$

which turns out to be a norm.

Theorem 6. *[18, Lemma 3.13] The space $L^1_w(\nu)$, with the norm $|\cdot|_\nu$, is a $|\nu|$-B.f.s. with the σ-Fatou property.*

As we have mentioned in the previous theorem $L^1_w(\nu)$ is a $|\nu|$-B.f.s. However, $|u|$ may not be a good local control measure for ν. On the other hand, remember that in the case of vector measures defined on σ-algebras there always exists a *Rybakov measure* μ, that is, there exists $x^*_0 \in X$ such that the positive finite measure $\mu := |\langle \nu, x^*_0\rangle| : \mathcal{R} \to [0, \infty)$ satisfies that $\mathcal{N}_0(\nu) = \mathcal{N}_0(\mu)$ [10], Ch. IX, Thm. 2.2.

2.4.2. ν-integrable functions

In this section we will recall when a function is integrable with respect to a vector measure in a δ-ring, as well as the fundamental results related to the space that these functions form. The vast majority of these were developed by D. R. Lewis in [15] and by P. R. Masani and H. Niemi in [18].

Definition 14. A function $f \in L^1_w(\nu)$ is ν-integrable, if for each $A \in \mathcal{R}^{loc}$ there exists a vector $x_A \in X$, such that

$$\langle x_A, x^*\rangle = \int_A fd\langle \nu, x^*\rangle, \ \forall \ x^* \in X^*. \tag{15}$$

The vector x_A is unique and well-defined, and will be denoted by $\int_A fd\nu$. The subset of $L^1_w(\nu)$ consisting of the functions which are integrable with respect to ν will be denoted by $L^1(\nu)$.

Let us observe that $\int_\Omega f\chi_A d\langle \nu, x^*\rangle = \int_A fd\langle \nu, x^*\rangle = \langle \int_A fd\nu, x^*\rangle$. Hence $f\chi_A \in L^1(\nu)$ and

$$\int_\Omega f\chi_A d\nu = \int_A fd\nu.$$

Theorem 7. *[18, Thm. 4.5] $L^1(\nu)$ is a closed subspace of $L^1_w(\nu)$ and the integral operator $I_\nu : L^1(\nu) \to X$ defined by*

$$I_\nu(f) = \int_\Omega fd\nu,$$

is linear and bounded with norm less than or equal to 1.

Proposition 12. Let $\varphi = \sum_{j=1}^{n} a_j \chi_{A_j} \in S(\mathcal{R})$. Then $\varphi \in L^1(\nu)$ and

$$\int_A \varphi d\nu = \sum_{j=1}^{n} a_j \nu(A_j \cap A), \ \forall A \in \mathcal{R}^{loc}.$$

Theorem 8. [15, Thm. 3.5] $S(\mathcal{R}) \subset L^1(\nu)$ is a dense set.

Theorem 9. [15, Thm. 5.1] If X does not contain a copy of c_0, then $L^1(\nu) = L^1_w(\nu)$.

Theorem 10. (Dominated Convergence Theorem) [15, Thm. 3.3] Let $f, f_n \in L^0(\mathcal{R}^{loc})$, $\forall \ n \in \mathbb{N}$ be such that $f_n \to f$ ν-a.e. and $g \in L^1(\nu)$ such that $|f_n| \leq |g|$, $\forall n \in \mathbb{N}$. Then $f \in L^1(\nu)$ and

$$\int_A f d\nu = \lim_{n \to \infty} \int_A f_n d\nu, \ \forall A \in \mathcal{R}^{loc}.$$

Lemma 4. [18, Thm. 4.9] Let $f \in L^1(\nu)$. Then there exist sets $N \in \mathcal{R}^{loc}$ and $B_n \in \mathcal{R}$, $\forall n \in \mathbb{N}$, such that:

i) $B_n \subset B_{n+1}$ and $N \cap B_n = \emptyset$ $\forall n \in \mathbb{N}$.
ii) $N \in \mathcal{N}_0(\nu)$.
iii) $supp f = N \cup \bigcup_{n=1}^{\infty} B_n$.

Theorem 11. The space $L^1(\nu)$ is a σ-order continuous $|\nu|$-Banach function space.

Remark 5. Let $f \in L^1(\nu)$ and $\{\varphi_n\} \subset S(\mathcal{R}^{loc})$ be such that $0 \leq \varphi_n \uparrow |f|$. From Lemma 4 we have that $supp f = N \cup R$, where $R = \bigcup_{n=1}^{\infty} B_n$, $B_n \in \mathcal{R}$, $B_n \subset B_{n+1}$, $\forall n \in \mathbb{N}$, $N \cap R = \emptyset$ and $N \in \mathcal{N}_0(\nu)$. Let us define

$$s_n = \varphi_n \chi_{B_n}.$$

Then $0 \leq s_n \leq s_{n+1}$ and $s_n \uparrow |f| \chi_R$. Since $|f| = |f| \chi_R$ ν-a.e. and $L^1(\nu)$ is σ-order continuous, it turns out that $s_n \to |f|$. As s_n is $\mathcal{R}$-simple, $\forall \ n \in \mathbb{N}$, we conclude that:

Given $f \in L^1(\nu)$ there exists a sequence $\{s_n\} \subset S(\mathcal{R})$ such that
$$0 \leq s_n \uparrow |f| \ \nu\text{-a.e.}, \ s_n \to |f| \ en \ L^1(\nu).$$

Lemma 5. [18, Lemma 4.16] Let $\nu : \mathcal{R} \to X$ be a vector measure, Z a Banach space and $T : X \to Z$ a linear operator.

i) *The function $T \circ \nu : \mathcal{R} \to Z$ is a vector measure such that $\mathcal{N}_0(\nu) \subset \mathcal{N}_0(T \circ \nu)$. Moreover, if $f \in L^1(\nu)$ then f is $T \circ \nu$-integrable and*

$$\int_A f d(T \circ \nu) = T \left(\int_A f d\nu \right), \ \forall \ A \in \mathcal{R}^{loc}.$$

ii) *If T is injective, then $\mathcal{N}_0(\nu) = \mathcal{N}_0(T \circ \nu)$. Hence $L^1(\nu) \subset L^1(T \circ \nu)$.*

Theorem 12. *[8, Prop. 2.3] A function $f \in L^0(\mathcal{R}^{loc})$ is ν-integrable if and only if there exists a sequence $\{s_n\} \in S(\mathcal{R})$ such that:*

i) *$\{s_n\}$ converges to f ν-a.e.*

ii) *$\left\{ \int_A s_n d\nu \right\}$ converges in X, $\forall \ A \in \mathcal{R}^{loc}$.*

Theorem 13. *[15, Thm. 3.2] Let $f \in L^1(\nu)$. The function $\tilde{\nu}_f : \mathcal{R}^{loc} \to X$ defined by*

$$\tilde{\nu}_f(A) := \int_A f d\nu, \tag{16}$$

is a vector measure, which will be called associate vector measure to f. Moreover,

$$|\tilde{\nu}_f|(A) = |f \chi_A|_\nu \ and \ |\tilde{\nu}_f|(A) = \int_A |f| d|\nu|, \ \forall A \in \mathcal{R}^{loc}.$$

It follows that $L^1(|\nu|) \subset L^1(\nu)$. Even more, a function $f \in L^1(\nu)$ is $|\nu|$-integrable if and only if the measure $\tilde{\nu}_f$ has finite variation, in which case

$$|f|_\nu = |\tilde{\nu}_f|(\Omega) \le |\tilde{\nu}_f|(\Omega) = |f|_{|\nu|} < \infty. \tag{17}$$

Definition 15. Let $\nu : \mathcal{R} \to X$ be a vector measure.

a) The measure ν is *σ-finite*, if there exists a sequence $\{B_n\} \subset \mathcal{R}$ and a set $N \in \aleph(\nu)$ such that $\Omega = \bigcup_{n=1}^{\infty} B_n \cup N$.

b) The measure ν is *strongly additive*, if $\lim_{n \to \infty} \nu(B_n) = 0$ holds for each collection of parwise disjoint sets $\{B_n\} \subset \mathcal{R}$.

Let us note that if $\mathcal{R}$ is a σ-algebra, then ν is strongly additive. Moreover, each strongly additive vector measure defined on a δ-ring is σ-finite [3, Lemma 1.1].

Theorem 14. *[8, Cor. 3.2] Let $\nu \to X$ be a vector measure. Then ν is strongly additive if and only if $\chi_\Omega \in L^1(\nu)$.*

L^1-space of Vector Measures Defined on δ-rings 237

Theorem 15. *[8, Prop. 3.4, Thm. 3.5] Let $\nu \to X$ be a vector measure σ-finite. Then:*

i) There exists a weak unit $g \in L^1(\nu)$.

ii) If $g \in L^1(\nu)$ is a weak unit, then $L^1(\nu)$ is lattice isometric to $L^1(\nu_g)$, where $\nu_g : \mathcal{R}^{loc} \to X$ is the associate vector measure to g.

3. Vector measures with scalar density f

In Theorem 13 we have that for each function f in $L^1(\nu)$ we can define the vector measure $\tilde{\nu}_f$. In this section we will extend this result to a more general class of functions, the locally ν-integrable functions.

3.1. The measure ν_f

Consider a δ-ring $\mathcal{R}$, a Banach space X, and a vector measure $\nu : \mathcal{R} \to X$. Thus, the space of ν-integrable functions $L^1(\nu)$ is obtained. As mentioned we will extend Theorem 13 by considering a more general class of functions. So let us start by defining the locally ν-integrable functions.

Definition 16. A function $f : \Omega \to \mathbb{K}$ is *locally ν-integrable* if $f\chi_B \in L^1(\nu)$, $\forall B \in \mathcal{R}$.

Let us denote by $L^1_{loc}(\nu)$ the set of (equivalence classes of) locally ν-integrable functions.

When $V \subset \mathbb{K}$, we have $f^{-1}(V) \cap B = (f|_B)^{-1}(V)$, $\forall B \in \mathcal{R}$. So, f is $\mathcal{R}^{loc}$-measurable if and only if $f|_B$ is $\mathcal{R}_B$-measurable, $\forall B \in \mathcal{R}$. On the other hand $f|_B = (f\chi_B)|_B$, $\forall B \in \mathcal{R}$. Hence

$$f \text{ es } \mathcal{R}^{loc}\text{-measurable if and only if } f\chi_B \text{ is } \mathcal{R}^{loc}\text{-measurable, } \forall B \in \mathcal{R}. \tag{18}$$

Observe that if $f \in L^1_{loc}(\nu)$, then $f\chi_B \in \mathcal{M}$, $\forall B \in \mathcal{R}$ and so $f \in \mathcal{M}$. It follows that $L^1_{loc}(\nu)$ is an ideal in $L^0(\nu)$. Moreover, $L^1(\nu) \subset L^1_{loc}(\nu)$.

Remark 6. Let Σ be a σ-algebra and $\mathcal{R} \subset \Sigma$ a δ-ring such that $A \cap B \in \mathcal{R}$, $\forall A \in \Sigma$ and $B \in \mathcal{R}$. Then $\Sigma \subset \mathcal{R}^{loc}$. Let us assume that $\Omega = \bigcup_{n=1}^{\infty} B_n$, where $B_n \in \mathcal{R}$, $\forall n \in \mathbb{N}$. Take $A \in \mathcal{R}^{loc}$. Then $A = \bigcup_{n=1}^{\infty} A \cap B_n$. Since $A \cap B_n \in \mathcal{R} \subset \Sigma$, $\forall n \in \mathbb{N}$, we have that $A \in \Sigma$. Therefore, in this case $\mathcal{R}^{loc} = \Sigma$.

Example 1. Let us consider $\Omega = (0,1)$, Σ the Lebesgue σ-algebra on Ω and $\mu : \Sigma \to \mathbb{R}$ the Lebesgue measure. Then the collection defined by

$$\mathcal{R} := \{A \in \Sigma : \exists\ a, b \in (0,1) \text{ such that } A \subset [a,b]\},$$

is a δ-ring on Ω. Since $\Omega = \bigcup_{n=2}^{\infty} [\frac{1}{n}, 1 - \frac{1}{n}]$, from the above remark we have $\mathcal{R}^{loc} = \Sigma$.

Let us consider the vector measure $\nu : \mathcal{R} \to \mathbb{R}$ defined by the restriction of μ to $\mathcal{R}$. Then by Proposition 3 i) we obtain that $|\nu| \leq \mu$. Otherwise for each $A \in \Sigma$,

$$\mu(A) = \lim_{n \to \infty} \mu(A \cap [\tfrac{1}{n}, 1 - \tfrac{1}{n}]) = \lim_{n \to \infty} \nu(A \cap [\tfrac{1}{n}, 1 - \tfrac{1}{n}])$$
$$\leq \lim_{n \to \infty} |\nu|(A \cap [\tfrac{1}{n}, 1 - \tfrac{1}{n}]) = |\nu|(A).$$

Hence $|\nu| = \mu$. So $L^1(\nu) = L^1(\mu)$.

Consider the function $f : \Omega \to \mathbb{R}$, given by $f(x) = \frac{1}{x}$, $\forall\ x \in \Omega$. Then $f \in L^1_{loc}(\nu)$, however $f \notin L^1(\nu)$. This example shows that the containment $L^1(\nu) \subset L^1_{loc}(\nu)$ can be proper.

Theorem 16. *Let $f \in L^1_{loc}(\nu)$. The function $\nu_f : \mathcal{R} \to X$ defined by*

$$\nu_f(B) := \int_B f\, d\nu,$$

is a vector measure. Moreover, for each $A \in \mathcal{R}^{loc}$,

$$\|\nu_f\|(A) = \sup_{x^* \in X} \int_A |f|\, d|\langle \nu, x^* \rangle| \quad \text{and} \quad |\nu_f|(A) = \int_A |f|\, d|\nu|. \tag{19}$$

Proof. Let $\{B_n\} \subset \mathcal{R}$ be a increasing collection of sets such that $B = \bigcup_{n=1}^{\infty} B_n \in \mathcal{R}$. Then $f\chi_B, f\chi_{B_n} \in L^1(\nu)$, $\forall\ n \in \mathbb{N}$, $f\chi_{B_n} \to f\chi_B$ and $|f\chi_{B_n}| \leq |f\chi_B|$, $\forall\ n \in \mathbb{N}$. From the Dominated Convergence Theorem in $L^1(\nu)$ we obtain that

$$\nu_f(B) = \int_B f\, d\nu = \lim_{n \to \infty} \int_{B_n} f\, d\nu = \lim_{n \to \infty} \nu_f(B_n).$$

By Lemma 2 it follows that ν_f is a vector measure.

Let us fix $x^* \in X$. Observe that for $B \in \mathcal{R}$, we have

$$\langle \nu_f, x^* \rangle(B) = \langle \int_B f\, d\nu, x^* \rangle = \int_B f\, d\langle \nu, x^* \rangle = \langle \nu, x^* \rangle_f(B).$$

From Theorem 4,

$$|\langle \nu_f, x^* \rangle|(A) = |\langle \nu, x^* \rangle_f|(A) = \int_A |f|\, d|\langle \nu, x^* \rangle|, \ \forall\ A \in \mathcal{R}^{loc}. \tag{20}$$

Hence
$$\|\nu_f\|(A) = \sup_{x^* \in X} \int_A |f| d|\langle \nu, x^* \rangle|.$$

Now let us fix $B \in \mathcal{R}$. Then $f\chi_B \in L^1(\nu)$, from Theorem 13 we have $|\nu_{f\chi_B}|(B) = \int_B |f| d|\nu|$. Besides observe that $|\nu_f|(B) = |\nu_{f\chi_B}|(B)$. So, from Propositions 7 ii) and (1), it follows that

$$|\nu_f|(A) = \sup_{B \in \mathcal{R}} \int_B |f| d|\nu| = \int_A |f| d|\nu|.$$

$\square$

Let $A \in \mathcal{R}^{loc}$. From (19) we obtain that $|\nu_f|(A) < \infty$ if and only if $f\chi_A \in L^1_w(\nu)$. In this case $|f\chi_A|_\nu = |\nu_f|(A)$. Futhermore if $N \in \mathcal{N}_0(\nu)$, then $f\chi_N = 0 \in L^1(\nu)$. So, $|\nu_f|(N) = 0$. Therefore

$$\mathcal{N}_0(\nu) \subset \mathcal{N}_0(\nu_f). \tag{21}$$

Example 2. Let Ω, $\mathcal{R}$ and ν be as in Example 1. We have seen that the function given by $f(x) = \frac{1}{x}$, $\forall\ x \in \Omega$ is a locally ν-integrable function. Consider its associated vector measure $\nu_f : \mathcal{R} \to \mathbb{R}$. Let us assume that $\chi_\Omega \in L^1(\nu_f)$ and define $B_n = [\frac{1}{n}, 1 - \frac{1}{n}]$, for $n = 3, 4, \ldots$. Notice that $\chi_{B_n} \uparrow \chi_\Omega$, hence by the Dominated Convergence Theorem

$$\int_\Omega \chi_\Omega d\nu_f = \lim_{n \to \infty} \int_\Omega \chi_{B_n} d\nu_f = \lim_{n \to \infty} \int_{\frac{1}{n}}^{1 - \frac{1}{n}} \frac{1}{x} dx = \infty.$$

This gives us a contradiction.

Therefore $\chi_\Omega \notin L^1(\nu_f)$. So, by Theorem 14 we have ν_f is not strongly additive.

The previous example shows that ν_f may not be strongly additive even though ν is. However from (2) we obtain the next result.

Proposition 13. Let $f \in L^1_{loc}(\nu)$. If ν is σ-finite, then ν_f is σ-finite.

3.2. The space $L^1(\nu_f)$

Notice that if $N \in \mathcal{N}_0(\nu_f)$, then $f\chi_N = 0$, ν-a.e. Let $g, h \in \mathcal{M}$ be such that $g = h\ \nu_f$-a.e. and $N \in \mathcal{N}_0(\nu_f)$ such that $g\chi_{N^c} = h\chi_{N^c}$. Then $fg\chi_{N^c} = fh\chi_{N^c}$ and $fg\chi_N = fh\chi_N = 0$ ν-a.e. Hence $fg = fh$, ν-a.e. So the operator $M_f : L^0(\nu_f) \to L^0(\nu)$, defined by

$$M_f(g) := fg, \forall\ g \in L^0(\nu_f), \tag{22}$$

is well defined and it is linear.

240   C. Avalos-Ramos

Theorem 17. *If $f \in L^1_{loc}(\nu)$ and $g \in L^0(\nu_f)$, then:*

i) *The function g is weakly ν_f-integrable if and only if $fg \in L^1_w(\nu)$. Moreover, the restriction $M_f : L^1_w(\nu_f) \to L^1_w(\nu)$ of the operator defined in (22) is a linear isometry.*

ii) *The function g is ν_f-integrable if and only if $fg \in L^1(\nu)$. Moreover, $M_f : L^1(\nu_f) \to L^1(\nu)$ is a linear isometry such that $I_{\nu_f} = I_\nu \circ M_f$.*

Proof. i) First note the following: if $\{g_n\} \subset L^1(\mathcal{R}^{loc})$ and $g \in L^1(\mathcal{R}^{loc})$ are such that $g_n \to g$, ν_f-a.e. and $N \in \mathcal{N}_0(\nu_f)$ such that $g_n \chi_{N^c} \to g\chi_{N^c}$. Then $fg_n\chi_{N^c} \to fg\chi_{N^c}$ and $fg_n\chi_N = fg\chi_N = 0$, ν-a.e. $\forall\, n \in \mathbb{N}$. So, $fg_n \to fg$ ν-a.e. Analogously if $g_n \to g$, $|\langle \nu_f, x^* \rangle|$-a.e., then $fg_n \to fg$ $|\langle \nu, x^* \rangle|$-a.e., for each $x^* \in X$.

Let us consider $s = \sum_{j=1}^{n} a_j \chi_{A_j} \in S(\mathcal{R}^{loc})$, by (20), for each $x^* \in X^*$,

$$\int_\Omega |fs|d|\langle \nu, x^* \rangle| = \sum_{j=1}^{n} |a_j| \int_{A_j} |f|d|\langle \nu, x^* \rangle|$$

$$= \int_\Omega |s|d|\langle \nu_f, x^* \rangle|.$$

Hence $s \in L^1_w(\nu_f)$ if and only if $fs \in L^1_w(\nu)$. Now take $g \in L^0(\nu_f)$ and consider $\{s_n\} \subset S(\mathcal{R}^{loc})$ such that $0 \le s_n \uparrow |g|$. By the Monotone Convergence Theorem for each $x^* \in X$,

$$\int_\Omega |fg|d|\langle \nu, x^* \rangle| = \lim_{n\to\infty} \int_\Omega |fs_n|d|\langle \nu, x^* \rangle| = \lim_{n\to\infty} \int_\Omega |s_n|d|\langle \nu_f, x^* \rangle|$$

$$= \int_\Omega |g|d|\langle \nu_f, x^* \rangle|.$$

It turns out that $g \in L^1_w(\nu_f)$ if and only if $fg \in L^1_w(\nu)$ and

$$||g||_{\nu_f} = \lim_{n\to\infty} ||s_n||_{\nu_f} = \lim_{n\to\infty} ||fs_n||_\nu = |fg|_\nu. \tag{23}$$

Which indicates that M_f is a linear isometry.

ii) Let us assume that $g \in L^1(\nu_f)$. If $s = \sum_{j=1}^{n} b_j \chi_{B_j} \in S(\mathcal{R})$, we have $fs \in L^1(\nu)$ and

$$\int_A sd\nu_f = \sum_{j=1}^{n} b_j \nu_f(A \cap B_j) = \sum_{j=1}^{n} b_j \int_A f\chi_{B_j} d\nu = \int_A fsd\nu, \ \forall\, A \in \mathcal{R}^{loc}.$$

$$\tag{24}$$

L^1-space of Vector Measures Defined on δ-rings 241

Now let $\{s_n\} \subset S(\mathcal{R})$ be such that $s_n \to g$ in $L^1(\nu_f)$ and ν_f-a.e. From the remark made at the beginning, $fs_n \to fg$ ν-a.e. Otherwise, from (23) we obtain that $\{fs_n\}$ is a Cauchy sequence in $L^1(\nu)$. Take $h \in L^1(\nu)$ such that $fs_n \to h$ in $L^1(\nu)$ and $\{s_{n_k}\}$ a subsequence of $\{s_n\}$ such that $fs_{n_k} \to h$ ν-a.e. Then $h = fg$ ν-a.e., hence $fg \in L^1(\nu)$ and $fs_n \to fg$ in $L^1(\nu)$.

Now assume that $fg \in L^1(\nu)$. Notice that $f|g| \in L^1(\nu)$. Let consider $\{B_n\} \subset \mathcal{R}$ such that $\sigma fg = \bigcup_{n=1}^{\infty} B_n$, ν-a.e. Take $\{\varphi_n\} \subset S(\mathcal{R}^{loc})$ such that $0 \le \varphi_n \uparrow |g|$ and define $s_n := \varphi_n \chi_{B_n}$. Then, $fs_n \to f|g|$, ν-a.e. By Dominated Convergence Theorem and (24), for each $A \in \mathcal{R}^{loc}$

$$\int_A |g| f d\nu = \lim_{n \to \infty} \int_A s_n f d\nu = \lim_{n \to \infty} \int_A s_n d\nu_f \in X.$$

By Theorem 12 we have that $|g| \in L^1(\nu_f)$, therefore $g \in L^1(\nu_f)$.

Finally if $\{s_n\} \subset S(\mathcal{R})$ is a sequence such that $s_n \to g$, ν_f-a.e., from (24) we obtain

$$I_\nu \circ M_f(g) = \int_A fg d\nu = \lim_{n \to \infty} \int_A fs_n d\nu = \int_A g d\nu_f = I_{\nu_f}(g).$$

$\square$

Corollary 1. *Let $f, h \in L^1_{loc}(\nu)$. If $|f| = |h|$ ν-a.e., then:*

i) $L^1_w(\nu_f) \equiv L^1_w(\nu_h)$ *and* $M_f(L^1_w(\nu_f)) = M_h(L^1_w(\nu_f))$.
ii) $L^1(\nu_f) \equiv L^1(\nu_h)$ *and* $M_f(L^1(\nu_f)) = M_h(L^1(\nu_f))$.

Proof. Take $g \in L^1(\mathcal{R}^{loc})$. Since $|f| = |h|$ ν-a.e., we have $|gf| = |gh|$ ν-a.e. Then $gf \in L^1_w(\nu)$ if and only if $gh \in L^1_w(\nu)$ and $gf \in L^1(\nu)$ if and only if $gh \in L^1(\nu)$. From previous theorem we obtain $L^1_w(\nu_f) \equiv L^1_w(\nu_h)$ and $L^1(\nu_f) \equiv L^1(\nu_h)$.

Now take $\tilde{f}, \tilde{h} \in L^0(\nu)$ such that $|\tilde{f}| = |\tilde{h}| = 1$, $f = |f|\tilde{f}$ and $|h| = h\tilde{h}$. Since $L^1(\nu_f)$ is a Banach lattice, it follows that $\tilde{f}g, \tilde{h}\tilde{f}g \in L^1(\nu_f)$, $\forall g \in L^1(\nu_f)$. Then

$$fg = |f|\tilde{f}g = h\tilde{h}\tilde{f}g, \ \forall g \in L^1(\nu_f).$$

This tells us that $M_f(L^1(\nu_f)) \subset M_h(L^1(\nu_f))$. For the other containments we proceed analogously. $\square$

As M_f is a linear isometry, it turns out that $M_f(L^1(\nu_f))$ and $M_f(L^1_w(\nu_f))$ are Banach subspaces of $L^1(\nu)$ and $L^1_w(\nu)$, respectively. Below we will study some properties of this spaces and of M_f.

242 C. Avalos-Ramos

Let us recall an ideal Y in a Banach lattice Z, is a *band*, if por each set $D \subset Y_\mathbb{R}$ such that there exists $g = \sup D \in Z_\mathbb{R}$, then $g \in Y_\mathbb{R}$.

Proposition 14. *If $f \in L^1_{loc}(\nu)$, then:*

i) $M_{|f|} : L^1_w(\nu_f) \to L^1_w(\nu)$ *is a lattice isometry and* $M_f(L^1_w(\nu_f)) \subset L^1_w(\nu)$ *is a band. So,* $M_f(L^1_w(\nu_f))$ *is a* $|\nu|$-*B.f.s. with the norm* $||\cdot||_\nu$.

ii) $M_{|f|} : L^1(\nu_f) \to L^1(\nu)$ *is a lattice isometry and* $M_f(L^1(\nu_f)) \subset L^1(\nu)$ *is a band. So,* $M_f(L^1(\nu_f))$ *is a* $|\nu|$-*B.f.s. wiht the norm* $||\cdot||_\nu$.

Proof. i) Let $g, h \in (L^1_w(\nu_f))_\mathbb{R}$. Since $|f| \geq 0$, we have $\sup\{|f|g, |f|h\} = |f|\sup\{g, h\}$, so $M_{|f|}$ is a lattice homomorphism. From Theorem 17 it follows that $M_{|f|}$ is a lattice isometry. Hence $M_{|f|}(L^1_w(\nu_f))$ is a Banach lattice. By the previous corollary $M_f(L^1_w(\nu_f))$ is Banach lattice.

Let $g \in L^1_w(\nu_f)$ and $h \in L^0(\nu)$ be such that $|h| \leq |fg|$, ν-a.e. Define $\hat{g} := \frac{h}{f}\chi_{supph}$, then $|\hat{g}| \leq |g|$. So, $\hat{g} \in L^1_w(\nu_f)$ and $h = \hat{g}f \in M_f(L^1_w(\nu_f))$. Hence $M_f(L^1_w(\nu_f))$ is an ideal in $L^0(\nu)$. It follows that $M_f(L^1_w(\nu_f))$ is a $|\nu|$-B.f.s.

Finally let $D \subset M_f(L^1_w(\nu_f))_\mathbb{R}$ be such that $\hat{h} = \sup D \in L^1_w(\nu)_\mathbb{R}$. Define

$$\hat{g} := \frac{\hat{h}}{|f|}\chi_{suppf} \in L^0(\nu). \text{ Then } |f|\hat{g} = \hat{h}\chi_{suppf} \in L^1_w(\nu)_\mathbb{R}. \text{ From Theorem 17}$$

we have $\hat{h} \in L^1_w(\nu_f)$. Otherwise as each $h \in D$ is expressed as $h = |f|g$ for some $g \in L^1_w(\nu_f)_\mathbb{R}$ it turns out that $supph \subset suppf$. Then $h \leq |f|\hat{g} \leq \hat{h}$, $\forall\, h \in D$. So, $\hat{h} = |f|\hat{g} \in M_{|f|}(L^1_w(\nu_f))_\mathbb{R}$. From previous corollary $\hat{h} \in M_f(L^1_w(\nu_f))_\mathbb{R}$.

ii) The proof is analogous. $\qquad\square$

Proposition 15. *If $L^1_w(\nu)$ has the Fatou property, then $L^1_w(\nu_f)$ has the Fatou property $\forall f \in L^1_{loc}(\nu)$.*

Proof. Let us assume that $L^1_w(\nu)$ has the Fatou property and $f \neq 0$. Notice that $\Omega \setminus suppf \in \mathcal{N}_0(\nu_f)$. Let $\{g_\tau\} \subset L^1_w(\nu_f)$ be such that $0 \leq g_\tau \uparrow$ and $\sup_\tau ||g_\tau||_{\nu_f} < \infty$. Then $g_\tau|f| \in M_f(L^1_w(\nu_f)) \subset L^1_w(\nu)$, $\forall\, \tau$, $0 \leq g_\tau|f| \uparrow$ and $\sup_\tau ||g_\tau f||_\nu < \infty$. Since $L^1_w(\nu)$ has the Fatou property, there exists $h \in L^1_w(\nu)$ such that $h = \sup_\tau g_\tau|f|$ and $||h||_\nu = \sup_\tau ||g_\tau f||_\nu$. As $M_f(L^1_w(\nu_f))$ is a band $h \in M_f(L^1_w(\nu_f))$. Hence there exists $g \in L^1_w(\nu_f)$ such that $h = |f|g$.

Moreover, since $g_\tau|f| \leq h$, we have $g_\tau \leq g$ ν_f-a.e. $\forall\, \tau$. Let us assume that there exists $g' \in L^1_w(\nu_f)$ such that $g_\tau \leq g'$ ν_f-a.e. $\forall\, \tau$.

Then $h \leq g'|f|$, and so, $g \leq g'$ ν_f-a.e. Hence $g = \sup_\tau g_\tau$. Accordingly

$$||g||_{\nu_f} = ||h||_\nu = \sup_\tau ||g_\tau f||_\nu = \sup_\tau ||g_\tau||_{\nu_f}. \qquad \square$$

By Theorem 17, the space $L^1(\nu_f)$ is isometrically isomorphic to a closed subspace of $L^1(\nu)$. Then $L^1(\nu_f)$ has all those properties that $L^1(\nu)$ inherits to its closed subspaces. Besides, since the operator I_{ν_f} is equal to the composition of the operator I_ν and a bouded operator it follows that the properties of I_{ν_f} are directly related to the properties of I_ν. For example, we have the following properties.

Corollary 2. *Let* $f \in L^1_{loc}(\nu)$.

  i) If $L^1(\nu)$ is reflexive, then $L^1(\nu_f)$ is reflexive.
  ii) If $L^1(\nu)$ is separable, then $L^1(\nu_f)$ is separable.
  iii) If $L^1(\nu)$ is weakly sequentially complete, then $L^1(\nu_f)$ is weakly sequentially complete.
  iv) If I_ν is a compact operator, then entonces I_{ν_f} is a compact operator.

Proposition 16. *Let* $f \in L^1_{loc}(\nu)$. *Then the following sentences are equivalent:*

  i) $f \neq 0$ ν-a.e.
  ii) $\mathcal{N}_0(\nu) = \mathcal{N}_0(\nu_f)$.
  iii) The linear isometry $M_f : L^1_w(\nu_f) \to L^1_w(\nu)$ is surjective.
  iv) The linear isometry $M_f : L^1(\nu_f) \to L^1(\nu)$ is surjective.

Proof. i) $\Rightarrow$ ii) Since $\mathcal{N}_0(\nu) \subset \mathcal{N}_0(\nu_f)$ it only remains to verify that $\mathcal{N}_0(\nu_f) \subset \mathcal{N}_0(\nu)$. Fix $A \in \mathcal{N}_0(\nu_f)$. By Theorem 13 $||f\chi_A||_\nu = 0$, then $f\chi_A = 0$ ν-a.e. As $f \neq 0$ ν-a.e. it follows that $A \in \mathcal{N}_0(\nu)$.
ii) $\Rightarrow$ i) We will establish the contrapositive. Let us assume that there exists $A \in \mathcal{R}^{loc}$ such that $A \notin \mathcal{N}_0(\nu)$ and $f\chi_A = 0$. Then $f\chi_A \in L^1_w(\nu)$ and $|\nu_f|(A) = |f\chi_A|_\nu = 0$. Hence $\mathcal{N}_0(\nu) \neq \mathcal{N}_0(\nu_f)$.
i) $\Rightarrow$ iii) Take $h \in L^1_w(\nu)$, then $\dfrac{h}{f} \in L^0(\mathcal{R}^{loc})$. Since $\mathcal{N}_0(\nu) = \mathcal{N}_0(\nu_f)$, we have that $\dfrac{h}{f} \in L^0(\nu) = L^0(\nu_f)$. Moreover, $\dfrac{h}{f}f = h \in L^1_w(\nu)$. By Theorem 17 i) we obtain $\dfrac{h}{f} \in L^1_w(\nu_f)$ and $M_f\left(\dfrac{h}{f}\right) = h$. Therefore M_f is surjective.
iii) $\to$ iv) Let $h \in L^1(\nu)$. Since $h \in L^1_w(\nu)$ and $M_f : L^1_w(\nu_f) \to L^1_w(\nu)$ is surjective, take $g \in L^1_w(\nu_f)$ such that $fg = h$. By Theorem 17 ii) $g \in L^1(\nu_f)$.
iv) $\to$ ii) Let us assume that there exists $A \in \mathcal{N}_0(\nu_f)$ such that $A \notin \mathcal{N}_0(\nu)$.

244   C. Avalos-Ramos

Take $B \in \mathcal{R}_A$ such that $||\nu||(B) > 0$. Since $\chi_B \in L^1(\nu) \subset L^1_w(\nu)$ and M_f is surjective there exists $g \in L^1(\nu_f) \subset L^1_w(\nu_f)$ such that $fg = \chi_B$. So, $f(x) \neq 0$ for almost every $x \in B$. Now as $B \in \mathcal{N}_0(\nu_f)$ we have that $f\chi_B = 0$, ν-a.e. which is a contradiction. $\qquad\square$

Example 3. If $A \in \mathcal{R}^{loc}$, then $\chi_A \in L^1_{loc}(\nu)$ and

$$\nu_{\chi_A}(B) = \nu(A \cap B), \ \forall \ B \in \mathcal{R}.$$

Moreover, $L^1(\nu_{\chi_A}) = \{g \in L^0(\nu) : g\chi_A \in L^1(\nu)\}$. We note that it is possible that $\chi_A \notin L^1(\nu)$, in this case $L^1(\nu) \subsetneq L^1_{loc}(\nu)$.

3.3. Characterizations of integrable functions in terms of $\mathcal{R}$

The definitions of the spaces of scalarly integrable and ν-integrable functions are given in terms of the σ-algebra $\mathcal{R}^{loc}$. In this section we will see how they can be expressed considering only the δ-ring $\mathcal{R}$.

Proposition 17. Let $f : \Omega \to \mathbb{K}$. Then:

i) $f \in L^1_w(\nu)$ if and only if for each $B \in \mathcal{R}$, $f\chi_B \in L^1_w(\nu)$ and $\sup_{B \in \mathcal{R}} ||f\chi_B||_\nu < \infty$. In this case $||f||_\nu = \sup_{B \in \mathcal{R}} ||f\chi_B||_\nu$.

ii) $f \in L^1(\nu)$ if and only if $f \in L^1_{loc}(\nu)$ and its associated vector measure $\nu_f : \mathcal{R} \to X$ is strongly additive.

Proof. i) Assume that $f \in L^1_w(\nu)$. Since $L^1_w(\nu)$ is a Banach lattice we have that $f\chi_B \in L^1_w(\nu)$ and $||f\chi_B||_\nu \leq ||f||_\nu, \ \forall \ B \in \mathcal{R}$. Then, $\sup_{B \in \mathcal{R}} ||f\chi_B||_\nu \leq ||f||_\nu < \infty$.

Now assume that $f\chi_B \in L^1_w(\nu), \ \forall \ B \in \mathcal{R}$ and $M := \sup_{B \in \mathcal{R}} ||f\chi_B||_\nu < \infty$, it follows that $f \in \mathcal{M}$. Let $x^* \in X$, then

$$\sup_{B \in \mathcal{R}} \int_B |f|d|\langle x^*, \nu\rangle| \leq M.$$

By Proposition 5, $f \in L^1(|\langle x^*, \nu\rangle|), \ \forall \ x^* \in X$. So $f \in L^1_w(\nu)$ and $||f||_\nu \leq M$.

ii) Let us assume that $f \in L^1(\nu)$. So $f \in L^1_{loc}(\nu)$. Moreover, ν_f is the restriction to $\mathcal{R}$ of the vector measure $\tilde{\nu}_f : \mathcal{R}^{loc} \to X$, defined on (24) hence ν_f is strongly additive.

Now assume that $f \in L^1_{loc}(\nu)$ and ν_f is strongly additive. From Theorem 14 it follows that $\chi_\Omega \in L^1(\nu_f)$. By Theorem 17, $f = f\chi_\Omega \in L^1(\nu)$. $\qquad\square$

3.4. Two δ-rings with the same associated σ-algebra

Let $\widehat{\mathcal{R}}$ be a δ-ring on Ω such that $\mathcal{R} \subset \widehat{\mathcal{R}}$ and $\mathcal{R}^{loc} = \widehat{\mathcal{R}}^{loc}$ and $\widehat{\nu} : \widehat{\mathcal{R}} \to X$ a vector measure such that $\nu(B) = \widehat{\nu}(B)$, $\forall B \in \mathcal{R}$. By Proposition 7 i) we have that $|\nu| \leq |\widehat{\nu}|$. Hence

$$\mathcal{N}_0(\widehat{\nu}) \subset \mathcal{N}_0(\nu).$$

Similarly

$$|\langle \nu, x^* \rangle| \leq |\langle \widehat{\nu}, x^* \rangle|, \ \forall \ x^* \in X^*.$$

Next example shows that it may possible that $\mathcal{N}_0(\widehat{\nu}) \subsetneq \mathcal{N}_0(\nu)$.

Example 4. Let $\Omega = \mathbb{N}_0 := \mathbb{N} \cup \{0\}$, $\widehat{\mathcal{R}} = 2^\Omega$. Define $\widehat{\nu} : \widehat{\mathcal{R}} \to \mathbb{R}$ by

$$\widehat{\nu}(A) := \sum_{n \in A} \frac{1}{2^n}.$$

Then $\widehat{\nu}$ is a vector measure such that $\mathcal{N}_0(\widehat{\nu}) = \{\emptyset\}$. Let us consider $\mathcal{R} := 2^{\mathbb{N}}$. Hence $\mathcal{R}$ is a δ-ring on Ω such that $\mathcal{R} \subset \widehat{\mathcal{R}}$ and $\mathcal{R}^{loc} = \widehat{\mathcal{R}}^{loc} = \widehat{\mathcal{R}}$. Define $\nu : \mathcal{R} \to \mathbb{R}$ by $\nu(B) := \widehat{\nu}(B)$. It turns out that $\mathcal{N}_0(\nu) = \{\emptyset, \{0\}\}$. Then $\mathcal{N}_0(\widehat{\nu}) \subsetneq \mathcal{N}_0(\nu)$.

Let us assume that $\mathcal{N}_0(\widehat{\nu}) \subsetneq \mathcal{N}_0(\nu)$. Take $A \in \mathcal{R}^{loc}$ such that $A \in \mathcal{N}_0(\nu)$ and $A \notin \mathcal{N}_0(\widehat{\nu})$. Then $\chi_A = 0$, ν-a.e. however $\chi_A \neq 0$, $\widehat{\nu}$-a.e. Hence, $L^0(\nu) \neq L^0(\widehat{\nu})$.

Proposition 18. *If* $\mathcal{N}_0(\widehat{\nu}) = \mathcal{N}_0(\nu)$, *then* $L^1(\widehat{\nu}) = L^1(\nu)$ *with equivalent norms and, for each* $A \in \mathcal{R}^{loc}$

$$\int_A f d\widehat{\nu} = \int_A f d\nu, \ \forall \ f \in L^1(\widehat{\nu}). \tag{25}$$

Proof. Since $\mathcal{N}_0(\widehat{\nu}) = \mathcal{N}_0(\nu)$ the spaces $L_w^1(\nu)$, $L_w^1(\widehat{\nu})$, $L^1(\nu)$ and $L^1(\widehat{\nu})$ are subspaces of $L^0(\nu)$. Fix $A \in \Sigma$ and $x^* \in X^*$. As $|\langle \nu, x^* \rangle| \leq |\langle \widehat{\nu}, x^* \rangle|$ we have that

$$\int_A |f| d|\langle \nu, x^* \rangle| \leq \int_A |f| |\widehat{\nu}, x^* \rangle|, \ \forall \ f \in \mathcal{M},$$

which tells us that $\|f\|_\nu \leq \|f\|_{\widehat{\nu}}$, $\forall \ f \in \mathcal{M}$. Hence, $L_w^1(\widehat{\nu}) \subset L_w^1(\nu)$.
Let us consider $f \in L^1(\widehat{\nu})$ and $B \in \mathcal{R}$, then

$$\left\langle \int_B f d\widehat{\nu}, x^* \right\rangle = \int_B f \langle \widehat{\nu}, x^* \rangle = \int_\Omega f \chi_B d\langle \nu, x^* \rangle.$$

Therefore $f\chi_B \in L^1(\nu)$ and $\int_B f d\nu = \int_B f d\hat{\nu}$. So, $f \in L^1_{loc}(\nu)$ and

$$\nu_f(B) := \int_B f d\nu = \int_B f d\hat{\nu} = \hat{\nu}_f(B), \ \forall \ B \in \mathcal{R}.$$

Since $f \in L^1(\hat{\nu})$, it turns out that $\hat{\nu}_f$ is a strongly additive vector measure. Hence ν_f is strongly additive. By Proposition 17, $f \in L^1(\nu)$.

For each $s = \sum_{j=1}^n b_j \chi_{B_j} \in S(\mathcal{R}) \subset S(\hat{\mathcal{R}})$ and each $A \in \mathcal{R}^{loc}$ we have

$$\int_A s d\hat{\nu} = \sum_{j=1}^n b_j \hat{\nu}(A \cap B_j) = \sum_{j=1}^n b_j \nu(A \cap B_j) = \int_A s d\nu. \qquad (26)$$

Now consider $f \in L^1(\nu)$ and take $\{s_n\} \subset S(\mathcal{R})^+$ such that $s_n \to |f|$ in $L^1(\nu)$ and $s_n \uparrow |f|$ ν-a.e. Then $s_n \uparrow |f|$ $\hat{\nu}$-a.e. By (24), $\left\{ \int_A s_n d\hat{\nu} \right\}$ converges in X. Since $S(\mathcal{R}) \subset S(\hat{\mathcal{R}})$, by Theorem 12 we have that $|f| \in L^1(\hat{\nu})$. Hence $f \in L^1(\hat{\nu})$.

Since the inclusion operator betwen the spaces $L^1(\nu)$ and $L^1(\hat{\nu})$ is positive we obtain the equivalence of norms.

Finally fix $A \in \mathcal{R}^{loc}$ and $f \in L^1(\hat{\nu})$. Let us take $\{s_n\} \subset S(\mathcal{R})$ such that $s_n \to f$ in $L^1(\nu)$, since $L^1(\hat{\nu}) = L^1(\nu)$ as Banach spaces it turns out that $s_n \to f$ in $L^1(\hat{\nu})$. From the continuity of I_ν and $I_{\hat{\nu}}$ and from (26), we have that

$$\int_A f d\hat{\nu} = \lim_{n \to \infty} \int_A s_n d\hat{\nu} = \lim_{n \to \infty} \int_A s_n d\nu = \int_A f d\nu.$$

$\square$

Corollary 3. If $\mathcal{N}_0(\langle \hat{\nu}, x^* \rangle) = \mathcal{N}_0(\langle \nu, x^* \rangle)$, $\forall \ x^* \in X^*$, then $L^1(\hat{\nu}) = L^1(\nu)$ and $L^1_w(\hat{\nu}) = L^1_w(\nu)$ with equivalent norms. In particular if $|\langle \nu, x^* \rangle| = |\langle \hat{\nu}, x^* \rangle|$, $\forall \ x^* \in X^*$, then $L^1(\hat{\nu}) \equiv L^1(\nu)$ and $L^1_w(\hat{\nu}) \equiv L^1_w(\nu)$.

Proof. Let us assume that $\mathcal{N}_0(\langle \hat{\nu}, x^* \rangle) = \mathcal{N}_0(\langle \nu, x^* \rangle)$, $\forall \ x^* \in X^*$. Then $\mathcal{N}_0(\hat{\nu}) = \mathcal{N}_0(\nu)$, by the previous proposition $L^1(\hat{\nu}) = L^1(\nu)$.

Let us note that $\langle \nu, x^* \rangle$ is the restriction to $\mathcal{R}$ of $\langle \hat{\nu}, x^* \rangle$. So, using the previous proposition for each $\langle \nu, x^* \rangle$, it turns out that $L^1(\langle \hat{\nu}, x^* \rangle) = L^1(\langle \nu, x^* \rangle)$, $\forall \ x^* \in X^*$. Thus

$$L^1_w(\hat{\nu}) = \bigcap_{x^* \in X^*} L^1(\langle \hat{\nu}, x^* \rangle) = \bigcap_{x^* \in X^*} L^1(\langle \nu, x^* \rangle) = L^1_w(\nu).$$

By the hypothesis we have that the norms $||\cdot||_{\hat{\nu}}$ and $||\cdot||_\nu$ are equivalent.

$\square$

Corollary 4. *Let Σ be a σ-algebra on Ω and $\widehat{\nu} : \Sigma \to X$ a vector measure. Let us assume that $\mathcal{R} \subset \Sigma$ is a δ-ring such that $\mathcal{R}^{loc} = \Sigma$ and let $\nu : \mathcal{R} \to X$ be the restriction to $\mathcal{R}$ of $\widehat{\nu}$. Then:*

i) *If $\mathcal{N}_0(\widehat{\nu}) = \mathcal{N}_0(\nu)$, then $L^1(\widehat{\nu}) = L^1(\nu)$.*
ii) *If $\mathcal{N}_0(\langle \widehat{\nu}, x^* \rangle) = \mathcal{N}_0(\langle \nu, x^* \rangle)$, $\forall \ x^* \in X^*$, then $L^1(\widehat{\nu}) = L^1(\nu)$ and $L^1_w(\widehat{\nu}) = L^1_w(\nu)$.*

Example 5. It is well known that $\chi_B \in L^1(\nu)$, $\forall \ B \in \mathcal{R}$. However the collection consisting of the sets $A \in \mathcal{R}^{loc}$ such that $\chi_A \in L^1(\nu)$ may be bigger. Following the notation used by P. R. Masani and H. Niemi in [18, Def. 5.1] let us define

$$\overline{\mathcal{R}} := \{A \in \mathcal{R}^{loc} : \chi_A \in L^1(\nu)\}.$$

They show that this collection is a δ-ring. Likewise they define the vector measure $\overline{\nu} : \overline{\mathcal{R}} \to X$

$$\overline{\nu}(A) := \int_\Omega \chi_A d\nu, \ \forall \ A \in \overline{\mathcal{R}}. \tag{27}$$

Then this measure is an extension of $\nu : \mathcal{R} \to X$ and $\mathcal{R}^{loc} = \overline{\mathcal{R}}^{loc}$ [18, Lemma 5.3]. They establish that the corresponding spaces of integrable functions to this measures are equal in [18, Thm. 5.8]. Moreover, they show that $|\langle \overline{\nu}, x^* \rangle| = |\langle \nu, x^* \rangle|$, $\forall \ x^* \in X^*$. Let us note that this result may be obtained from Corollary 3 and Proposition 18 directly:

Corollary 5. *Let $\overline{\mathcal{R}} := \{A \in \mathcal{R}^{loc} : \chi_A \in L^1(\nu)\}$ and $\overline{\nu} : \overline{\mathcal{R}} \to X$ the vector measure defined in (27). Then $L^1_w(\overline{\nu}) \equiv L^1_w(\nu)$ and $L^1(\overline{\nu}) \equiv L^1(\nu)$.*

Next we will give conditions to the equality of the collections of null sets holds.

Remark 7. Let $B \in \mathcal{R}$. Since $\widehat{\mathcal{R}} \subset \mathcal{R}^{loc}$, it turns out that $A \cap B \in \mathcal{R}$, $\forall \ A \in \widehat{\mathcal{R}}$. Thus, $A = A \cap B \in \mathcal{R}_B$, $\forall \ A \in \widehat{\mathcal{R}}_B$. Hence $\widehat{\mathcal{R}}_B = \mathcal{R}_B$, $\forall \ B \in \mathcal{R}$.

Lemma 6. *Let $\widehat{\mathcal{R}}$ be a δ-ring on Ω such that $\mathcal{R} \subset \widehat{\mathcal{R}}$ and $\mathcal{R}^{loc} = \widehat{\mathcal{R}}^{loc}$, in addition let $\widehat{\nu} : \widehat{\mathcal{R}} \to X$ such that ν is the restriction of $\widehat{\nu}$ to $\mathcal{R}$.*

i) *$\mathcal{N}_0(\widehat{\nu}) = \mathcal{N}_0(\nu)$ if and only if for each $A \in \mathcal{R}^{loc} \setminus \mathcal{N}_0(\widehat{\nu})$ there exists $B \in \mathcal{R}_A$ such that $B \notin \mathcal{N}_0(\widehat{\nu})$.*
ii) *If there exist $\{B_n\} \subset \mathcal{R}$ and $N \in \mathcal{N}_0(\widehat{\nu})$ such that $\Omega = \bigcup_{n=1}^\infty B_n \cup N$, then $|\langle \nu, x^* \rangle| = |\langle \widehat{\nu}, x^* \rangle|$, $\forall \ x^* \in X^*$ and $|\nu| = |\widehat{\nu}|$. Hence $\mathcal{N}_0(\widehat{\nu}) = \mathcal{N}_0(\nu)$.*

Proof. i) Assume that $\mathcal{N}_0(\widehat{\nu}) = \mathcal{N}_0(\nu)$. Let us take $A \in \mathcal{R}^{loc} \setminus \mathcal{N}_0(\widehat{\mathcal{R}})$. Thus $|\nu|(A) \neq 0$. Hence there exists $B \in \mathcal{R}_A$ such that $\widehat{\nu}(B) = \nu(B) \neq 0$.

To establish the other implication only remains to check $\mathcal{N}_0(\nu) \subset \mathcal{N}_0(\widehat{\nu})$. Thus take $A \in \mathcal{R}^{loc} \setminus \mathcal{N}_0(\widehat{\nu})$ and $B \in \mathcal{R}_A$ such that $B \notin \mathcal{N}_0(\widehat{\nu})$. Then $\widehat{\nu}(C) \neq 0$ for some $C \in \widehat{\mathcal{R}}_B = \mathcal{R}_B$. However $\nu(C) = \widehat{\nu}(C)$, then $A \notin \mathcal{N}_0(\nu)$.

ii) Observe that since $\widehat{\mathcal{R}}_B = \mathcal{R}_B$ it follows that if $B \in \mathcal{R}$, then $|\widehat{\nu}|(B) = |\nu|(B)$.

Now let $\{B_n\} \subset \mathcal{R}$ and $N \in \mathcal{N}_0(\widehat{\nu})$ such that $\Omega = \bigcup_{n=1}^{\infty} B_n \cup N$ and take $A \in \mathcal{R}^{loc}$. Thus $A \cap B_n \in \mathcal{R}$, $\forall\, n \in \mathbb{N}$ and $A \cap N \in \mathcal{N}_0(\widehat{\nu}) \subset \mathcal{N}_0(\nu)$. Hence

$$|\widehat{\nu}|(A) = \sum_{n=1}^{\infty} |\widehat{\nu}|(A \cap B_n) = \sum_{n=1}^{\infty} |\nu|(A \cap B_n) = |\nu|(A).$$

Therefore $|\widehat{\nu}| = |\nu|$. Analogously it is shown that $|\langle \nu, x^* \rangle| = |\langle \widehat{\nu}, x^* \rangle|$, $\forall\, x^* \in X^*$. $\qquad\square$

Remark 8. Under the context in which we are working J. M. Calabuig, O. Delgado and E. A. Sánchez Pérez considered in [4] the following condition on the measure $\widehat{\nu}$. Let $x^* \in X^*$.

$$\text{If } C \in \widehat{\mathcal{R}} \text{ is such that } \sup_{B \in \mathcal{R}_C} |\langle \widehat{\nu}, x^* \rangle(B)| = 0, \text{ then } \langle \widehat{\nu}, x^* \rangle(C) = 0. \quad (28)$$

Let us see that this condition is equivalent to $\mathcal{N}_0(|\langle \widehat{\nu}, x^* \rangle|) = \mathcal{N}_0(|\langle \nu, x^* \rangle|)$. Take $x^* \in X^*$ and assume that $\mathcal{N}_0(|\langle \widehat{\nu}, x^* \rangle|) = \mathcal{N}_0(|\langle \nu, x^* \rangle|)$. Let $C \in \widehat{\mathcal{R}}$ such that $\sup_{B \in \mathcal{R}_C} |\langle \widehat{\nu}, x^* \rangle(B)| = 0$. Since ν is the restriction of $\widehat{\nu}$ to $\mathcal{R}$, we have

$$\sup_{B \in \mathcal{R}_A} |\langle \nu, x^* \rangle(B)| = \sup_{B \in \mathcal{R}_C} |\langle \widehat{\nu}, x^* \rangle(B)| = 0.$$

Thus $\langle \nu, x^* \rangle(B) = 0$, $\forall\, B \in \mathcal{R}_C$. So $C \in \mathcal{N}_0(|\langle \nu, x^* \rangle|) = \mathcal{N}_0(|\langle \widehat{\nu}, x^* \rangle|)$. Therefore $\langle \widehat{\nu}, x^* \rangle(C) = 0$.

Now assume that (28) holds. It is only necesary establish that $\mathcal{N}_0(\nu) \subset \mathcal{N}_0(\widehat{\nu})$. Let us consider $C \in \mathcal{N}_0(|\langle \nu, x^* \rangle|)$ such that $C \in \widehat{\mathcal{R}}$. Then $\langle \nu, x^* \rangle(B) = 0$, $\forall\, B \in \mathcal{R}_C$. Thus,

$$\sup_{B \in \mathcal{R}_C} |\langle \widehat{\nu}, x^* \rangle(B)| = \sup_{B \in \mathcal{R}_C} |\langle \nu, x^* \rangle(B)| = 0.$$

By the hypothesis $\langle \widehat{\nu}, x^* \rangle(C) = 0$. It follows that $C \in \mathcal{N}_0(|\langle \widehat{\nu}, x^* \rangle|)$.

Take $A \in \mathcal{N}_0(|\langle \nu, x^* \rangle|)$. Then $C \in \mathcal{N}_0(|\langle \nu, x^* \rangle|)$ holds for each $C \in \widehat{\mathcal{R}}_A$. It turns out that $C \in \mathcal{N}_0(|\langle \widehat{\nu}, x^* \rangle|)$. Hence, $|\langle \widehat{\nu}, x^* \rangle|(A) = \sup_{C \in \widehat{\mathcal{R}}} |\langle \widehat{\nu}, x^* \rangle|(C) = 0$.

In [4, Lemma 2] it is shown that if condition (28) holds for each $x^* \in X^*$, then $L^1(\widehat{\nu}) \subset L^1(\nu)$. Since $\mathcal{N}_0(|\langle \widehat{\nu}, x^* \rangle|) = \mathcal{N}_0(|\langle \nu, x^* \rangle|)$, $\forall\, x^* \in X^*$ implies $\mathcal{N}_0(\widehat{\nu}) = \mathcal{N}_0(\nu)$, this result can be obtained as a consequence of Proposition 18.

Moreover, in [4, Lemma 3] it is proved that under certain conditions $L^1(\widehat{\nu}) \equiv L^1(\nu)$. What follows from Lemma 6 iii) and Proposition 18.

3.5. When f is a ν-integrable function

We have seen that given a function $f \in L^1_{loc}(\nu)$ it is posible that the vector measure ν_f is not strongly additive. However by Proposition 17 ii), we have that ν_f is strongly additive when $f \in L^1(\nu)$. Hence if $f \in L^1(\nu)$, then ν_f is σ-finite.

Given a function $f \in L^1(\nu)$ we may associate the vector measures

$$\nu_f : \mathcal{R} \to X \quad \text{and} \quad \tilde{\nu}_f : \mathcal{R}^{loc} \to X.$$

By Theorems 13 and 16 we obtain

$$\|\nu_f\|(A) = \|\tilde{\nu}_f\|(A), \ \forall\, A \in \mathcal{R}^{loc}.$$

Hence, from Propositions 8 and 10 it is follows that

$$\|\tilde{\nu}_f\|(A) = \sup \left\{ \left\| \sum_{j=1}^{n} a_j \nu_f(A_j) \right\|_X : \{a_j\}_1^n \subset B_{\mathbb{K}}, \ \{A_j\} \in \pi_A \right\}$$

$$= \sup\{\|\nu_f\|(B) : B \in \mathcal{R}_A\}.$$

Next result follows from Theorems 13 and 16.

Lemma 7. Let $f \in L^1(\nu)$. Then

$$|\tilde{\nu}_f|(A) = |\nu_f|(A) = \int_A |f| d|\nu| = \sup_{B \in \mathcal{R}_A} \int_B |f| d|\nu|, \ \forall\, A \in \mathcal{R}^{loc}.$$

Lemma 8. Let $f \in L^1(\nu)$. Then

$$\|\|\tilde{\nu}_f\|\|(A) = \|\|\nu_f\|\|(A) = \sup\{\|\|\nu_f\|\|(B) : B \in \mathcal{R}_A\}, \ \forall\, A \in \mathcal{R}^{loc}.$$

Proof. Take $A \in \mathcal{R}^{loc}$. By definition

$$\|\|\tilde{\nu}_f\|\|(A) := \sup\{\|\nu_f(C)\|_X : C \in (\mathcal{R}^{loc})_A\}.$$

Since $\mathcal{R}_A \subset \mathcal{R}_A^{loc}$, it is follows that $||\nu_f||(A) \leq ||\tilde{\nu}_f||(A)$. Let us define

$$a := \sup_{B \in \mathcal{R}_A} |||\nu_f|||(B).$$

As the quasi-variation is increasing it turns out that $a \leq ||\nu_f||(A)$.

Let us consider $C \in (\mathcal{R}^{loc})_A$. Since $f \in L^1(\nu)$, by Lemma 4 there exist $N \in \mathcal{N}_0(\nu)$ and an increasing sequence of sets $\{B_n\} \subset \mathcal{R}$ such that $\sigma f = N \cup \bigcup_{n=1}^{\infty} B_n$. Hence $f\chi_{B_n} \to f$ ν-a.e. Moreover, $|f\chi_{B_n}| \in L^1(\nu)$, $\forall$ $n \in \mathbb{N}$. By the Dominated Convergence Theorem we have

$$\tilde{\nu}_f(C) = \int_C f d\nu = \lim_{n \to \infty} \int_C f\chi_{B_n} d\nu = \lim_{n \to \infty} \tilde{\nu}_f(C \cap B_n).$$

As $C \cap B_n \in \mathcal{R}$, $\forall n \in \mathbb{N}$, it follows that

$$|\tilde{\nu}_f(C)|_X = \lim_{n \to \infty} |\nu_f(C \cap B_n)|_X \leq ||\nu_f||(A). \tag{29}$$

Taking supremum over $C \in (\mathcal{R}^{loc})_A$ it turns out that $||\tilde{\nu}_f||(A) \leq ||\nu_f||(A)$.

Now let us fix $B \in \mathcal{R}_A$. From equality in (29) we have that

$$||\nu_f(B)||_X = \lim_{n \to \infty} ||\nu_f(B \cap B_n)||_X \leq a.$$

Taking supremun over $B \in \mathcal{R}_A$ we obtain $|||\nu_f|||(A) \leq a$. $\qquad \square$

Observe that if $f \in L^1(\nu)$, then for each $A \in \mathcal{R}^{loc}$ and each $x^* \in X^*$

$$|\langle \nu_f, x^* \rangle|(A) = \int_A |f| d|\langle \nu, x^* \rangle| = |\langle \tilde{\nu}_f, x^* \rangle|(A). \tag{30}$$

Hence, from Proposition 18 and Corollary 3 we obtain next result.

Corollary 6. *If $f \in L^1(\nu)$, then $L_w^1(\tilde{\nu}_f) \equiv L_w^1(\nu_f)$ and $L^1(\tilde{\nu}_f) \equiv L^1(\nu_f)$.*

4. Generalization to δ-rings

In this section we will establish for δ-rings, results that are well known for vector measures defined on σ-algebras [20, Section 3.1]. In most cases the proofs are done in an analogous way, yet we include them so that the paper is self-contained. We will also discuss some examples that illustrate the development.

4.1. An equivalent norm on $L^1(\nu)$

Let $\nu : \mathcal{R} \to X$ be a vector measure defined on a δ-ring whose values are in a Banach space. Let us consider $f \in L^1(\nu)$. As is known for measures defined on σ-algebras [20, p. 112], we can get an equivalent norm in $L^1(\nu)$ by definition of the quasi-variation of ν_f. Namely, we define the function $||| \cdot ||| : L^1(\nu) \to [0, \infty)$ as

$$|||f|||_\nu := |||\nu_f|||(\Omega) = \sup_{B \in \mathcal{R}} \left\| \int_B f d\nu \right\|_X . \tag{31}$$

From Lemma 8 we have that $||\nu_f|| = ||\tilde{\nu}_f||$. Since the vector measure $\tilde{\nu}_f$ is defined on the σ-algebra $\mathcal{R}^{loc}$ it follows that the quasi-variation is finite. Then the function $||| \cdot |||_\nu$ is well defined.

Proposition 19. *The function $||| \cdot |||_\nu$ defines an equivalent norm to $|| \cdot ||_\nu$ on $L^1(\nu)$.*

Proof. Let $f \in L^1(\nu)$. It is enough to prove that $||f||_\nu = 0$ implies that $f = 0$. The rest of the properties follow from linearity of the integral operator and the properties of the norm on X.

Notice that $|||f|||_\nu = |||\nu_f|||(\Omega)$. From inequality ii) in Proposition 9 it follows that

$$|||f|||_\nu \le ||f||_\nu \le 4|||f|||_\nu, \tag{32}$$

in the complex case. And in the real case

$$|||f|||_\nu \le ||f||_\nu \le 2|||f|||_\nu. \tag{33}$$

It turns out that if $||f||_\nu = 0$ we have that $|f|_\nu = 0$, and so $f = 0$. The equivalence of both norms is also followed. $\square$

We have said that $|| \cdot ||_\nu$ is a lattice norm, however even when $||| \cdot |||_\nu$ is an equivalent norm, it may not be a lattice norm. We will provide conditions for $|| \cdot ||_\nu$ to be a lattice norm, as happens for classical vector measures [20, Ex. 3.10]. It is necessary to note first that since $||\nu_f||(\Omega) = ||\tilde{\nu}_f||(\Omega) = ||f||_\nu$, from Proposition 8 it follows that

$$||f||_\nu = \sup \left\{ \left\| \int_\Omega s f d\nu \right\|_X : s \in S(\mathcal{R}), ||s||_\infty \le 1 \right\}, \ \forall \ f \in L^1(\nu). \tag{34}$$

Proposition 20.

i) The function $||| \cdot |||_\nu$ is a lattice norm if and only if $|||f|||_\nu = ||f||_\nu$, $\forall$ $f \in L^1(\nu)$.

ii) If $\left\lVert \int_\Omega f d\nu \right\rVert_X = ||f||_\nu$, $\forall\, f \in L^1(\nu)$, *then* $||| \cdot |||_\nu$ *is a lattice norm.*

Proof. i) Let us assume that $|| \cdot ||_\nu$ is a lattice norm. Let us fix $f \in L^1(\nu)$ and take $s \in S(\mathcal{R})$ such that $\sup |s(\omega)| \leq 1$. As $|sf| \leq |f|$, from the hypothesis it follows that

$$\left\lVert \int_\Omega s f d\nu \right\rVert_X \leq |||sf|||_\nu \leq |||f|||_\nu.$$

For (34) we have that $||f||_\nu \leq |||f|||_\nu$. The remaining inequality follows from (32) and (33) in Proposition 19. The other implication is clear.

ii) Let $f \in L^1(\nu)$. Assume that $\left\lVert \int_\Omega f d\nu \right\rVert_X = ||f||_\nu$. By definition of $|| \cdot ||_\nu$ and Proposition 19 it is true that

$$||f||_\nu = || \int_\Omega f d\nu ||_X \leq |||f|||_\nu \leq ||f||_\nu.$$

Therefore $|||f|||_\nu = ||f||_\nu$, $\forall\, f \in L^1(\nu)$. From i) the result follows. $\qquad\square$

It is well known that the integral operator is linear and bounded with norm less or equal to 1. In the case that $L^1(\nu) \neq \{0\}$ the norm is exactly equal to 1, as we can see in the next proposition.

Proposition 21. *If* $\nu : \mathcal{R} \to X$ *is a vector measure such that* $\nu \neq 0$, *then* $|I_\nu| = 1$.

Proof. Take $f \in L^1(\nu)$. From (34) and (7) we get that

$$\begin{aligned}
||f||_\nu &= \sup\{||I_\nu(sf)||_X : s \in S(\mathcal{R}),\ ||s||_\infty \leq 1\} \\
&\leq ||I_\nu|| \sup\{||f||_\nu ||s||_\infty : s \in S(\mathcal{R}),\ ||s||_\infty \leq 1\} \leq ||I_\nu|| ||f||_\nu.
\end{aligned}$$

$\qquad\square$

4.2. *Positive vector measure*

In the present section we present interesting results obtained when the vector measure is positive.

Definition 17. Let X be a Banach lattice. A vector measure $\nu : \mathcal{R} \to X$ is *positive* if $\nu(A) \geq 0$, $\forall\, A \in \mathcal{R}$.

The next example of a positive vector measure will be useful later.

Example 6. Given a vector measure $\nu : \mathcal{R} \to X$, define the function $[\nu] : \mathcal{R} \to L^1(\nu)$ as

$$[\nu](A) := \chi_A. \tag{35}$$

Since $L^1(\nu)$ has σ-order continuos norm we have that $[\nu]$ is a vector measure; clearly it is positive. As $\nu(B) = 0$ if and only if $\chi_B = 0$, $\forall\, B \in \mathcal{R}$, it turns out that $\mathcal{N}_0(\nu) = \mathcal{N}_0([\nu])$. If $s \in S(\mathcal{R})$, then $s \in L^1([\nu])$ and $\int_\Omega sd[\nu] = s$. Now take $f \in L^1([\nu])$. There exists a sequence $\{s_n\} \subset S(\mathcal{R})$ such that $s_n \to f$ $[\nu]$-a.e. and $\int_\Omega s_n d[\nu] \to \int_\Omega fd[\nu]$. Being as $\int_\Omega s_n d[\nu] = s_n$, $\forall\, n \in \mathbb{N}$, we have that $s_n \to \int_\Omega fd[\nu]$ in $L^1(\nu)$. Then there exists a subsequence $\{s_{n_k}\} \subset \{s_n\}$ such that $s_{n_k} \to \int_\Omega fd[\nu]$ ν-a.e. Therefore $\int_\Omega fd[\nu] = f$. We may conclude that

$$|||f|||_{[\nu]} = \sup_{B \in \mathcal{R}} \left\{ \left\| \int_B fd[\nu] \right\|_\nu \right\} = \sup_{B \in \mathcal{R}} ||f\chi_B||_\nu = ||f||_\nu.$$

It follows that $||| \cdot |||_{[\nu]}$ is a lattice norm. From Proposition 20, we get that

$$||f||_{[\nu]} = |||f|||_{[\nu]} = ||f||_\nu, \ \forall\, f \in L^1([\nu]).$$

Since $S(\mathcal{R})$ is dense both $L^1(\nu)$ and $L^1([\nu])$, we have that $L^1(\nu) \equiv L^1([\nu])$.

Even in the case of σ-algebras, the semi-variation can be not σ-additive, as we may see in [20, p. 104], however we have

$$||\nu|| \text{ is } \sigma\text{-additive if and only if } ||\nu|| = |\nu|.$$

From Propositions 7 and 10 i) it follows that if $|\nu|$ is σ-additive then $|\nu| = |\nu|$ on $\mathcal{R}^{loc}$. Reciprocally, it is clear that if $|\nu| = |\nu|$ on $\mathcal{R}^{loc}$, then $|\nu|$ is σ-additive. In this case we have $|\nu|(A) < \infty$, $\forall\, A \in \mathcal{R}$.

Recall that a Banach lattice X is an *abstract L^1-space* when

$$||x + y||_X = ||x||_X + ||y||_X, \ \ \forall\, x, y \in X^+.$$

An example of an abstract L^1-space is any space $L^1(\mu)$, where μ is a positive measure.

When a positive vector measure take its values on an abstract L^1-space we have that $||\nu|| = |\nu|$ en $\mathcal{R}^{loc}$, [20, Ex. 3.1].

Example 7. Let X be an abstract L^1-space and $\nu : \mathcal{R} \to X$ a positive vector measure. Let consider $A \in \mathcal{R}^{loc}$ and $\{A_j\}_1^n \in \pi_A$, then

$$\sum_{j=1}^{n} ||\nu(A_j)||_X = \left|\left|\sum_{j=1}^{n} \nu(A_j)\right|\right|_X = \left|\left|\nu\left(\bigcup_{j=1}^{n} A_j\right)\right|\right|_X \le ||\nu||(A).$$

Taking the supremum over π_A we obtain that $|\nu|(A) \le |\nu|(A)$. Since the other inequality is always true, it follows that

$$|\nu| = ||\nu||.$$

In this case $||\nu||$ is σ-additive.

Moreover, for $B \in \mathcal{R}$ and $\{B_j\}_1^n \in \pi_B$, we have

$$\left|\left|\sum_{j=1}^{n} \nu(B_j)\right|\right|_X = \left|\left|\nu\left(\bigcup_{j=1}^{n} B_j\right)\right|\right|_X \le ||\nu(B)||_X.$$

So $|\nu|(B) \le ||\nu(B)||_X \le ||\nu||(B) = |\nu|(B)$. Hence

$$|\nu|(B) = ||\nu(B)||_X = ||\nu||(B) < \infty, \quad \forall \, B \in \mathcal{R}. \tag{36}$$

Even though the following Remark and Lemma are mentioned in [14, p. 7], we consider it pertinent to present the proofs.

Remark 9. Assume that X is a Banach lattice. Clearly if the integral operator I_ν is positive, then the vector measure ν is positive. On the other hand whether ν is positive, it follows that $I_\nu(s) \in X^+, \forall \, s \in S(\mathcal{R})^+$. Consider $f \in L^1(\nu)^+$. From Remark 5, there exists $s_n \in S(\mathcal{R})^+$ such that $s_n \uparrow f$, ν-a.e. From the Dominated Convergence Theorem we have that $I_\nu(f) = \lim_{n\to\infty} I_\nu(s_n)$. Then $I_\nu(f) \in X^+$. That is to say that I_ν is a positive operator. So, by the inequality (3)

$$\left|\left|\int_\Omega f d\nu\right|\right| \le \int_\Omega |f| d\nu, \quad \forall \, f \in L^1(\nu).$$

Lemma 9. *Let X be a Banach lattice. If $\nu : \mathcal{R} \to X$ is a positive vector measure, then*

$$||f||_\nu = \left|\left|\int_\Omega |f| d\nu\right|\right|_X, \quad \forall \, f \in L^1(\nu).$$

Proof. Take $f \in L^1(\nu)$. As ν is a positive vector measure and $|f| \ge 0$, it follows that $\tilde{\nu}_{|f|} : \mathcal{R}^{loc} \to X$ is a positive vector measure. From Theorem 17

$$L^1\text{-space of Vector Measures Defined on }\delta\text{-rings} \qquad 255$$

and the corresponding case for σ-algebras, it turns out that

$$||fg||_\nu = ||g||_{\tilde{\nu}_{||f||}} = \left|\left|\int_\Omega |g| d\tilde{\nu}_{|f|}\right|\right|_X = \left|\left|\int_\Omega |fg| d\nu\right|\right|_X, \ \forall\, g \in L^1(\nu_{|f|}).$$

Since $\chi_\Omega \in L^1(\tilde{\nu}_{|f|})$, we have the equality. $\qquad\qquad\qquad\square$

4.3. $|\nu|$-integrable function

As we have mentioned in the Theorem 13, D. R. Lewis proved that for every ν-integrable function f, the variation of measure $\tilde{\nu}_f$ is given as the integral of $|f|$ with respect to $|\nu|$. From which it follows that $L^1(|\nu|) \subseteq L^1(\nu)$. Next we will see some cases in which the equality holds.

Proposition 22. *Let X be an abstract L^1-space and $\nu : \mathcal{R} \to X$ a positive vector measure. Then:*

i) $L^1(|\nu|) \equiv L^1(\nu)$. Moreover,

$$\left|\left|\int_\Omega |f| d\nu\right|\right|_X = \int_\Omega |f| d|\nu|.$$

ii) There exists $x_0^* \in (X^*)^+$ such that

$$||\nu(B)||_X = ||\nu||(B) = |\nu|(B) = \langle \nu, x_0^* \rangle(B), \quad B \in \mathcal{R}.$$

That is, the measures $|\nu|$ and $\langle \nu, x_0^* \rangle$ are equal on $\mathcal{R}$ and so

$$L^1(\nu) \equiv L^1(|\nu|) \equiv L^1(\langle \nu, x_0^* \rangle).$$

Proof. i) Take $f \in L^1(\nu)$. Let us consider the case $f \in L^1(\nu)^+$. Then $\tilde{\nu}_f : \mathcal{R}^{loc} \to X$ is a positive vector measure defined on a σ-algebra. Therefore the semi-variation of $\tilde{\nu}_f$ is finite. Since X is an abstract L^1-space, from the Example 7 it follows that $|\tilde{\nu}_f|(\Omega) < \infty$. By Theorem 13 we have that $f \in L^1(|\nu|)$. Then $f \in L^1(|\nu|), \forall\, f \in L^1(\nu)$.

As $\tilde{\nu}_{|f|}$ is a positive vector measure From Theorem 13, Example 7 and Lemma 8 i) it turns out that

$$||f||_\nu = |\tilde{\nu}_{|f|}|(\Omega) = ||\tilde{\nu}_{|f|}||(\Omega) = ||f||_{|\nu|}.$$

Finally form Lemma 9 and the above

$$\left|\left|\int_\Omega |f| d\nu\right|\right|_X = ||f||_\nu = ||f||_{|\nu|} = \int_\Omega |f| d|\nu|.$$

ii) Since X is an abstrac L^1-space, then it is order isometric to the space $L^1(\eta)$ for some positive measure $\eta : \mathcal{S} \to [0, \infty]$, defined on a measurable space $(\Lambda, \mathcal{S})$. Then we can assume that $X = L^1(\eta)$, whit its usual norm $||\cdot||_\eta$.

Let $B \in \mathcal{R}$. Then

$$||\nu(B)||_\eta = \int_\Lambda \nu(B)d\eta = \langle \nu(B), \chi_\Lambda \rangle = \langle \nu, \chi_\Lambda \rangle(B).$$

If $x_0^* = \chi_\Lambda \in X^*$, we obtain the desired assertion.

By (36) we have that for $B \in \mathcal{R}$

$$||\nu(B)||_X = ||\nu||(B) = |\nu|(B) = \langle \nu, x_0^* \rangle(B).$$

It is follows that $|\nu|$ and $\langle \nu, x_0^* \rangle$ coincide on $\mathcal{R}$. Therefore $|\nu|(A) = |\langle \nu, x_0^* \rangle|(A)$, $\forall A \in \mathcal{R}^{loc}$. From i) we get the equalities. $\square$

Proposition 23. *The space $L^1(\nu)$ is lattice isomorphic to an abstract L^1-space if and only if $L^1(|\nu|) = L^1(\nu)$. In that case $L^1(\nu) = L^1_w(\nu)$.*

Proof. Let us assume that $L^1(\nu)$ is lattice isomorphic to an abstract L^1-space Z. Take $T : L^1(\nu) \to Z$ the lattice isomorphism and consider $f \in L^1(\nu)^+$ and $A \in \mathcal{R}^{loc}$. Notice that $f\chi_A \in L^1(\nu)^+$ and $T(f\chi_A) \in Z^+$. Let us defined the function η_f on $\mathcal{R}^{loc}$ as $\eta_f(A) := T(f\chi_A)$. Therefore η_f is a positive vector measure. By Example 7 we have $|\eta_f|(\Omega) < \infty$. Since T is an isomorphism it turns out that

$$||\tilde{\nu}_f(A)||_X = \left|\left|\int_\Omega f\chi_A d\nu\right|\right|_X \leq ||f\chi_A||_\nu$$
$$\leq ||T^{-1}|| \, ||T(f\chi_A)||_Z = ||T^{-1}|| \, ||\eta_f(A)||_Z \leq ||T^{-1}|| \, |\eta_f|(A).$$

From Proposition 7 we obtain that $|\tilde{\nu}_f|(A) \leq ||T^{-1}|| \, |\eta_f|(A)$, $\forall A \in \mathcal{R}^{loc}$. So $|\tilde{\nu}_f|(\Omega) < \infty$. By Theorem 13 we conclude that $f \in L^1(|\nu|)$. Then $L^1(\nu)^+ \subset L^1(|\nu|)$, and so $L^1(\nu) \subset L^1(|\nu|)$. The other containment is always true. Accordingly $L^1(\nu) = L^1(|\nu|)$.

As $L^1(|\nu|)$ is an abstract L^1-space the other implication holds.

Since $L^1(|\nu|)$ is σ-order continuous and has the σ-Fatou property, it turns out that $L^1(\nu)$ is too. Hence $L^1(\nu) = L^1_w(\nu)$ [5, Prop. 5.4]. $\square$

Proposition 24. *The space $L^1(\nu)$ is an abstract L^1-space if and only if $L^1(|\nu|) \equiv L^1(\nu)$.*

Proof. Let us assume that $L^1(\nu)$ is an abstract L^1-space. From the previous proposition we have that $L^1(\nu) = L^1(|\nu|)$. Now consider the vector measure $[\nu] : \mathcal{R} \to L^1(\nu)$ introduced in the Example 6. From this and the Example 7 it follows that $|\chi_B|_\nu = |\chi_B|_{[\nu]} = |[\nu]|(B)$. On the other hand $||\nu(B)||_X \leq ||\nu||(B) = ||\chi||_\nu$, by Proposition 7 it turns out that

$|\nu|(B) \le \|\chi_B\|_\nu$. Therefore $\|\chi_B\|_{|\nu|} = |\nu|(B) = \|\chi_B\|_\nu, \forall B \in \mathcal{R}$. Since $L^1(\nu)$ is an abstract L^1-space we have

$$|s|_{\||\nu\||} = \|s\|_\nu, \ \forall \ s \in S(\mathcal{R}). \tag{37}$$

As $S(\mathcal{R})$ is a dense subspace in both $L^1(\nu)$ and $L^1(|\nu|)$, from (37) the desired assertion is obtained. $\qquad\square$

Corollary 7. *If* $L^1(\nu) = L^1(|\nu|)$, *then* $L^1(\nu_f) = L^1(|\nu_f|), \ \forall \ f \in L^1_{loc}(\nu)$.

Proof. Since $L^1(|\nu_f|) \subset L^1(\nu_f)$, it only remains to prove the other containment. Take $s \in S(\mathcal{R})$, from (19) we have that

$$\int_A sd|\nu_f| = \int_A |f|sd|\nu|, \ \forall \ A \in \mathcal{R}^{loc}. \tag{38}$$

Now take $g \in L^1(\nu_f)$ and a sequence $\{s_n\} \in S(\mathcal{R})$ such that $0 \le s_n \uparrow |g|$, ν_f-a.e. By Theorem 17 ii) $fg \in L^1(\nu)$; Moreover, $|f|s_n \uparrow |fg|$, ν-a.e. By the Monotone Convergence Theorem and (38),

$$\int_\Omega |g|d|\nu_f| = \lim_{n\to\infty} \int_\Omega s_n d|\nu_f| = \lim_{n\to\infty} \int_\Omega |f|s_n d|\nu| = \int_\Omega |fg|d|\nu|.$$

From the hypothesis we obtain that $L^1(\nu_f) \subset L^1(|\nu_f|)$. $\qquad\square$

In contraction to the integral operators associated with vector measures defined in σ-algebras, for the integral operators with respect to measures defined in δ-rings, the condition that the variation is finite is not necessary in order to the integral operator is compact [19, Thm. 1], as we can see in the following example.

Example 8. Let us consider the measure space (Ω, Σ, μ) where $\Omega = \mathbb{N}$, $\Sigma = 2^\Omega$ and μ the counting. Let $\mathcal{R} := \{B \subset \mathbb{N} : \mu(B) < \infty\}$ and define $\nu : \mathcal{R} \to \mathbb{R}$ by

$$\nu(A) := \mu(A), \ \forall \ A \in \mathcal{R}.$$

Then $\mathcal{R}^{loc} = \Sigma$ and $L^1(\nu) = l^1$. Observe that $|\nu(B)| = \mu(B), \ \forall \ B \in \mathcal{R}$. Which implies that $\||\nu\||(\Omega) = \infty$, so $|\nu|(\Omega) = \infty$. On the other hand $I_\nu : l^1 \to \mathbb{R}$ is a finite range operator and so it is a compact operator.

However, even without imposing conditions on the variation of ν, the following result still holds.

Theorem 18. *Let* $\nu : \mathcal{R} \to X$ *be a vector measure. If* $I_\nu : L^1(\nu) \to X$ *is a compact operator, then* $L^1(\nu) = L^1(|\nu|)$.

Proof. Since always we have $L^1(|\nu|) \subset L^1(\nu)$, it only remains to prove the other containment. Take $f \in L^1(\nu)$ and consider its associated vector measure $\nu_f : \mathcal{R} \to X$. By Theorem 17 we have $I_{\nu_f} = I_\nu \circ M_f$, where the operator $M_f : L^1(\nu_f) \to L^1(\nu)$, defined by $M_f(g) = fg$, is continuous. As I_ν is a compact operator it turns out that I_{ν_f} is compact. From Lemma 7, $|\nu_f| = |\tilde{\nu}_f|$. Since $\tilde{\nu}_f$ is defined on a σ-algebra, its variation is finite [19, Thm. 4]. Having in mind Theorem 13, we conclude that $f \in L^1(|\nu|)$. $\square$

Corollary 8. *Let $\nu : \mathcal{R} \to X$ be a vector measure. If $L^1(\nu)$ is reflexive and $\dim L^1(\nu) = \infty$, then the integral operator I_ν is not compact.*

Proof. Since $L^1(|\nu|)$ is reflexive only in the case of its dimension is finite [12, Ch. IV Ex. 13.2], from the hypothesis we have $L^1(\nu) \neq L^1(|\nu|)$. Then by previous theorem we obtain that I_ν is not compact. $\square$

From the theory of operators we know that if a linear operator is compact, then it is completely continuous. When ν is defined in a σ-algebra, in [20, p. 153] it is proved that completely continuous I_ν implies that the range $\mathbf{R}(\nu)$ of ν is relatively compact. In the following example we see that this does not necessarily happen when ν is defined in a δ-ring.

Example 9. Let $\mathcal{R}$ and $\nu : \mathcal{R} \to \mathbb{R}$ be as in Example 8. Since $||\nu||(\Omega) = \infty$, $\mathbf{R}(\nu) \subset \mathbb{R}$ is an unbounded set. Hence $\mathbf{R}(\nu)$ is not relatively compact set. However I_ν is a compact operator.

4.4. *When measure ν is σ-finite*

Considering the Propositions 7 iii) and 11 it follows that if $|\nu|$ is σ-finite, in particular if $|\nu|(\Omega) < \infty$, then the measure ν is σ-finite. However, the fact that the semi-variation of a vector measure ν is bounded does not guarantee that ν is σ-finite [8, Ex. 2.2].

It is well known that if X does not contain a copy of c_0 and $||\nu||$ is bounded, then ν is strongly additive and, consequently, ν is σ-finite [9, p. 36]. On the other hand, even though $c_0 \subset X$, if ν is bounded and X^* is separable, the same conclusion is obtained, as stated below.

Proposition 25. *Let $\nu : \mathcal{R} \to X$ be a vector measure. If $||\nu||(\Omega) < \infty$ and X^* is separable, then ν is σ-finite.*

Proof. Since X^* is separable, we have that X is too. So, let us take a dense subset $\{x_n^*\} \subset X$. As $||\nu||(\Omega) < \infty$, then $|\langle \nu, x_n^* \rangle|(\Omega) < \infty$,

$\forall\ n \in \mathbb{N}$. Having in mind the Proposition 7 iii), for each $n \in \mathbb{N}$, take $\{A_{n,k}\} \subset \mathcal{R}$ and $N_n \in \mathcal{N}_0(|\langle \nu, x_n^* \rangle|)$ such that $\Omega = \bigcup_{k=1}^{\infty} A_{n,k} \cup N_n$. Then $\Omega = \bigcup_{n=1}^{\infty} \bigcup_{k=1}^{\infty} A_{n,k} \cup N$, where $N := \bigcap_{n=1}^{\infty} N_n$.

In order to establish the conclusion it only remainds to prove that $N \in \mathcal{N}_0(\nu)$. Fix $x^* \in X$, $\{x_{n_j}^*\} \subset \{x_n^*\}$ such that $x_{n_j}^* \to x^*$ and $B \in \mathcal{R}_N$. Since

$$\langle \nu, x^* \rangle(B) = \lim_{j \to \infty} \langle \nu, x_{n_j}^* \rangle(B),$$

and $\langle \nu, x_{n_j}^* \rangle(B) = 0$, $\forall\ j \in \mathbb{N}$, we have $\langle \nu, x^* \rangle(B) = 0$. It follows that $N \in \mathcal{N}_0(\langle \nu, x^* \rangle)$, $\forall\ x^* \in X$. Then $||\nu||(N) = 0$. $\qquad\square$

Proposition 26. Let $\nu : \mathcal{R} \to X$ be a vector measure. If $L^1(\nu) = L^1(|\nu|)$, then ν is σ-finite if and only if its variation is σ-finite.

Proof. Let us assume that ν is a σ-finite vector measure and take $\{B_n\} \subset \mathcal{R}$ and $N \in \mathcal{N}_0(\nu)$ such that $\Omega = \bigcup_{n=1}^{\infty} B_n \cup N$. Then $\chi_{B_n} \in S(\mathcal{R}) \subset L^1(\nu) = L^1(|\nu|)$. This implies that $|\nu|(B_n) < \infty$, $\forall\ n \in \mathbb{N}$. By Proposition 11 we have $\mathcal{N}_0(|\nu|) = \mathcal{N}_0(\nu)$. So, $|\nu|(N) = 0$. Therefore $|\nu|$ is a σ-finite measure.

As we have observed the other implication always holds. $\qquad\square$

The following example shows that it is not enough the equality $L^1(\nu) = L^1(|\nu|)$ to guarantee that the vector measure ν is σ-finite.

Example 10. Let Γ be an uncountable set and $\mu : 2^\Gamma \to [0, \infty]$ the counting measure. Let us consider the δ-ring $\mathcal{R} := \{B \subset \Gamma : \mu(B) < \infty\}$ and the Banach space $X := l^1(\Gamma)$. Define $\nu : \mathcal{R} \to X$ by $\nu(A) = \chi_A$, $\forall\ A \in \mathcal{R}$. It turns out that $\mathcal{R}^{loc} = 2^\Gamma$ and $\mathcal{N}_0(\nu) = \{\emptyset\}$. Moreover, since $||\nu(B)||_{l^1(\Gamma)} = \mu(B)$, $\forall\ B \in \mathcal{R}$, Proposition 7 tells us that $|\nu| = \mu$. Then $L^1(|\nu|) = l^1(\Gamma) = L^1(\nu)$ [8, Ex. 2.2]. However ν is not σ-finite.

Recall that a Banach space X is *weakly compactly generated* if there exists $K \subset X$ weakly compact such that the space spanned by K is dense in X.

By considering a classical vector measure G. P. Curbera proved that the space $L^1(\nu)$ is a weakly compactly generated space [6, Thm. 2]. In the following example we will see that this can no longer be valid when considering a vector measure defined in a δ-ring.

Example 11. Let us consider the hypothesis of the Example 10 again. We have seen that $L^1(\nu) = l^1(\Gamma)$. We will show that $l^1(\Gamma)$ is not weakly compactly generated space.

Since Γ is an uncountable set, we have that $l^1(\Gamma)$ is not separable. Now consider a weakly compact subset $K \subset l^1(\Gamma)$. As in $l^1(\Gamma)$ weak convergence and convergence in norm are equivalents [7, Cor. II.2.2], it follows that K is a compact set. Then K is separable. Hence K cannot be total. Since K is arbitrary, we conclude that $L^1(\nu) = l^1(\Gamma)$ is not weakly compactly generated.

On the other hand, the following is true.

Proposition 27. *If ν is a σ-finite vector measure, then $L^1(\nu)$ is a weakly compactly generated space.*

Proof. Let assume that ν is a σ-finite vector measure. By Theorem 15, $L^1(\nu)$ is lattice isometric to $L^1(\nu_g)$, where ν_g is a vector measure defined on a σ-algebra. Since $L^1(\nu_g)$ is weakly compactly generated space it turns out that $L^1(\nu)$ is weakly compactly generated space. $\qquad\square$

Acknowledgments

The reviewers are thanked for their invaluable comments.

References

[1] Y. A. Abramovich and C. D. Aliprantis, *An Invitation to Operator Theory.* Providence, Graduate Studies in Math., Vol. 50, AMS, 2002.

[2] C. Bennett and R. Sharpley, *Interpolation of operators.* Boston, Academic Press, 1988.

[3] J. K. Brooks and N. Dinculeanu, 'Strong additivity, absolute continuity and compactness in spaces of measures', *J. Math. Anal. Appl.* **45**(1974), 156-175.

[4] J. M. Calabuig, O. Delgado and E. A. Sánchez Pérez, 'Factorizing operators on Banach function spaces through spaces of multiplication operators'. *J. Math. Anal. Appl.* **364**(2010), 88-103.

[5] J. M. Calabuig, O. Delgado, M. A. Juan and E. A. Sánchez-Pérez, 'On the Banach lattice structure of L_w^1 of a vector measure on a δ-ring'. *Collect. Math.* **65**(2014), 67-85.

[6] G. P. Curbera, 'Operators into L^1 of a vector measure and applications to Banach lattices', *Math. Ann.* **293**(1992), 317-330.

[7] M. M. Day, *Normed Linear Spaces.* Berlin-Heidelberg-New York, Springer-Verlag, 1973.

L^1-space of Vector Measures Defined on δ-rings   261

[8] O. Delgado, 'L^1-spaces of vector measures defined on δ-rings'. *Arch. Math* **84**(2005), 432-443.

[9] O. Delgado, *Further Developments on L^1 of a Vector Measure*. Ph. D. Thesis, University of Sevilla, 2004.

[10] J. Diestel and J. J. Uhl Jr., *Vector Measures*. Providence, R.I. Amer. Math. Soc., 1977.

[11] N. Dinculeanu, *Vector measures*. Oxford, Pergamon Press, 1953.

[12] N. Dunford and J. T. Schwartz, *Linear Operators, Part I General Theory*. New York, Interscience Publishers, Inc., 1958. Pure and Applied Mathematics, Vol. VII.

[13] E. Jiménez Fernández, M. A. Juan and E. A. Sánchez-Pérez, 'The Köthe dual of an Abstract Banach Lattice', *J. Funct. Spaces Appl.* (2013).

[14] M. A. Juan, *Vector measures on δ-rings and representations theorems of Banach lattices*. Ph. D. Thesis, Univ. Politec. of Valencia, 2011.

[15] D. R. Lewis, 'On integrability and summability in vector spaces'. *Illinois J. Math* **16**(1972), 294-207.

[16] J. Lindenstrauss and L. Tzafriri, *Classical Banach Spaces II*. Berlin-Heidelberg-New York, Springer-Verlag, 1979.

[17] P. R. Masani and H. Niemi, 'The integration theory of Banach space valued measures and the Tonelli-Fubini theorems. I. Scalar-valued measures on δ-rings'. *Adv. Math.* **73**(1989), 204-241.

[18] P. R. Masani and H. Niemi, 'The integration theory of Banach space valued measures and the Tonelli-Fubini theorems. II. Pettis integration'. *Adv. Math.* **75**(1989), 121-167.

[19] S. Okada, W. J. Ricker and L. Rodríguez-Piazza, 'Compactness of the integration operator associated with a vector measure'. *Studia Math.* **120** (**2**)(2002), 133-149.

[20] S. Okada, W. J. Ricker and E. A. Sánchez-Pérez, *Optimal Domain and Integral Extension of Operators*. Basel-Boston-Berlin, Birkhauser, 2008.

[21] H. H. Schaefer, *Banach Lattices and Positive Operators*. Berlin-Heidelberg-New York, Springer-Verlag, 1974.

[22] A. Zaanen, *Integration*. Amsterdam, North-Holland Publishing Company, 1967.

[23] A. Zaanen, *Riesz Spaces II* Amsterdam-New York-Oxford, North-Holland Publishing Company, 1983.

© 2025 World Scientific Publishing Company
https://doi.org/10.1142/9789819812202_0013

Chapter 13

More on Unified Approach to Integration

Mangatiana A. Robdera

*Department of Mathematics, University of Botswana,
Private Box 0022, Gaborone, Botswana*

1. Introduction

Integral theory is a broad subject, which has many influences on the development of many mathematical areas of research and their applications. It has a long history that can be traced back to two thousand years ago. The modern form of the integration theory began with Newton and Leibniz in the seventeenth century. A foundation of the classical modern integration theory was laid by Riemann in the nineteenth century. A systematic study of the theory really started in early twentieth century with the work of Lebesgue who established the well known Lebesgue integral. Such a theory has since dominated the mathematics arena. The Lebesgue's approach is a difficult one and the definition of the Lebesgue integral requires a considerable amount of measure theory. In addition, the Lebesgue integral does not integrate the derivatives as the Newton calculus does. The (special) Denjoy (1912) and the Perron (1914) integrals gave generalizations of the Lebesgue integral that can recover a function from its pointwise finite derivative. Unfortunately, both the Denjoy and Perron integrals are difficult to handle. In 1957/1958, Henstock and Kurzweil independently gave a Riemann-type approach to the Denjoy-Perron integral which is simpler and easier to work with.

There are very sound reasons to want to generalize the theory of integration to vector valued functions and (even) vector valued integrators. The comparison among the generalizations of the Lebesgue integral to the vector valued settings is one of the most fruitful areas of research in the modern theory of integration. The Lebesgue-Bochner integral and the Pettis integral have been the preeminent and dominant notions in this context [1, 3, 5, 6, 8, 9, 17]. These kinds of integral integrate vector valued functions against a scalar valued integrator. On the other hand, the Bartle-Dunford-Schwartz integral [5], subsequently generalized by Dobrakov (see for example [12]), integrates scalar functions with respect to vector measure. The traditional formulations of the above-mentioned approaches to integration run into practical difficulties mainly because they all require the initial development of a workable measure theory before any useful results for the integral can be obtained. Alternative approaches avoiding such difficulties are available: each one of the Daniell [4], the Birkohoff [2, 8], the McShane [7] and the Henstock-Kurzweil [10] integrals, is a type of integration that more or less generalizes the simpler approach of the Riemann integral.

The classical integrals have a multitude of different properties; some of which are remarkably complex with lots of intricate details. The integrals devised for special purposes frequently are designed so as to be stronger than the classical integration procedures with regard to some chosen properties. In emphasizing one property heavily, it frequently happens that other properties are lost. However, the saving feature is that all of these assorted approaches have strong resemblances. According to E. H. Moore dictum, the existence of strong similarities in the central part of different theories indicates the existence of a more general theory of which these different theories are all special cases. A relatively satisfactory approach in trying to unify the different integration theories was introduced in [13] (see also [15]) for integrals of normed vector space valued functions with respect to normed vector space valued integrators.

My purpose in this note is to show that with a little effort, one can provide a treatment of the subject that is not only unifying but also more intuitive and technically straightforward. Our approach is based on a slight alteration of the ineffable mathematical ingenuous of the Riemann approach which still remains, in my view, the quintessence of the integration theory. Our approach, not only strengthen the content of integration theory, but also provides a thread tying the subject matter together. There is no attempt to be encyclopedic. We will only give a glimpse overview of our approach, and we have tried to make the subject accessible for an audience which is primarily interested in and familiar in Riemann integration theory in the more general setting of topological vector spaces.

As an application, we show that such an approach is flexible enough to further generalize the spectral theorem on Hilbert spaces. Likewise, this approach can be used to push stochastic integration towards the more general setting of integrable processes taking values in topological vector spaces, further strengthening and simplifying the results obtained in [14].

Throughout this note, $\mathbb{F}$ will be one of the scalar fields $\mathbb{R}$ or $\mathbb{C}$. All vector spaces mentioned here are over $\mathbb{F}$. All unexplained notation and terminology can be found in our standard references [1, 5, 6, 11].

2. Classical Integrals

In this section, we give a brief review of the classical integrals. We shall denote by $I(a, b)$ any interval with end points $a < b$ and by $\ell(I(a, b)) = b - a$ its length. A partition of $[a, b]$ is a collection $P = \{I(a_i, b_i) : i = 1, \ldots, n\}$ of finitely many pairwise disjoint subintervals $I(a_i, b_i)$ whose union is $[a, b]$. A partition P is said to be tagged if for each i, a number t_i is chosen from $I(a_i, b_i)$. Given a function $f : [a, b] \to X$, where X is a normed vector space, and a tagged partition P of $[a, b]$, the Riemann sum of f at P is the number $Sf(P) := \sum_{i=1}^{n} \ell(I(a_i, b_i)) f(t_i)$.

Riemann Integral. A vector $A \in X$ is the Riemann-integral of $f :$ $[a, b] \to X$ if for every $\varepsilon > 0$, there is a constant $\delta_\varepsilon > 0$ such that if P is any tagged partition of $[a, b]$ satisfying $[a_i, b_i] \subset (t_i - \delta_\varepsilon, t_i + \delta_\varepsilon)$, $\|Sf(P) - A\| < \varepsilon$.

A more general definition of the integral is obtained if one allows δ_ε to be any continuous positive function $\delta : [a, b] \to (0, \infty)$ called a *gauge function*. A tagged partition P of $[a, b]$ is said to be δ-fine if $0 < b_i - a_i < \delta(t_i)$ for $i = 1, \ldots, n$. The existence of a δ-fine partition is guaranteed by the *Nested Intervals Theorem*.

Henstock-Kurzweil Integral. A vector $A \in X$ is the Henstock-Kurzweil integral of a function $f : [a, b] \to X$ if for every $\varepsilon > 0$, there is a function $\delta : [a, b] \to (0, \infty)$ such that if P is a δ-fine tagged partition of $[a, b]$ satisfying $[a_i, b_i] \subset (t_i - \delta(t_i), t_i + \delta(t_i))$, one has $\|Sf(P) - A\| < \varepsilon$ (see for example [1]).

McShane Integral. If we drop the requirement that for $(I(a_i, b_i), t_i) \in P$, $t_i \in I(a_i, b_i)$ in the Henstock-Kurzweil integral, then the above defined vector A is the McShane integral of f over the interval $[a, b]$ (see for example [1]).

Given a σ-algebra Σ of subsets of a non-empty set Ω, a set function $\mu : \Sigma \to [0, \infty]$ is called a *measure* if it satisfies: $\mu(\emptyset) = 0$ and $\mu(\bigcup_{k=1}^{\infty} E_k) = \sum_{k=1}^{\infty} \mu(E_k)$ for every collection $\{E_k : k \in \mathbb{N}\}$ of pairwise disjoint sets in Σ.

A function $f : \Omega \to X$, where X is a vector space, is said to be *simple* if it has the form $f = \sum_{k=1}^{n} 1_{S_k} a_k$, where the a_k are distinct vectors in

X, $S_k \in \Sigma$, and $\mu(S_k) < \infty$ whenever $a_k \neq 0$, then the Lebesgue-Bochner integral of f over a set $E \in \Sigma$ is defined by $\int_E f d\mu = \sum_{k=1}^n \mu(S_k \cap E) a_k$ (see for example [5]).

A function $f : \Omega \to X$ is said to be Σ-*measurable*, if there exists a sequence of Σ-simple functions $\{s_n\}$ such that $s_n \to f$ and $\|s_n\| \nearrow \|f\|$ pointwise μ-almost everywhere.

Lebesgue-Bochner integral. A Σ-measurable function $f : \Omega \to X$ is Lebesgue-Bochner integrable over a set $E \in \Sigma$ if there exists a sequence of simple functions $\{s_n\}$ such that $\lim_{n\to\infty} \int_E \|f - s_n\| d\mu = 0$, where the integral is an ordinary Lebesgue integral. In this case, the Lebesgue-Bochner integral of f is defined by $\int_E f d\mu = \lim_{n\to\infty} \int_E s_n d\mu$ (see for example [5]).

Given an algebra $\mathcal{A}$ of subsets Ω and a Banach space X, a *additive vector measure* (or a *vector measure*, for short) is a function $\mu : \mathcal{A} \to X$ such that for any two disjoint sets A and B in $\mathcal{A}$, one has $\mu(A \cup B) = \mu(A) + \mu(B)$.

A vector measure $\mu : \mathcal{A} \to X$ is called *countably additive (or σ-additive)* if for any sequence $\{A_k\}$ of disjoint sets in $\mathcal{A}$ such that $\bigcup_k A_k \in \mathcal{A}$, it holds that $\mu(\bigcup_{k=1}^\infty A_k) = \sum_{k=1}^\infty \mu(A_k)$, with the series on the right-hand side convergent in the norm of the Banach space X.

It can be proved that an additive vector measure $\mu : \mathcal{A} \to X$ is countably additive if and only if for any sequence $\{A_k\}$ as above one has $\lim_{n\to\infty} \|\mu(\bigcup_{k=1}^\infty A_k)\| = 0$.

The integral of Σ-simple functions of the form $f = \sum_{k=1}^n a_k 1_{S_k}$ where the a_k are scalars and $S_k \in \Sigma$ is defined by $\int_E f d\mu = \sum_{k=1}^n a_k \mu(S_k \cap E)$.

Integration with respect to vector measures. A measurable function $f : \Omega \to \mathbb{F}$ or $[-\infty, \infty]$ is integrable over $E \in \Sigma$ if there exists a sequence $\{s_n\}$ of simple functions that converges μ-almost everywhere to f and the sequence $\{\int_E s_n d\mu\}$ converges in norm in X. In this case, the integral of f over E is defined by $\int_E f d\mu = \lim_{n\to\infty} \int_E s_n d\mu$ (see for example [5, 6]).

It should be clear that the McShane integrability implies the Henstock-Kurzweil integrability. It is known that for functions taking values in a finite dimensional vector space, the McShane integral is equivalent to the Lebesgue-Bochner integral.

3. Definition of Convergence

In this section, we will discuss a notion that is central to any approach to integrals. An essential and integral component of the different approaches to classical integrals is the limiting process. The central concept behind any notion of limit is the idea of arbitrary closeness, nearness, or convergence.

Unified Approach to Integration 267

We are used to talking about convergence of sequences, convergence of a functions, or convergence of nets. We believe that actually, the notion of convergence could and should be defined solely in terms of the topology of the range space.

Definition 1. An infinite subset A of a topological space X *converges to* $x_0 \in X$, which we denote $A \to x_0$, (or $\lim A = x_0$) if every neighborhood of x_0 contains all but finitely many elements of A.

Such a definition is motivated by the definition of another topological notion: a cluster point. Recall that in a topological space, a point x_0 is said to be a *cluster point* (or an *accumulation point*) of an infinite subset A if every neighborhood of x_0 contains infinitely many elements of A. Clearly, a point of convergence of a set A is an accumulation point of A, but there is a clear distinction between these two notions as evidenced by the classical example: 1 is an accumulation point for the subset $\left\{ (-1)^n \frac{n}{n+1} : n \in \mathbb{N} \right\}$ of the real numbers $\mathbb{R}$ but is not a point of convergence. It is easy to see that if the topological space is Hausdorff, an infinite convergent set can only converge to a unique element.

Definition 1 puts in a conspicuous position the topological nature of the notion of convergence. Such a treatment of convergence appeals to us as one which can readily fit to any course in analysis. For example, the above definition of convergence is closely related to the all-important topological concept of compactness. Indeed, it can be shown that a subset A of a topological space is compact if and only if every infinite subset of A admits a subset that converges to an element in A. Many other results related to compactness can be expressed in terms of the above definition of convergence [16], but at the moment we are not going into that direction.

A well known criterion for convergence that has many virtuous applications in metric spaces is the so-called *Cauchy criterion*. The Cauchy criterion can be naturally extended to the more general setting of topological vector spaces. Recall that a topological vector space is a vector space assigned a Hausdorff topology with respect to which the vector operations are continuous.

Definition 2. An infinite subset A of a Hausdorff topological vector space X *is Cauchy* if for every neighborhood $\mathcal{O}$ of 0, there exists a finite subset $N_{\mathcal{O}}$ of A such that $x - y \in \mathcal{O}$ for all $x, y \in A \setminus N_{\mathcal{O}}$. We then say that a closed infinite subset A is *(topologically) complete* if every Cauchy subset of A is convergent to an element in A.

Note that the above definition is slightly different from the conventional definition of completeness: *A subset A of a metric space is complete if every Cauchy sequence in A converges to an element in A.* In the setting of a topological vector space, we refer to this latest as the *sequential completeness*. Obviously, (topological) completeness implies sequential completeness. It turns out that in fact, the two notions are equivalent.

Theorem 1. *In a Hausdorff topological vector space, (topological) completeness and sequential completeness are equivalent properties.*

Proof. Assume that a Hausdorff topological vector space X is sequentially complete. Let A be an infinite Cauchy subset of X. Fix a neighborhood $\mathcal{O}$ of 0 in X. Then, there exists a finite subset N_1 of A such that $x - y \in \mathcal{O}$ for all $x, y \in A \setminus N_1$. Fix $x_1 \in A \setminus N_1$ and consider $\mathcal{O}_1 = \frac{1}{2}\mathcal{O}$. Then, there exists a finite subset N_2 of $A \setminus N_1$ such that $x - y \in \mathcal{O}_1$ for all $x, y \in A \setminus N_1 \cup N_2$. Fix $x_2 \in A \setminus N_1 \cup N_2$ and consider $\mathcal{O}_2 = \frac{1}{3}\mathcal{O}$. Continuing in this manner, we choose successively $x_1, x_2, \ldots$ in A, and a sequence of neighborhoods of 0: $\mathcal{O}_1 \supset \mathcal{O}_2 \supset \ldots$ such that $x_n \in A \setminus \bigcup_{k=1}^{n} N_k$ and that $x_n - x_{n+1} \in \mathcal{O}_n$ for all $n \in \mathbb{N}$. Clearly, $\{x_n : n \in \mathbb{N}\}$ form a Cauchy sequence of elements of A. So it converges to some element $a \in A$. Now, pick n large enough so that $x_n - a \in \frac{1}{2}\mathcal{O}$. If $x \in A \setminus \bigcup_{k=1}^{n} N_k$, then $x - x_n \in \frac{1}{n}\mathcal{O}$. So $x - a = x - x_n + x_n - a \in \mathcal{O}$, completing the proof. $\square$

4. Elements of Integration

In this section, we establish the main framework of our approach to integration and try to separate the essential parts of the different approaches.

Semirings of subsets. In what follows, we shall denote by 2^{Ω} the power set of a nonempty set Ω, that is, the collection of all subsets of Ω. We say that a collection $\mathcal{S} \subset 2^{\Omega}$ of sets is a *semiring* if it satisfies the following properties:

(s1) $\emptyset \in \mathcal{S}$;
(s2) If $E, F \in \mathcal{S}$, then $E \cap F \in \mathcal{S}$;
(s3) If $E, F \in \mathcal{S}$, then there exists a finite number of mutually disjoint elements $C_1, C_2, \ldots, C_n \in \mathcal{S}$ such that $E \setminus F = \bigcup_{k=1}^{n} C_k$.

An example of a semiring of sets is the collection of half-open, half-closed real intervals $[a, b) \subset \mathbb{R}$. The power set 2^{Ω} of any non-empty set Ω is obviously a semiring. It turns out that the simplest family of sets on which a "reasonable" theory of integration can be built, is a semiring. However, there are natural collections of sets that satisfy other properties that are stronger than those of a semiring. They provide us with more examples of semirings. For example, it easily to check that:

- An *algebra of sets,* that is, a collection $\mathcal{A}$ of subsets of a nonempty sets Ω that is closed under finite intersection and under complementation, is a semiring.
- A *ring of sets,* that is, a collection $\mathcal{R}$ of subsets of a nonempty sets Ω that closed under finite union, and under complementation, is a semiring.

Another useful collection of sets is that of σ-*algebra*. A collection $\mathcal{A}$ of sets is said to be a σ-*algebra* if it is an algebra and if it satisfies: $\bigcup_{n\in\mathbb{N}} A_n \in \mathcal{A}$ for every sequence of sets $\{A_n\}$ in $\mathcal{A}$. Thus, any σ-algebra is also an example of a ring, and hence a semiring.

We have noticed that topological or metric properties in the domain space Ω are quite irrelevant in our definition of the integral. However, when there is a topology on Ω, another interesting example of a semiring can be given.

We generalize the idea of intervals of real numbers to the more general setting of arbitrary topological spaces as follows.

Definition 3. An *interval* in a topological space Ω is any connected set with nonempty interior. We shall denote by $\mathcal{I}(\Omega)$ the subset of the powerset 2^Ω consisting of all intervals in Ω.

It is easy then to check that if E and F are two intervals in Ω, then either $E \cap F = \emptyset$ or $E \cap F$ is also an interval. Thus, a collection of intervals that satisfies property 4 is indeed a semiring. The collection of left-open, right-closed real intervals $[a, b) \subset \mathbb{R}$ is an example of such a semiring.

Integrands and Integrators. There are two main parts in an integral: the *integrand,* that is, the part that has to be integrated, and the *integrator,* the part against which we do the integration. Both parts may take (extended) real numbers, or complex numbers, or even vector values.

In our approach, we shall simply agree with the following.

Definition 4. By an *integrand,* we are referring to a function $f : \Omega \to X$ from a nonempty set Ω taking values in a given vector space X.

Note that in the above definition, we do not require any topology neither on the domain space Ω nor on the vector space X.

Recall that a set function μ from a semiring $\mathcal{S}$ of subsets of a nonempty set Ω into a vector space Y is said to be *additive* if for every pairwise disjoint finite sequence $A_1, A_2, ..., A_n \in \mathcal{S}$, with $\bigcup_{i=1}^{n} A_i \in \mathcal{S}$, we have $\mu(\bigcup_{i=1}^{n} A_i) = \sum_{i=1}^{n} \mu(A_i)$.

One of the most conspicuous advantage of the Lebesgue integral, as contrasted with the Riemann integral, is the superiority of the convergence theorems associated with it. These theorems in turn rest on the countable additivity of the Lebesgue measure (integrator) λ, that is, the property that for every pairwise disjoint infinite sequence $A_1, A_2, ... \in \mathcal{S}$, with $\bigcup_{i=1}^{\infty} A_i \in \mathcal{S}$, we have $\lambda(\bigcup_{i=1}^{\infty} A_i) = \sum_{i=1}^{\infty} \lambda(A_i)$. It would not have been unreasonable to believe that without the invention of the Lebesgue integral, finitely additive set functions would have lost all interest and importance. Our approach reinstates the usefulness of the finite additivity property.

Definition 5. Given a vector space Y, by a Y-valued *integrator* we mean a finitely additive set function $\mu : \mathcal{S} \to Y$, where $\mathcal{S}$ is a semiring of a nonempty set Ω. The triplet $(\Omega, \mathcal{S}, \mu)$ is then referred to as a Y-valued *integrator space*.

Here again, we make a point that no topology is required for the vector space Y.

Example 1. The length function is a real valued integrator on the semiring of intervals in $\mathbb{R}$. A nonnegative real valued Lebesgue measure defined on a σ-algebra is a integrator. More generally, an outer measure defines an integrator on the power set of a given nonempty set.

Subpartitions, Multiplicators and Riemann sums. Let $(\Omega, \mathcal{S}, \mu)$ be a given Y-valued integrator space, where Y is a vector space.

Definition 6. We shall call an $\mathcal{S}$-*subpartition* of Ω any finite collection π of pairwise disjoint elements of $\mathcal{S}$. We denote by $\bigsqcup \pi$ the subset of Ω obtained by taking the union of all E in π.

It is worth noticing that $\bigsqcup \pi$ is a subset of Ω. an $\mathcal{S}$-subpartition π is simply called an $\mathcal{S}$-*partition* if $\bigsqcup \pi = \Omega$. We denote by $\Pi(\Omega, \mathcal{S})$ the set of all $\mathcal{S}$-subpartitions of Ω.

An $\mathcal{S}$-subpartition π is said to be *tagged* if for each $E \in \pi$, one associates a point $t_E \in \bigsqcup \pi$ in such a way that $t_E \neq t_F$ whenever $E \neq F$. We use the notation $T(\pi) := \{t_E : E \in \pi\}$. In what follows, $\mathcal{S}$-subpartitions are always supposed to be tagged. If one requires each tagging point t_E to be an element of the tagged set E, then we say that the tagging of the S-subpartition is *conditional.*

One more bit of terminology is called for.

Definition 7. By *a multiplicator,* we simply mean a bilinear mapping $m : X \times Y \to Z$, where X, Y, Z are vector spaces.

Example 2. If $X = Y$ is an inner product space and $Z = \mathbb{F}$, then the inner product $m(u, v) = \langle u, v \rangle$ is a multiplicator on $X \times Y$. If X,Y are real normed spaces, let $\mathcal{L}(Y, X)$ be the normed space of bounded linear operators from Y to X. Then the mapping $m : \mathcal{L}(Y, X) \times Y \to X : m(A, x) = Ax$ is a multiplicator. If X,Y,Z are real normed spaces, the composition $m : \mathcal{L}(X, Z) \times \mathcal{L}(Y, X) \to \mathcal{L}(Y, Z)) : m(A, B) = AB$ is a multiplicator.

Definition 8. Let (Ω, S, μ) be Y-valued integrator space. Let X, Z be two vector spaces, and $f : \Omega \to X$ a function. Given a multiplicator $m : X \times Y \to Z$, an expression of the form

$$\sum_{E \in \pi, t_E \in E} m\left(\mu(E), f(t_E)\right) =: R_\mu^m(f, \pi) \tag{1}$$

is called an *m-Riemann sums* of f with respect to $\pi \in \Pi(\Omega, S)$. We shall call the set

$$R_\mu^m(f) := \left\{ R_\mu^m(f, \pi) : \pi \in \Pi(\Omega, S) \right\}$$

as the $\mu, \Updownarrow$-*Riemann set* of f.

Clearly, $R_\mu^m(f)$ is an infinite subset of Z.

Now that we have achieved our preparatory works, we are in a position to introduce our definition of the integral.

5. Definition of the integral

Definition 9. Let Y be a vector space and (Ω, S, μ) a Y-valued integrator space. Let $m : X \times Y \to Z$ be a multiplicator, where X is a linear space and Z a Hausdorff topological vector space. We say that a function $f : \Omega \to X$ is S, μ, m-*integrable* if $R_\mu^m(f)$, the μ, m-Riemann set of f, is convergent to an element in Z. The limit $\lim R_\mu^m(f)$ is then a vector in Z, called the S, μ, m-*integral* of f over the set Ω, and shall be denoted by $\int m(\mu, f)$.

In other words:

A function $f : \Omega \to X$ is $\mathcal{S}, \mu, m$-integrable if there is a vector $\int m(\mu, f)$ in Z with the property that for every neighborhood $\mathcal{N}$ of 0 in Z, there exists a finite subset $N(f)$ of $R_\mu^m(f)$ such that

$$R_\mu^m(f) \setminus N(f) \subset \int m(\mu, f) + \mathcal{N}.$$

Since the topology on Z is Hausdorff, the vector $\int m(\mu, f)$ is unique whenever it exists.

For the particular cases where either the integrand f, or the integrator μ, is scalar-valued, we always consider the multiplicator defined by either of the vector scalings $m(\mu(E), f(t_E)) = f(t_E)\mu(E)$ or $m(f(t_E), \mu(E)) = \mu(E)f(t_E)$. In either case, we simply denote the integral $\int m(\mu, f)$ by the standard notation $\int f d\mu$.

It is transparent that our approach does not require any knowledge of measure theory. In addition, it broadens the class of integrable functions large enough to contain most of the classical notions of integral.

Example 3. Consider $\Omega = [a, b]$ a bounded interval in the real number system. Let $\mathcal{S}$ be the semiring consisting of left-closed, right-open subintervals of $[a, b]$. Then, the length function ℓ is an integrator on $\mathcal{S}$. Since partitions and δ-fine partitions are subpartitions, the $\mathcal{S}, \ell$-integral extends the classical Riemann and Henstock-Kurzweil integrals over a bounded interval $[a, b]$.

We would like to highlight the following facts.

(1) One of the main virtues of our approach is that it does not require any topological or metric property on the domain space Ω.
(2) The use of subpartitions allows our approach to cover the case of improper integrals.
(3) Another benefit of our approach is the fact that the $\mathcal{S}, \mu, m$-integral is able to integrate functions that are not necessarily measurable, and on sets that are not necessarily measurable.
(4) Furthermore, our approach has the advantage that many desired classical properties of the integral follow immediately from the properties of limit and therefore their proofs are obtained at not cost at all.

Some useful characterizations of integrability are in order.

Proposition 1. *If f is $\mathcal{S}, \mu, m$-integrable, then for every neighborhood $\mathcal{N}$ of 0 in Z, there exists a finite subset $N(f)$ of $R_\mu^m(f)$ such that*

$$R_\mu^m(f) \setminus N(f) - R_\mu^m(f) \setminus N(f) \subset \mathcal{N}.$$

Proof. Fix a balanced neighborhood $\mathcal{N}$ of 0 in Z and let $\mathcal{U}$ a balanced neighborhood of 0 in Z such that $\mathcal{U}+\mathcal{U} \subset \mathcal{N}$. Let $N_0(f)$ be a finite subset of $R_\mu^m(f)$ such that $R_\mu^m(f) \setminus N_0(f) \subset \int m(\mu, f)+\mathcal{U}$. Let $R_\mu^m(f, \pi), R_\mu^m(f, \rho) \in R_\mu^m(f) \setminus N_0(f)$. Then,

$$R_\mu^m(f, \pi) - R_\mu^m(f, \rho) = R_\mu^m(f, \pi) - \int m(\mu, f) + \int m(\mu, f) - R^m(f, \rho)$$

$$\in \mathcal{U} + \mathcal{U} \subset \mathcal{N}.$$

The proof is complete. $\qquad\qquad\qquad\qquad\qquad\qquad\qquad\qquad\qquad\square$

The above proposition justifies the introduction of the following definition.

Definition 10. We say that a function $f : \Omega \to X$ satisfies the *Cauchy criterion for* $\mathcal{S}, \mu, m$-*integrability* if for every neighborhood $\mathcal{O}$ of 0 in Z, there exists a finite subset $N_\mathcal{O}(f) \subset R_\mu^m(f)$ such that for every $R_\mu^m(f, \pi), R_\mu^m(f, \rho) \in R_\mu^m(f) \setminus N_\mathcal{O}(f)$,

$$R_\mu^m(f, \pi) - R_\mu^m(f, \rho) \in \mathcal{O}.$$

Clearly, the Cauchy criterion for $\mathcal{S}, \mu, m$-integrability corresponds exactly to the Cauchy-ness of the set $R_\mu^m(f)$. Thus, if Z is complete, then $R_\mu^m(f)$ converges to an element $\int m(\mu, f)$ in Z; that is, f is $\mathcal{S}, \mu, m$-integrable. This can be stated formally as such:

Proposition 2. *Assume that* X, Y *are vector spaces and* Z *a complete Hausdorff topological vector space and let* $(\Omega, \mathcal{S}, \mu)$ *is a* Y-*valued integrator space. A function* $f : \Omega \to X$ *satisfies the Cauchy criterion for* $\mathcal{S}, \mu, m$-*integrability if and only if it is* $\mathcal{S}, \mu, m$-*integrable.*

6. Spaces of Integrable Functions

One of the disadvantages of the classical Henstock-Kurzweil integral is that, unlike the space of Lebesgue-Bochner integrable functions, there seems to be no known natural topology for the space of Henstock-Kurzweil integrable functions. In this section, we shall see that in fact, a linear topology can indeed be naturally defined for the space of integrable functions.

Let $(\Omega, \mathcal{S}, \mu)$ denote a fixed Y-valued integrator space where Y is given vector space, and $m : X \times Y \to Z$ a multiplicator where X is a vector space, and Z is Hausdorff a topological vector space. We notice that being a limit operator, the $\mathcal{S}, \mu, m$-integral is linear. Consequently, the set of all

$\mathcal{S}, \mu, m$-integrable functions $f : \Omega \to X$ is quickly seen to be a linear space. We shall denote such a space by $\mathcal{I}^m (\Omega, \mathcal{S}, \mu, X)$.

If $A \subset \Omega$, we say that a function $f : \Omega \to X$ is $\mathcal{S}, \mu, m$-integrable over A if $f \cdot 1_A$ is $\mathcal{S}, \mu, m$-integrable. For simplicity, we denote by $\mathcal{I}^m (A, \mathcal{S}, \mu, X)$ the subset of $\mathcal{I}^m (\Omega, \mathcal{S}, \mu, X)$ consisting of functions $\mathcal{S}, \mu, m$-integrable over A, and we define

$$\int_A m(\mu, f) := \int m(\mu, f \cdot 1_A).$$

Standard argument shows that a function $f \in \mathcal{I}^m (\Omega, \mathcal{S}, \mu, X)$ if and only if it for every $A \subset \Omega$ $f \in \mathcal{I}^m (A, \mathcal{S}, \mu, X)$.

If $\mu(A) = 0$, then for any $f : \Omega \to X$, $R^m(f \cdot 1_A) = \{0\}$. In this case, we agree to set that $\int_A m(\mu, f) = 0$. It follows that $\int m(\mu, f) = \int m(\mu, g)$ whenever

$$\mu (\{\omega \in A : f(\omega) \neq g(\omega)\}) = 0.$$

We say that two functions $f, g : \Omega \to X$ are μ-*equivalent* and we write $f \overset{\mu}{\sim} g$ if $\mu (\{\omega \in A : f(\omega) \neq g(\omega)\}) = 0$. It is quickly seen that the relation $\overset{\mu}{\sim}$ is an equivalence relation on $\mathcal{I}^m (\Omega, \mathcal{S}, \mu, X)$. We denote by $I^m (\Omega, \mathcal{S}, \mu, X)$ the quotient space $\mathcal{I}^m (\Omega, \mathcal{S}, \mu, X) / \overset{\mu}{\sim}$.

Let $\mathcal{B}_0(Z)$ be a base of neighborhoods of 0 in the topological vector space Z. For each $\mathcal{O} \in \mathcal{B}_0(Z)$, let $B(0, \mathcal{O})$ denote the subset of $\mathcal{I}^m (\Omega, \mathcal{S}, \mu)$ consisting of integrable functions g such that all but finitely many elements of $R_\mu^m(g)$ are contained in $\mathcal{O}$. Set a subset $\mathcal{U} \subset \mathcal{I}^m (\Omega, \mathcal{S}, \mu, X)$ to be open if it is the union of translates of $B(0, \mathcal{N})$ where $\mathcal{O} \in \mathcal{B}_0(Z)$. Stated more formally:

Definition 11. We say that a subset $\mathcal{U}$ in $\mathcal{I}^m (\Omega, \mathcal{S}, \mu, X)$ is open if for every $f \in \mathcal{U}$, there exists $\mathcal{O} \in \mathcal{B}_0(Z)$ such that $f + B(0, \mathcal{O}) \subset \mathcal{U}$.

Let τ be the set of all open sets as defined above. Clearly, $\emptyset, \mathcal{I}^m (\Omega, \mathcal{S}, \mu, X) \in \tau$; τ is stable under arbitrary union. Assume that $\mathcal{U}, \mathcal{V} \in \tau$ and let $f \in \mathcal{U} \cap \mathcal{V}$. Then, there exist $\mathcal{O}_\mathcal{U}$ and $\mathcal{O}_\mathcal{V}$ in $\mathcal{B}_0(Z)$ such that $f + B(0, \mathcal{O}_\mathcal{U}) \subset \mathcal{U}$ and $f + B(0, \mathcal{O}_\mathcal{V}) \subset \mathcal{V}$. Clearly, $\mathcal{O}_\mathcal{U} \cap \mathcal{O}_\mathcal{V} \in \mathcal{B}_0(Z)$ and $f + B(0, \mathcal{N}_\mathcal{U} \cap \mathcal{N}_\mathcal{V}) \subset \mathcal{U} \cap \mathcal{V}$. Thus, τ is stable under finite intersection. Hence, τ is a topology on the vector space $\mathcal{I}^m (\Omega, \mathcal{S}, \mu, X)$.

To see that τ is compatible with the vector space structure of $\mathcal{I}^m (\Omega, \mathcal{S}, \mu, X)$, let $B(0, \mathcal{N}) \in \tau$, and choose a neighborhood $\mathcal{O}$ of 0 in Z such that $\mathcal{O} + \mathcal{O} \subset \mathcal{N}$. Then,

$$B(0, \mathcal{O}) + B(0, \mathcal{O}) \subset B(0, \mathcal{O} + \mathcal{O}) \subset B(0, \mathcal{N}.)$$

This proves that the addition is continuous. Also if $g \in B(0, \mathcal{N})$, then all but finitely many elements $R^m(g)$ are in $\mathcal{N}$. Then, for a scalar λ, $R^m_\mu(\lambda g) = \lambda R^m_\mu(g)$. Thus, we also have all but finitely many elements of $R^m_\mu(\lambda g)$ are in $\lambda \mathcal{N}$. This show that the scalar multiplicator is also continuous. Hence, the topology τ is linear and $(\mathcal{I}^m(\Omega, \mathcal{S}, \mu, X), \tau)$ is indeed a topological vector space.

The collection $\mathcal{B}_0 := \{B(0, \mathcal{N}) : \mathcal{N} \in \mathcal{B}_0(Z)\}$ is a base of neighborhood of 0 in $\mathcal{I}^m(\Omega, \mathcal{S}, \mu, X)$. In what follows, $\mathcal{I}^m(\Omega, \mathcal{S}, \mu, X)$ is always endowed with such a topology τ. The space $I^m(\Omega, \mathcal{S}, \mu, X)$ will be equipped with the quotient topology that is obviously a Hausdorff topology. We shall continue to use τ to denote such a topology.

We finish this section with the following interesting result.

Theorem 2. *Let $(\Omega, \mathcal{S}, \mu)$ be a real-valued integrator space, and X a complete Haudorff topological vector space. Then, the vector space $\mathcal{I}(\Omega, \mathcal{S}, \mu, X)$ of X valued integrable functions is topologically complete.*

Proof. In view of Theorem 3, it suffices to prove that $\mathcal{I}(\Omega, \mathcal{S}, \mu, X)$ is sequentially complete. So let (f_n) be a cauchy sequence of X valued functions in $\mathcal{I}(\Omega, \mathcal{S}, \mu, X)$. Fix a balanced neighborhood $\mathcal{O}$ of zero in X. There exists $N_\mathcal{O} \in \mathbb{N}$ such that for $h, k > N_\mathcal{O}$, one has $f_h - f_k \in B(0, \mathcal{O})$. That is, all but finitely many elements of the Riemann set $R_\mu(f_h - f_k)$ are in $\mathcal{O}$. We can choose a nonempty element A of $\mathcal{S}$ such that $\mu(A)(f_h(\omega) - f_k(\omega)) \in \mathcal{O}$. This implies that the sequence $(f_n(\omega))$ is Cauchy in X. The completeness of the Hausdorff topological space X allows us to define a function by $\omega \mapsto f(\omega) = \lim_{n \to \infty} f_n(\omega)$.

On the other hand, for $h, k > N_\mathcal{O}$, since f_h (respectively, f_k) is an element of $\mathcal{I}(\Omega, \mathcal{S}, \mu, X)$, all but finitely many elements of the Riemann set $R_\mu(f_h)$ (respectively, $R_\mu(f_k)$) are in $\mathcal{O}$. Let $R_\mu(f_h, \pi)$ (respectively, $R_\mu(f_k, \pi)$) be such an element.

$$\int f_h d\mu - \int f_k d\mu = \int f_h d\mu - R_\mu(f_h, \pi)$$
$$+ R_\mu(f_h, \pi) - R_\mu(f_k, \pi)$$
$$+ R_\mu(f_k, \pi) - \int f_k d\mu \in 3\mathcal{O}.$$

This proves that the sequence $(\int f_n d\mu)$ is Cauchy, and hence converges to say a in Z.

Now since $f(\omega) = \lim_{n \to \infty} f_n(\omega)$, there exists $N_\omega > N_\mathcal{O}$ such that if $h, k > N_\omega$, $f_h(\omega) - f_k(\omega) \in \mu(\Omega)^{-1}\mathcal{O}$. Therefore, for every subpartition

$\pi = \{(E_i, \omega_i) : i = 1, ..., n\}$ and for every $h, k > \max\{N_{\omega_i} : i = 1, ..., n\} =: N_\pi$, we have

$$R_\mu(f_h, \pi) - R_\mu(f_k, \pi) = \sum_{i=1}^{n} \mu(E_i)(f_h(\omega_i) - f_k(\omega_i)) \in \mathcal{O}.$$

If we let $h \to \infty$, we have

$$R_\mu(f, \pi) - R_\mu(f_k, \pi) \in \mathcal{O}.$$

Since $\int f_n d\mu \to a$, there exists $N > N_\pi$ such that $\int f_n d\mu - a \in \mathcal{O}$ whenever $n > N$. It follows that if $h, k > N$,

$$R_\mu(f, \pi) - a = R_\mu(f, \pi) - R_\mu(f_k, \pi) + R_\mu(f_k - f_h, \pi)$$
$$+ R_\mu(f_h, \pi) - \int f_h d\mu + \int f_h d\mu - a \in 4\mathcal{O}.$$

Since $\mathcal{O}$ was chosen arbitrarily, this shows that $f \in \mathcal{I}(\Omega, \mathcal{S}, \mu, X)$ and $\int f d\mu = a$. The proof is complete. $\qquad\square$

7. Extended Lebesgue-Nikodým Theorem

The Lebesgue-Nikodým Theorem states that for a Lebesgue measure $\lambda : \Sigma \to [0, \infty)$, an additive set function $F : \Sigma \to \mathbb{R}$ which is λ-absolutely continuous, is the integral of a Lebesgue integrable function $f : \Omega \to \mathbb{R}$; that is, for all measurable sets A, $F(A) = \int_A f d\lambda$. Such a property is not shared by vector valued set functions in general. In this section, we show that such a property can be recovered if one uses our approach to integrals.

Let $\mu : \mathcal{S} \subset 2^\Omega \to [0, \infty)$ be an integrator and let X be a Hausdorff topological vector space. A set function $F : \mathcal{S} \to \mathbb{R}$ is μ-*absolutely continuous* if for every neighborhood $\mathcal{N}$ of 0 in X, there exists $\delta > 0$ such that $F(A) \in \mathcal{N}$ whenever $\mu(A) < \delta$. Our next result is an extension of the Lebesgue-Nikodým Theorem.

Theorem 3. *Let X be a complete Hausdorff topological vector space, and $\mu : \mathcal{S} \subset 2^\Omega \to [0, \infty)$ an integrator. Assume that $F : \mathcal{S} \to X$ is a μ-absolutely continuous additive set function. Then there exists $f \in \mathcal{I}(\Omega, \mathcal{S}, \mu, X)$ such that*

$$F(A) = \int f \cdot 1_A d\mu$$

for all $A \in \mathcal{S}$.

Proof. For every $\pi \in \Pi(\Omega, \mathcal{S})$, consider the function defined on Ω by

$$F_\pi(\omega) = \sum_{E \in \pi} \frac{F(E \cap A)}{\mu(E \cap A)} 1_E(\omega).$$

Clearly, $F_\pi \in \mathcal{I}(\Omega, \mathcal{S}, \mu, X)$. The μ-absolute continuity of F ensures that

$$\int f \cdot 1_A d\mu = \sum_{E \in \pi} F(E \cap A) = F\left(\bigcup_{E \in \pi} (E \cap A)\right) \to F(A) \qquad (2)$$

as $\mu\left(\bigcup_{E \in \pi}(E \cap A)\right) \to \mu(A)$.

We claim that the set $\{F_\pi : \pi \in \Pi(\Omega, \mathcal{S})\}$ is Cauchy in $\mathcal{I}(\Omega, \mathcal{S}, \mu, X)$. Fix a balanced neighborhood $\mathcal{N}$ of 0 in X. There exists $\delta > 0$ and a finite subset $N_1 \subset \Pi(\Omega, \mathcal{S})$ such that for π_1 and π_2 in $\Pi(\Omega, \mathcal{S}) \setminus N_1$, whenever $\mu(A) < \delta$,

$$F\left(\bigcup_{E \in \pi_1} (E \cap A)\right) - F\left(\bigcup_{E \in \pi_2} (E \cap A)\right) = F\left(\bigcup_{E \in \pi_1} (E \cap A) \setminus \bigcup_{E \in \pi_2} (E \cap A)\right) \in \mathcal{N}.$$

Fix such a π_1 and π_2 in $\Pi(\Omega, \mathcal{S}) \setminus N_1$. There exists $N_2 \subset \Pi(\Omega, \mathcal{S}) \setminus N_1$ such that for $\eta \in \Pi(\Omega, \mathcal{S}) \setminus N_2$

$$R_\mu(F_{\pi_1}, \eta) - \int F_{\pi_1} \cdot 1_A d\mu, \ R_\mu(F_{\pi_2}\eta) - \int F_{\pi_2} \cdot 1_A d\mu \in \mathcal{N}.$$

It follows that for $\eta \in \Pi(\Omega, \mathcal{S}) \setminus N_2$

$$R_\mu(F_{\pi_1}, \eta) - R_\mu(F_\rho, \eta) = R_\mu(F_{\pi_1}, \eta) - \int F_{\pi_1} \cdot 1_A d\mu$$

$$+ F\left(\bigcup_{E \in \pi_1} (E \cap A)\right) - F\left(\bigcup_{E \in \pi_2} (E \cap A)\right)$$

$$+ \int F_{\pi_2} \cdot 1_A d\mu - R_\mu(F_{\pi_2}, \eta) \in 3\mathcal{N}.$$

This proves our claim.

By Theorem 2, there exists $f \in \mathcal{I}(\Omega, \mathcal{S}, \mu, X)$ such that $\{F_\pi : \pi \in \Pi(\Omega, \mathcal{S})\}$ converges to f. So given a neighborhood $\mathcal{N}$ of 0 in X, there exists a finite subset U_1 of $\Pi(\Omega, \mathcal{S}) \setminus N_2$ such that for $\eta \in \Pi(\Omega, \mathcal{S}) \setminus U_1$,

$$R_\mu(f - F_\pi, \eta) \in \mathcal{N}. \qquad (3)$$

On the other hand, it follows from 2 that there exists a finite subset U_2 of $\Pi(\Omega, \mathcal{S}) \setminus U_1$ such that for $\eta \in \Pi(\Omega, \mathcal{S}) \setminus U_2$,

$$R_\mu(F_\pi, \eta) - F(A) \in \mathcal{N}. \qquad (4)$$

Finally, by definition of the integral there exists a finite subset U_3 of $\Pi(\Omega, \mathcal{S}) \setminus U_2$ such that for $\eta \in \Pi(\Omega, \mathcal{S}) \setminus U_3$,

$$\int f \cdot 1_A d\mu - R_\mu(f, \eta) \in \mathcal{N}.$$

Combining (2), (3), and (4) we have for $\eta \in \Pi(\Omega, \mathcal{S}) \setminus U_3$

$$\int f \cdot 1_A d\mu - F(A) = \int f \cdot 1_A d\mu - R_\mu(f, \eta)$$
$$+ R_\mu(f - F_\pi, \eta)$$
$$+ R_\mu(F_\pi, \eta) - F(A) \in 3\mathcal{N}.$$

The desired result follows since $\mathcal{N}$ is arbitrary chosen. The proof is complete. $\qquad\square$

8. Further Notions of Integrability

Again in this section, we fix a Y-valued integrator space $(\Omega, \mathcal{S}, \mu)$, where Y is a given vector space.

Unconditional Integrability. In the definition of $\mu, \Updownarrow$-Riemann sums (Definition 8) if we associate to each E of the partition π, a point $t_E \in \bigcup \pi$ that is not necessarily in E, we say that we have an *unconditional μ, m-Riemann sums* of f with respect to $\pi \in \Pi(\Omega, \mathcal{S})$, that is,

$$\overline{R}_\mu^m(f, \pi) := \sum_{E \in \pi, t_E \in \bigcup \pi} m\left(\mu(E), f(t_E)\right).$$

The set

$$\overline{R}_\mu^m(f) := \left\{ \overline{R}_\mu^m(f, \pi) : \pi \in \Pi(\Omega, \mathcal{S}) \right\}$$

will be called the *unconditional $\mu, \Updownarrow$-Riemann set* of f.

Definition 12. Let Y be a topological vector space and $(\Omega, \mathcal{S}, \mu)$ a Y-valued integrator space. Let $m : X \times Y \to Z$ be a multiplicator, where X is a linear space and Z a Hausdorff topological vector space. We say that a function $f : \Omega \to X$ is *unconditionally $\mathcal{S}, \mu, m$-integrable* if the set $\overline{R}_\mu^m(f)$ is convergent in Z. That is, if there is a vector $m(\mu, f)$ with the property that for every neighborhood $\mathcal{N}$ of 0 in Z, there exists a finite subset $N(f)$ of $\overline{R}_\mu^m(f)$ such that

$$\overline{R}_\mu^m(f) \setminus N(f) \subset m(\mu, f) + \mathcal{N}.$$

We clearly have $R_\mu^m(f) \subset \overline{R}_\mu^m(f)$. In other words, unconditional $\mathcal{S}, \mu, m$-integrability of a function f implies its $\mathcal{S}, \mu, m$-integrability.

Example 4. Since a σ-algebra $\mathcal{S}$ is a semiring, a Lebesgue measure or a vector measure λ on $\mathcal{S}$ is an integrator, we see that the Lebesgue-Bochner integral [5], the Birkhoff integral [2] and the Bartle-Dunford-Schwartz integral [1] are all particular cases of the unconditional $\mathcal{S}, \lambda$-integral. If ℓ is the length function of the semiring of left-closed and right-open subintervals on an interval I_0 in $\mathbb{R}$, then one can see that the McShane integrable functions are unconditional $\mathcal{S}, \ell$-integrable.

We denote by $\mathcal{U}^m(\Omega, \mathcal{S}, \mu, X)$ the vector space of unconditionally $\mathcal{S}, \mu, m$-integrable functions. It should be clear from the definitions that:

Proposition 3. $\mathcal{U}^m(\Omega, \mathcal{S}, \mu, X) \subset \mathcal{I}^m(\Omega, \mathcal{S}, \mu, X)$ and $m(\mu, f) = \int m(\mu, f)$ for every $f \in \mathcal{U}^m(\Omega, \mathcal{S}, \mu, X)$.

Let Γ be a nonempty set and $\varpi : \Gamma \to \Omega$ be an injection. Then,

$$\varpi^{-1}(\mathcal{S}) = \left\{ \varpi^{-1}(E) : E \in \mathcal{S} \right\}$$

is a semiring of subsets of Γ, and one can naturally define an integrator $\eta : \varpi^{-1}(\mathcal{S}) \to Y$ by setting $\eta(\varpi^{-1}(E)) = \mu(E)$.

Theorem 4. *Let* $m : X \times Y \to Z$ *be a multiplicator where* X *is a vector space and* Z *a Hausdorff topological vector space. The following statements are equivalent for a function* $f : \Omega \to X$.

- $f \in \mathcal{U}^m(\Omega, \mathcal{S}, \mu, X)$.
- *For every nonempty set* Γ *and for every injection* $\varpi : \Gamma \to \Omega$, *the function* $f \circ \varpi \in \mathcal{U}^m(\varpi^{-1}(\Omega), \varpi^{-1}(\mathcal{S}), \eta)$.

In such a case, $\int_{\varpi^{-1}(\Omega)} m(f \circ \varpi, \eta) = m(\mu, f)$.

Proof. Clearly, $2 \Rightarrow 1$. To see $1 \Rightarrow 2$, suppose $f \in \mathcal{U}^m(\Omega, \mathcal{S}, \mu, X)$ and let there is a vector $m(\mu, f)$ be its integral. Let $\mathcal{N}$ be a neighborhood of 0 in Z. Then, there exists a finite subset $N(f)$ of $\overline{R}_\mu^m(f)$ such that $\overline{R}_\mu^m(f) \setminus N(f) \subset m(\mu, f) + \mathcal{N}$. Let $\overline{R}_\mu^m(f, \pi) \in \overline{R}_\mu^m(f)$ be such that

$$\overline{R}_\mu^m(f, \pi) - \int m(\mu, f) \in \mathcal{N},$$

and let $\varpi : \Gamma \to \Omega$ be an injection. Then,

$$R_\eta^m(f \circ \varpi, \varpi^{-1}(\pi)) = \sum_{\varpi^{-1}(E) \in \varpi^{-1}(\pi), \varpi(t_E) \in \varpi^{-1}(E)} m\left(\eta(\varpi^{-1}(E)), f \circ \varpi(\varpi^{-1}(t_E))\right)$$

$$= \sum_{E \in \pi, t_E \in E} m\left(\mu(E), f(t_E)\right) \in \overline{R}_\mu^m(f, (\pi)).$$

Thus,

$$R_\eta^m(f \circ \varpi, \varpi^{-1}(\pi)) - \overline{\int} m(\mu, f) \in \mathcal{N}.$$

This shows that $f \circ \varpi \in \mathcal{I}^m\left(\varpi^{-1}(\Omega), \varpi^{-1}(\mathcal{S}), \eta\right)$ and

$$\int_{\varpi^{-1}(\Omega)} m\left(f \circ \varpi, \eta\right) = \overline{\int} m(\mu, f).$$

The proof is complete. □

It follows that unconditionally $\mathcal{S}, \mu, m$-integrability is an hereditary property, namely:

Corollary 1. *A function $f : \Omega \to X$ is unconditionally $\mathcal{S}, \mu, m$-integrable if and only if it is unconditionally $\mathcal{S}, \mu, m$-integrable over all of its subsets.*

Such a property sets apart the unconditional $\mathcal{S}, \mu, m$-integrability from the classical integrals such as Lebesgue-Bochner integral, the Birkhoff integral or the Bartle-Dunford-Schwartz integral which all lack the hereditary property due to, among other things, the measurability requirement.

Norm and Weak Integrability. For a semiring $\mathcal{S}$ of subsets of a nonempty set Ω, we denote by $\mathcal{A}(\mathcal{S})$ the Banach space of scalar valued integrators on $\mathcal{S}$ with the variation norm, $\|\sigma\| = |\sigma|(\Omega)$, where $|\sigma|(E)$ denotes the variation of μ on the set E.

In what follows, we shall denote by X^* the continuous dual of a topological vector space X.

Definition 13. Let $(\Omega, \mathcal{S}, \mu)$ be a Y-valued integrator space, X a Hausdorff topological vector space.

(1) We say that a function $f : \Omega \to X$ is *scalarly $\mathcal{S}, \mu$-integrable* if for every $x^* \in X^*$, the scalar function

$$x^* f : \Omega \to \mathbb{F} : \omega \mapsto x^* f(\omega)$$

belongs to $\mathcal{I}(\Omega, \mathcal{S}, \mu, X)$. We denote by $s\mathcal{I}(\Omega, \mathcal{S}, \mu, X)$ the space of all scalarly $\mathcal{S}, \mu$-integrable functions $f : \Omega \to X$.

(2) If in addition, the vector space X is a normed space, and $0 < p < \infty$, then we say that a function $f : \Omega \to X$ is $\mathcal{S}, \mu, p$-*integrable* if the non-negative function

$$\|f\|_X^p : \Omega \to \mathbb{R} : \omega \mapsto \|f(\omega)\|_X^p$$

belongs to $\mathcal{I}(\Omega, \mathcal{S}, \mu, X)$. We denote by $\mathcal{I}^p(\Omega, \mathcal{S}, \mu, X)$ the space of all $\mathcal{S}, \mu, p$-integrable functions $f : \Omega \to X$.

For example, if Ω is a topological space, Σ the σ-algebra containing the Borel σ-algebra, and λ is the Lebesgue measure on Σ, we see that the space $\mathcal{L}^p(\Omega, \Sigma, \lambda, X)$ of the Lebesgue-Bochner p-integrable functions is contained in $\mathcal{I}^p(\Omega, \Sigma, \lambda, X)$. Again, unlike $\mathcal{L}^p(\Omega, \Sigma, \lambda, X)$, the space $\mathcal{I}^p(\Omega, \Sigma, \lambda, X)$ can include non-measurable functions, thus

$$\mathcal{L}^p(\Omega, \Sigma, \lambda, X) \subsetneq \mathcal{I}^p(\Omega, \Sigma, \lambda, X).$$

In particular, if X is an infinite dimensional Banach space, we have

$$\mathcal{L}^1(\Omega, \Sigma, \lambda, X) \subset \mathcal{I}^1(\Omega, \Sigma, \lambda, X) \subset \mathcal{I}(\Omega, \Sigma, \lambda, X).$$

The reverse of the second inclusion does not hold as evidenced by the classical example of the real valued function $f(x) = \frac{\sin x}{x}$ defined on $\Omega = [0, \infty)$.

In fact, if we denote by $m(\Omega, \Sigma, \lambda, X)$ the vector space of all λ-measurable functions, and if A is a λ-measurable function, then we have

$$\mathcal{L}^1(A, \Sigma, \lambda, X) \subset \mathcal{I}^1(A, \Sigma, \lambda, X) \cap m(\Omega, \Sigma, \lambda, X).$$

There are known example of non-measurable functions that are norm-integrable. For example, the function $t \mapsto \delta_t$ that associates to every $t \in [a, b]$, the Dirac measure δ_t in the space of Radon measures on $[a, b]$. Other examples of non-measurable operator-valued function $t \mapsto U_t$ are also given in [19], for which $\int \|U_t\| \, dt < \infty$.

On the other hand, it can be shown that if X is a finite dimensional vector space, then $\mathcal{U}(\Omega, \Sigma, \lambda, X) = \mathcal{I}^1(\Omega, \Sigma, \lambda, X)$. The situation is different for infinite dimensional space. The classical *Dvoretski-Rogers theorem* asserts that:

If X is an infinite dimensional Banach space, and if $\{\alpha_n\}$ is a sequence of positive real numbers such that $\sum_n \alpha_n^2 < \infty$, then there exists an unconditionally summable sequence $\{x_n\}$ in X such that $\|x_n\| = \alpha_n$, for every $n \in \mathbb{N}$.

Thus, if X is an infinite dimensional Banach space, then there always exists an unconditional but not norm-summable sequence in X. Our next result is an extension of such a result.

Theorem 5. *Let X be an infinite dimensional Banach space. Let λ be the Caratheodory extension of a regular measure μ defined on a σ-algebra Σ containing the Borel subsets of a topological space Ω. Let $A \subset \Omega$ such that $\lambda(A) < \infty$. Then, $\mathcal{U}(A, \Sigma, \lambda, X) \setminus \mathcal{I}^1(A, \Sigma, \lambda, X)$ is not empty.*

Proof. Since X is infinite dimensional, there exists a sequence $\{x_n\}$ of elements of X that is unconditionally summable but not norm summable. Let $\{U_n\}$ be a pairwise disjoint sequence of open subsets of A. Clearly, $\sum_n \lambda(U_n) < \lambda(A) < \infty$. Furthermore, by removing sets of λ-measure zero, we may assume that $\lambda(U_n) > 0$ for all $n \in \mathbb{N}$. Set $y_n = \frac{1}{\lambda(U_n)} x_n$ for each $n \in \mathbb{N}$. Consider the function

$$f(t) = \begin{cases} y_n, & t \in U_n \\ 0, & \text{otherwise.} \end{cases}$$

The unconditional summability of the sequence $\{x_n\}$ corresponds exactly to the unconditional integrability of the function f. On the other hand, the partial sums of the series $\sum_n \lambda(U_n) \|y_n\|$ are elements of Riemann set $R_\lambda(\|f\|)$ of $\|f\|$ which diverges. Thus, $f \in \mathcal{U}(A, \Sigma, \lambda, X) \setminus \mathcal{I}^1(A, \Sigma, \lambda, X)$.

On the other hand, given $\epsilon > 0$, there exists a finite subset $N_\epsilon \in \mathbb{N}$ such that for every finite subset K of $\mathbb{N}$, $\qquad\qquad\qquad\qquad\qquad\square$

Theorem 6. *Let $(\Omega, \mathcal{S}, \mu)$ be an $\mathbb{F}$-valued integrator space, and X a topological vector space. If $f \in s\mathcal{I}(\Omega, \mathcal{S}, \mu, X)$, then there exists a vector $\int^{**} f d\mu \in X^{**}$ such that for every $x^* \in X^*$,*

$$\left\langle x^*, \int^{**} f d\mu \right\rangle = \int x^* f d\mu.$$

Proof. We notice that for each $x^* \in X^*$, the map $\mu_{x^*} : A \mapsto \int_A x^* f d\mu$ for $A \in 2^\Omega$ defines an element of the Banach space $\mathcal{A}(\mathcal{S})$. A straight forward application of the closed graph theorem shows that the operator $T : X^* \to \mathcal{A}(\mathcal{S}) : x^* \mapsto \mu_{x^*}$ is bounded. Its adjoint maps $\mathcal{A}(\mathcal{S})^*$ into X^{**}. The indicator function 1_Ω defines an element of $\mathcal{A}(\mathcal{S})^*$ as follows: $\langle \nu, 1_\Omega \rangle = \nu(\Omega)$ for every $\nu \in \mathcal{A}(\mathcal{S})$. It follows in particular that $T^* 1_\Omega \in X^{**}$ and we have for every $x^* \in X^*$,

$$\langle x^*, T^* 1_\Omega \rangle = \langle T x^*, 1_\Omega \rangle = \langle \mu_{x^*}, 1_\Omega \rangle = \int x^* f d\mu.$$

Hence, $\int^{**} f d\mu = T^* 1_\Omega \in X^{**}$ as desired. $\qquad\qquad\qquad\qquad\square$

The following definition clearly extends the notion of Pettis integrability. For details on Pettis integral, see for example [5].

Definition 14. Under the hypothesis of Theorem 6, if for $f \in s\mathcal{I}(\Omega, \mathcal{S}, \mu, X)$, the vector $\int^{**} f d\mu$ actually belongs to X, then we say that f is *weakly $\mathcal{S}, \mu$-integrable*. We denote by $w\mathcal{I}(\Omega, \mathcal{S}, \mu, X)$ the vector space of weakly $\mathcal{S}, \mu$-integrable functions.

For the case where $\mu : \mathcal{S} \to [0, \infty)$ is a nonnegative integrator and X is a Banach space, we have

$$\mathcal{I}^1(\Omega, \mathcal{S}, \mu, X) \subset w\mathcal{I}(\Omega, \mathcal{S}, \mu, X) \subset s\mathcal{I}(\Omega, \mathcal{S}, \mu, X);$$

and if $\mathcal{P}^1(\Omega, \mathcal{S}, \mu, X)$ denotes the space of Pettis $\mathcal{S}, \mu$-integrable functions, that is, the space of all μ-measurable functions $f : \Omega \to X$ such that for every $x^* \in X^*$ the function $|x^* f| : \Omega \to \mathbb{F}$ is $\mathcal{S}, \mu$-integrable, then

$$\mathcal{P}^1(\Omega, \mathcal{S}, \mu, X) \subset w\mathcal{I}(\Omega, \mathcal{S}, \mu, X).$$

Note that if Ω is a topological space and $\mathcal{S}$ is an algebra containing the open sets, we have

$$\mathcal{L}^1(\Omega, \mathcal{S}, \mu, X) \subset \mathcal{P}^1(\Omega, \mathcal{S}, \mu, X).$$

Here again, no measurability is required. If Ω is a topological space, $\mathcal{S}$ the σ-algebra containing the Borel σ-algebra, and μ is the Lebesgue measure on $\mathcal{S}$, we have

$$\mathcal{L}^1(\Omega, \mathcal{S}, \mu, X) \subsetneq \mathcal{P}^1(\Omega, \mathcal{S}, \mu, X) \subsetneq w\mathcal{I}(\Omega, \mathcal{S}, \mu, X).$$

9. Spectral Integrals

As an example of application of our approach to the definition of the integral, we put forward a generalization of spectral theorem to infinite dimension Hilbert spaces. We shall establish the integral representation of self-adjoint bounded operator A on a Hilbert Space H.

We denote by $\mathcal{L}(H)$ the vector space of all bounded linear operators on a Hilbert H. Let $\mathcal{S}$ be the a semiring of sets in $\mathbb{C}$ and $\mathcal{P}(H)$ the vector subspace of $\mathcal{L}(H)$ consisting of orthogonal projections in H. We define a *spectral integrator* as an additive set function $\sigma : \mathcal{S} \to \mathcal{P}(H)$ satisfying $\sigma(\Omega) = I$ the identity on H. The scaling operator $m : \mathbb{C} \times \mathcal{P}(H) \to \mathcal{L}(H)$, $(\lambda, T) \mapsto \lambda T$ is a well-defined multiplicator. For $\pi \in \Pi(\mathbb{C}, \mathcal{S})$, the *spectral $\sigma, \updownarrow$-Riemann sum* of a function $f : \mathbb{C} \to \mathbb{C}$ is then given by

$$R_\sigma(f, \pi) = \sum_{E \in \pi, t_E \in E} f(t_E)\sigma(E).$$

Definition 15. We say that a function $f : \mathbb{C} \to \mathbb{C}$ is *spectral integrable over* Ω if it is $\mathcal{S}, \sigma$-integrable. The element $\int f d\sigma = \lim R_\sigma(f)$ is then called the *spectral integral* of f.

The set $\Lambda(\sigma) := \mathbb{C} \setminus \bigcup U_i$ where the union is taken over all sets in $\mathcal{S}$ for which $\sigma(U_i) = 0$ is called the *spectrum* of the spectral integrator σ. The integrator σ is said to be *compact* if $\Lambda(\sigma)$ is compact.

Theorem 7. *If σ is a compact spectral integrator, there exists a unique normal operator $A : H \to H$ such that for all $h, k \in H$,*

$$\left\langle \int f d\sigma h, k \right\rangle = \langle Ah, k \rangle$$

where $f : \mathbb{C} \to \mathbb{C}$ is the identity $f(z) = z$.

Proof. Since $\Lambda(\sigma)$ is compact, we have $\varphi(h, k) := \left\langle \int z d\sigma h, k \right\rangle$ is finite for all $h, k \in H$. Clearly, φ is linear with respect to the first variable and conjugate linear with respect to the second variable. We also have

$$|\varphi(h, h)| = \left| \left\langle \int z d\sigma h, h \right\rangle \right| \leq M \|h\|^2$$

where $M = \sup \{|z| : z \in \Lambda(\sigma)\}$. By the parallelogram law, we have

$$|\varphi(h, k)| \leq \frac{1}{4} M \left(\|h + k\|^2 + \|h - k\|^2 + \|h + ik\|^2 + \|h - ik\|^2 \right)$$
$$\leq M \left(\|h\|^2 + \|k\|^2 \right).$$

This show that $\|\varphi\| \leq 2M$.

For fixed $h \in H$, let $\phi_h(k) = \overline{\varphi(h, k)}$. By the Riesz representation theorem, there exists a vector, say Ah such that $\phi_h(k) = \langle k, Ah \rangle$. It then follows that $\varphi(h, k) = \langle Ah, k \rangle$. The linearity of the operator A easily follows from the linearity of φ. Moreover,

$$\|Ah\|^2 = |\varphi(h, Ah)| \leq \|\varphi\| \|h\| \|Ah\|,$$

which implies that $\|A\| \leq \|\varphi\|$. Similarly, we have a bounded linear operator $\overline{A}$ such that $\left\langle \int \overline{z} d\sigma h, k \right\rangle = \langle \overline{A}h, k \rangle$.

We now show that A is normal. We have

$$\langle h, A'k \rangle = \overline{\langle \overline{A}k, h \rangle} = \overline{\left\langle \int \overline{z} d\sigma k, h \right\rangle} = \int z d\sigma \overline{\langle k, h \rangle} = \left\langle \int z d\sigma h, k \right\rangle = \langle Ah, k \rangle.$$

Therefore, $A^* = \overline{A}$.

Unified Approach to Integration 285

Let $F \in \mathcal{S}$. We have for arbitrary $h, k \in H$

$$\langle A^*h, \sigma(F)k \rangle = \left\langle \int \bar{z} d\sigma h, \sigma(F)k \right\rangle = \lim \langle R(\bar{z}, \sigma)h, \sigma(F)k \rangle .$$

For every $\pi \in \Pi\left(\mathbb{C}, \mathcal{S}, \sigma\right)$, we have

$$\langle R_\sigma(\bar{z}, \pi)h, \sigma(F)k \rangle = \left\langle \sum_{E \in \pi, t_E \in E} t_E \sigma(E)h, \sigma(F)k \right\rangle$$

$$= \left\langle \sum_{E \in \pi, t_E \in E} t_E \sigma(F)\sigma(E)h, k \right\rangle$$

$$= \left\langle \sum_{E \cap F \in \pi, t_{E \cap F} \in E \cap F} t_E \sigma(F \cap E)h, k \right\rangle$$

$$= \langle R_\sigma(\bar{z} 1_F, \pi)h, k \rangle$$

It follows from the continuity of the inner product that

$$\langle A^*h, \sigma(F)k \rangle = \left\langle \int \bar{z} 1_F d\sigma h, k \right\rangle .$$

This implies that

$$\langle AA^*h, k \rangle = \langle A^*h, A^*k \rangle = \overline{\langle A^*k, A^*h \rangle}$$

$$= \overline{\int \bar{z} \langle (d\sigma) k, A^*h \rangle} = \int z \langle (d\sigma) A^*h, k \rangle$$

$$= \int z\bar{z} \langle (d\sigma) h, k \rangle = \left\langle \int |z|^2 d\sigma h, k \right\rangle .$$

Similarly and by symmetry, we also have $\langle A^*Ah, k \rangle = \left\langle \int |z|^2 d\sigma h, k \right\rangle$. This shows that $AA^* = A^*A$, i.e. A is a normal operator. $\square$

10. Stochastic Integrals

The standard theory of stochastic integration is unfortunately developed only generally enough for limited scope of applications. In this section, we attempt to show that it is possible to describe stochastic integration for larger, more general classes of integrand and integrator processes.

In what follows, V is a Hausdorff topological vector space. We fix an integrator space $\left(\Omega, 2^\Omega, \mu\right)$ where $\mu : 2^\Omega \to \mathbb{R}$ is a nonnegative scalar integrator and T an well-ordered index set. It is clear that the restriction of μ to any semiring in 2^Ω is an integrator. It is easily verified that if $\mathcal{S}_1 \subset \mathcal{S}_2$

are two semiring in 2^Ω, then clearly, $\Pi(\Omega, \mathcal{S}_1) \subset \Pi(\Omega, \mathcal{S}_2)$. The following proposition is then easily verified.

Prop 1. Let $\mathcal{S}_1 \subset \mathcal{S}_2$ be two semiring in 2^Ω. Then for every $A \in 2^\Omega$,

(1) $I(A, \mathcal{S}_1, \mu, V) \subset I(A, \mathcal{S}_2, \mu, V)$ and
(2) $\int_A f d\mu_{|\mathcal{S}_1} = \int_A f d\mu_{|\mathcal{S}_2}$ for $f \in I(A, \mathcal{S}_1, \mu, V)$.

For simplicity of notation, in lieu of $\mu_{|\mathcal{S}}$, we sometimes simply write μ.

Conversely, assume that $f \in I(A, \mathcal{S}_2, \mu, V)$. Then, the map $A \mapsto \int_A f d\mu$ is an additive set function from $\mathcal{S}_1$ into V. By the extended Lebesgue-Nikodým Theorem 3, there exists a μ-almost unique function $E(f \mid \mathcal{S}_1) \in I(A, \mathcal{S}_1, \mu, V)$ such that $\int_A f d\mu = \int_A E(f \mid \mathcal{S}_1) d\mu$. Such a function is called the *conditional expectation* of f given $\mathcal{S}_1$.

Note that as opposed to the classical definition of conditional expectation, here the measurability of the function $E(f \mid \mathcal{S}_1)$ is not required. However, it is easily checked that basic properties of the classical conditional expectations are preserved, namely:

(1) If $f \in I(A, \mathcal{S}_1, \mu, V)$, then $E(f \mid \mathcal{S}_1) \sim f$.
(2) If $\mathcal{S}_1 \subset \mathcal{S}_2 \subset 2^\Omega$ are semirings, $f \in I(A, \mathcal{S}_2, \mu, V)$, then
$$E\left(E\left(f \mid \mathcal{S}_1\right) \mid \mathcal{S}_2\right) \sim E\left(f \mid \mathcal{S}_1\right).$$
(3) If $\mathcal{S}_1 \subset \mathcal{S}_2$ are semirings, $f, g \in I(A, \mathcal{S}_2, \mu, V)$, and α, β scalars, then
$$E\left(\alpha f + \beta g \mid \mathcal{S}_1\right) \sim \alpha E\left(f \mid \mathcal{S}_1\right) + \beta E\left(g \mid \mathcal{S}_1\right).$$

Definition 16. Let V be a topological vector space. A V-valued *stochastic process* is any function $X : T \to I(\Omega, \mathcal{S}, \mu, V)$ where T is a well-ordered set.

A stochastic process is said to be *discrete* if T is the set of all natural numbers, and *continuous* if T is an interval. The usual examples are $T = \mathbb{N}$, $T = [0, \infty)$, and $T = [a, b] \subset \mathbb{R}$.

Definition 17. By a *filtration*, we mean a family $\{\mathcal{S}_t : t \in T\}$ of semirings in 2^Ω satisfying $\mathcal{S}_s \subset \mathcal{S}_t$ whenever $s < t$. A stochastic process $X : T \to I(\Omega, 2^\Omega, \mu, V)$ is said to be *adapted* to a filtration $\{\mathcal{S}_t : t \in T\}$ if for each $t \in T$, $X(t) \in I(\Omega, \mathcal{S}_t, \mu, V)$.

The theory of martingales plays a very important and useful role in the study of stochastic processes.

Definition 18. Let V be a topological vector space. A *martingale* $X : T \to I(\Omega, 2^\Omega, \mu, V)$ is a stochastic process adapted to some filtration $\{\mathcal{S}_t : t \in T\}$ that satisfies $E\left(X(t) \mid \mathcal{S}_s\right) = X(s)$ for all $s \leq t$.

Here again, neither the absolute integrability nor the measurability of the functions involved are required.

Let $P = \{a = t_0 < t_1 < \cdots < t_n = b\}$ be a partition of an interval $[a, b]$ in T. A stochastic process $X : T \to I(\Omega, \mathcal{S}, \mu, V)$ defines an integrator (still denoted by X) on the semiring of half-open intervals $[s, t)$ of T, by $X([s, t)) = X(t) - X(s)$ for $s \leq t$.

Given a multiplicator $m : U \times V \to W$, we define

$$\mathcal{M} : I(\Omega, \mathcal{S}, \mu, U) \times I(\Omega, \mathcal{S}, \mu, V) \to I(\Omega, \mathcal{S}, \mu, W)$$

by $m(f, g)(\omega) = m(f(\omega), g(\omega))$ for every $\omega \in \Omega$. Clearly, m is a multiplicator.

Definition 19. Let $m : U \times V \to W$ be a multiplicator. Let $X : T \to I(\Omega, \mathcal{S}, \mu, U)$ and $Y : T \to I(\Omega, \mathcal{S}, V)$ be two stochastic processes. Given a partition $P = \{a \leq t_0 < t_1 < \cdots < t_n \leq b\}$ of an interval $[a, b]$ in T, we define the *stochastic $\mu, \updownarrow$-Riemann sum* of Y with respect to X at P to be:

$$Y_X(P) = \sum_{i=1}^{n} m\left[Y(t_{i-1}), (X(t_i) - X(t_{i-1}))\right].$$

It is clear that for every partition P of $[a, b] \subset T$, $Y_X(P) \in I(\Omega, \mathcal{S}, \mu, W)$. We shall denote by $\pi([a, b])$ the set of all partitions of $[a, b] \subset T$, and by $R(Y_X, [a, b])$ the set of all stochastic $\mu, \updownarrow$-Riemann sums over $[a, b]$.

We are now ready to give our definition of stochastic integral.

Definition 20. Let $m : U \times V \to W$ be a multiplicator, where W is a topological vector space. Let $X : T \to I(\Omega, \mathcal{S}, \mu, U)$ and $Y : T \to I(\Omega, \mathcal{S}, \mu, V)$ be two stochastic processes. We say that Y is *stochastically integrable with respect to X over the interval* $[a, b] \subset T$ if the $\lim R(Y_X, [a, b])$ exists in $I(\Omega, \mathcal{S}, \mu, W)$.

Such a limit is called the *stochastic integral of Y with respect to X over* $[a, b]$ and will be denoted by $\int_a^b \updownarrow(Y, X)$.

In other words:

The stochastic process Y is stochastically integrable with respect to stochastic process X over $[a, b]$ if there exists an element $\int_a^b m(Y, X)$ in $I(\Omega, \mathcal{S}, \mu, W)$ such that for every $\mathcal{N} \in \mathcal{B}_0(W)$, all but finitely many $Y_X(P)$ are in $\int_a^b \updownarrow(Y, X) + B(0, \mathcal{N})$.

The process Y is called *the integrand process* while the process X is the *integrator process.*

It follows immediately from the linearity of the limit that the stochastic integration is a linear operator. That is to say, if $Y, Z : T \to I(\Omega, \mathcal{S}, \mu, V)$

are both stochastic integrable with respect to $X : T \to I(\Omega, \mathcal{S}, \mu, U)$ over $[a, b]$ and α, β are any pair of scalars then

$$\int_a^b \updownarrow ((\alpha Y + \beta Z), X) = \alpha \int_a^b \updownarrow (Y, X) + \beta \int_a^b \updownarrow (Z, X).$$

We notice that if $Y : T \to I(\Omega, \mathcal{S}, \mu, V)$ is stochastically integrable with respect to $X : T \to I(\Omega, \mathcal{S}, \mu, U)$ over $[a, b] \subset T$, then it is stochastically integrable over any subinterval $[c, d] \subset [a, b]$. Hence, if $Y : T \to I(\Omega, \mathcal{S}, \mu, V)$ is stochastically integrable with respect to $X : T \to I(\Omega, \mathcal{S}, \mu, U)$ over $[a, b]$ then

$$Z : [a, b] \to I(\Omega, \mathcal{S}, \mu, W) \tag{5}$$
$$t \mapsto \int_a^t \updownarrow (Y, X)$$

defines another stochastic process that we shall call the *indefinite stochastic integral* of Y with respect to X.

As in the classical theory, the case of martingales presents a particular feature.

Theorem 8. *The indefinite stochastic integral of a martingale with respect to a martingale adapted to the same filtration is a martingale adapted to the same filtration.*

Proof. Let $t \mapsto Z(t) = \int_{a_0}^t \updownarrow (Y, X)$ be the indefinite stochastic integral of $Y : T \to I(\Omega, \mathcal{S}, \mu, V)$ with respect to a martingale $X : T \to I(\Omega, \mathcal{S}, \mu, W)$ adapted to the same filtration as Y. We note that for $P = \{s = t_0 < t_1 < \cdots < t_n = t\}$, $Y(t_{i-1}) \in I(\Omega, \mathcal{S}_s, \mu, V)$ and thus $E(Y(t_{i-1})|\mathcal{S}_s) = Y_{t-1}$. It follows that

$$E[Y_X(P)|\mathcal{S}_s] = E\left(\sum_{i=1}^n [\updownarrow (Y(t_{i-1}), (X(t_i) - X(t_{i-1})) |\mathcal{S}_s)]\right)$$

$$= \sum_{i=1}^n E[\updownarrow (Y(t_{i-1}), (X(t_i) - X(t_{i-1})) |\mathcal{S}_s)]$$

$$= \sum_{i=1}^n m(Y(t_{i-1}), E[(X(t_i) - X(t_{i-1})) |\mathcal{S}_s]) = 0$$

and

$$E\left(\left(\int_s^t \updownarrow (Y, X)\right)|\mathcal{S}_s\right) = E\left[\left(\int_s^t \updownarrow (Y, X) - Y_X(P)\right)|\mathcal{S}_s\right].$$

Fix $\mathcal{N} \in \mathcal{B}_0(Z)$. For all but finitely many $P \in \pi([s,t])$, we have $\int \updownarrow(Y,X) - Y_X(P) \in B(0,\mathcal{N})$. We infer that for $s < t$,

$$E\left((Z(t) - Z(s))|\mathcal{S}_s\right) = E\left(\left(\int_{a_0}^t \updownarrow(Y,X) - \int_{a_0}^s \updownarrow(Y,X)\right)|\mathcal{S}_s\right)$$

$$= E\left(\left(\int_s^t \updownarrow(Y,X)\right)|\mathcal{S}_s\right)$$

$$= E\left(\left(\int_s^t \updownarrow(Y,X)\right) - Y_X(P)|\mathcal{S}_s\right) \in B(0,\mathcal{N}).$$

Since $\mathcal{N} \in \mathcal{B}_0(W)$ is arbitrary, we conclude that $E\left(Z(t)|\mathcal{S}_s\right) = Z(s)$. This completes the proof. $\square$

I wish to express my sincere appreciation to the reviewers for the useful and pertinent remarks/comments that greatly contributed to the improvement of the writing of this chapter.

References

[1] R. G. Bartle, A Modern Theory of Integration, Graduate Studies in Math. AMS, 2001, Providence, RI.

[2] G. Birkhoff, Integration of Functions with Values in a Banach Space. In: Transactions of the American Mathematical Society, Vol. 38, No. 2, 1935, pp. 357-378.

[3] Z. M. Bogdanovich, Vector and Operator Valued Measures and Applications, 1973, ISBN:978-0-12-702450-9, https://doi.org/10.1016/C2013-0-11628-5.

[4] P. J. Daniell, A General Form of Integral, Annals of Mathematics, Second Series, Annals of Mathematics, 1918, 19 (4): 279–294, doi:10.2307/1967495.

[5] J. Diestel and J. J. Uhl , *Vector Measures*, AMS Mathematical Survey 15, 1977, Providence, RI.

[6] N. Dinculeanu, *Vector Integration and Stochastic Integration in Banach spaces*, 2000, John Wiley & Sons Inc.

[7] D. H. Fremlin, *The Generalized McShane Integral*, Illinois Jour. of Math, Volume 39, Number 1, Spring 1995.

[8] J. Friedrich, *Integration of Banach space-valued functions: Bochner and Birkhoff integration*. Diplomica Verlag, Hamburg 2013, pp. 28-46, ISBN 978-3-8428-4043-0.

[9] R. A. Gordon, *Riemann integration in Banach spaces*, Rocky Mt. J. Math. 21 (3) (1991) 923–949.

[10] Guoju Ye, *On Henstock–Kurzweil and McShane integrals of Banach space-valued functions*, J. Math. Anal. Appl. 330, 2007, pp. 753-765.

[11] E. J. McShane, *Partial Orderings and Moore-Smith Limits*, Amer. Math. Monthly 59, 1952, pp. 1-11.

[12] T. V. Panchapagesan, *On the distinguishing features of the Dobrakov integral*, Divulg. Mat. 3 1995, no. 1, 79–114. Zbl 0883.28011.

[13] M. A. Robdera, *A Unified Approach to Integration Theory*, IJAPM, Vol.9(1), 2019, pp. 21-28, doi: 10.17706/ijapm.2019.9.1.21-28.

[14] M. A. Robdera, *A New General Approach to Vector Valued Stochastic Integration*, International Journal of Modeling and Optimization, Vol. 4, No. 4, 2014, pp. 299-304.

[15] M. A. Robdera and Kagiso D.N., *On the differentiability of vector valued additive set functions*, Adv. in Pure Math., 3, 2013, pp. 653-659.

[16] M. A. Robdera, *Compactness Principles for Topological Vector spaces*, Topological Algebra and its Applications (to appear).

[17] J. Rodriguez,, *On integration of vector functions with respect to vector measures*, Czechoslovak Mathematical Journal. 56(131), 2006, pp. 805-825, MR 2261655 (2007j:28019).

[18] D. K. Sari, L. Peng-Yee, D. Zhao, *A New Topology on the space of Primitives of Henstock-Kurzweil Integrable Functions*. Southeast Asian Bull. of Math. 42(5) 2018, pp. 719-728.

[19] E. G. T. Thomas, *Vector-valued integration with application to operator-valued H^∞-space*, IMA J. Math. Control Inform. 14 (1997), 109-136.

© 2025 World Scientific Publishing Company
https://doi.org/10.1142/9789819812202_0014

Chapter 14

Some Unified Integral Representations of the Four-parameter Mittag-Leffler Functions

Ankit Pal

Division of Mathematics, School of Advanced Sciences & Languages, VIT Bhopal University, Kothrikalan, Sehore-466 114, Madhya Pradesh, India
Email: ankit.pal@vitbhopal.ac.in

Kiran Kumari

Department of Mathematics and Humanities, Sardar Vallabhbhai National Institute of Technology, Surat, Gujarat, 395 007, India
Email: kiran.kumariprajapati@gmail.com

> In this work, we propose some unified integral representations for the four parameter Mittag-Leffler function, and evaluated our findings in terms of generalized special functions. Additionally, a special case of the four-parameter Mittag-Leffler function has been corollarily presented.

1. Introduction and Preliminaries

In the fields of science and technology, special functions are constantly important in expressing diverse concepts and theories. Many special functions are included in this list, such as *Bessel's function, generalized hypergeometric function, generalized Mittag-Leffler functions, Wright hypergeometric function, H-function, I-function, Aleph function, generalized M-series, R-function, Legendre polynomials, Laguerre polynomials, Hermite polynomials, Srivastava polynomials*, and many more. Science and technical fields are well aware of the applicability of these specific functions. In the theory of applied analysis, generalized functions and symmetric functions are very common, and they have been applied to a variety of fields such as group theory, Lie algebras, algebraic geometry, and probability theory. The Mittag-Leffler function, which was introduced in connection with a method of summation of some divergent series, has received much interest from scientists due to its wide applications in pure as well as applied mathematics.

The Mittag-Leffler function naturally occurs as a solution of fractional-order differential equations or fractional-order integral equations.

The Mittag-Leffler function, a generalization of the exponential function, was introduced by the Swedish mathematician Gösta Mittag-Leffler [9] in 1903 is defined as,

$$E_\alpha(z) = \sum_{n=0}^{\infty} \frac{z^n}{\Gamma(\alpha n + 1)}, \quad \Re(\alpha) > 0. \tag{1}$$

In 1905, Wiman [22] generalized the Mittag-Leffler function $E_\alpha(z)$ and came up with the following definition:

$$E_{\alpha,\beta}(z) = \sum_{n=0}^{\infty} \frac{z^n}{\Gamma(\alpha n + \beta)}, \tag{2}$$

where $\alpha, \beta \in \mathbf{C}, \Re(\alpha) > 0, \Re(\beta) > 0$.

Later, Prabhakar [17] defined the three-parameter Mittag-Leffler function $E_{\alpha,\beta}^\rho(z)$ in 1971 which can be represented as,

$$E_{\alpha,\beta}^\rho(z) = \sum_{n=0}^{\infty} \frac{(\rho)_n}{\Gamma(\alpha n + \beta)} \frac{z^n}{n!}, \tag{3}$$

where $\alpha, \beta, \rho \in \mathbf{C}$ and $\Re(\alpha) > 0, \Re(\beta) > 0, \Re(\rho) > 0$.

Wright [23–25] introduced the Fox-Wright function, an extension of the generalized hypergeometric function, which can be represented in the following form:

$$_p\Psi_q(z) = {}_p\Psi_q \left[\begin{matrix} (r_i, R_i)_{1,p} \\ (s_j, S_j)_{1,q} \end{matrix} \middle| z \right] = \sum_{n=0}^{\infty} \frac{\Gamma(r_1 + R_1 n) \cdots \Gamma(r_p + R_p n)}{\Gamma(s_1 + S_1 n) \cdots \Gamma(s_q + S_q n)} \frac{z^n}{n!}, \tag{4}$$

where $r_i, s_j \in \mathbf{C}, i = 1, 2, \ldots, p; j = 1, 2, \ldots, q$, and the coefficients $R_1, \ldots, R_p \in \mathbf{R}^+$ and $S_1, \ldots, S_q \in \mathbf{R}^+$ satisfying the condition

$$\sum_{j=1}^{q} S_j - \sum_{i=1}^{p} R_i > -1. \tag{5}$$

In particular, when $R_i = S_j = 1$ $(i = 1, 2, \ldots, p; j = 1, 2, \ldots, q)$, (4) reduces to

$$_p\Psi_q \left[\begin{matrix} (r_1, 1), \ldots, (r_p, 1) \\ (s_1, 1), \ldots, (s_q, 1) \end{matrix} \middle| z \right] = \frac{\prod_{i=1}^{p} \Gamma(r_i)}{\prod_{j=1}^{q} \Gamma(s_j)} {}_pF_q \left[\begin{matrix} r_1, \ldots, r_p \\ s_1, \ldots, s_q \end{matrix} \middle| z \right], \tag{6}$$

where $_pF_q(\cdot)$ is the generalized hypergeometric function [4, 19].

Many researchers have looked at the various characteristics of generalized Mittag-Leffler functions and their extensions over the past few decades. Prabhakar [17] gave diverse properties of three-parameter Mittag-Leffler function $E^{\rho}_{\alpha,\mu}(z)$. In order to demonstrate the existence and uniqueness of the solution to the corresponding integral equation of the first kind, he also examined the integral operator with the Mittag-Leffler function as the kernel. Later, using generalized fractional calculus operators, Kilbas et al. [7] examined the several compositions of the three-parameter Mittag-Leffler function.

In this paper, we recall the four-parameter Mittag-Leffler function due to Srivastava and Tomovski [21]:

$$E^{\beta,\rho}_{\alpha,\mu}(z) = \sum_{m=0}^{\infty} \frac{(\beta)_{\rho m}}{\Gamma(\alpha m + \mu)} \frac{z^m}{m!}, \tag{7}$$

where $z, \mu, \beta \in \mathbf{C}$ and $\Re(\alpha) > \max\{0, \Re(\rho) - 1\}; \Re(\rho) > 0$. It is interesting to point out that the series representation of (7) yields the following relationships:

i. Setting $\alpha = 1, \rho = 1$, yields

$$E^{\beta,1}_{1,\mu}(z) = \sum_{m=0}^{\infty} \frac{(\beta)_m}{\Gamma(m + \mu)} \frac{z^m}{m!} = \frac{1}{\Gamma(\mu)} \, {}_1F_1\left(\beta; \mu; z\right), \tag{8}$$

where ${}_1F_1\left(\cdot\right)$ is the well-known confluent hypergeometric function [19].

ii. Setting $\rho = 1$, gives

$$E^{\beta,1}_{\alpha,\mu}(z) = \sum_{m=0}^{\infty} \frac{(\beta)_m}{\Gamma(\alpha m + \mu)} \frac{z^m}{m!} = E^{\beta}_{\alpha,\mu}(z), \tag{9}$$

where $E^{\beta}_{\alpha,\mu}(z)$ is the three-parameter Mittag-Leffler function (3).

iii. Substituting $\beta = 1, \rho = 1$, yields

$$E^{1,1}_{\alpha,\mu}(z) = \sum_{m=0}^{\infty} \frac{(1)_m}{\Gamma(\alpha m + \mu)} \frac{z^m}{m!} = E_{\alpha,\mu}(z), \tag{10}$$

where $E_{\alpha,\mu}(z)$ is the two-parameter Mittag-Leffler function (2).

iv. Substituting $\mu = 1, \beta = 1, \rho = 1$, yields

$$E^{1,1}_{\alpha,1}(z) = \sum_{m=0}^{\infty} \frac{(1)_m}{\Gamma(\alpha m + 1)} \frac{z^m}{m!} = E_{\alpha}(z), \tag{11}$$

where $E_{\alpha,\mu}(z)$ is the well-known Mittag-Leffler function (1).

Many authors have recently studied various unified integral formulas involving certain special functions. The list of such authors includes Chandak et al. [2], Choi and Agarwal [3], Jain et al. [5], Khan et al. [6], Nisar et al. [10], Nisar et al. [11], Singh et al. [20] etc. In the past few years, some generalizations of Mittag-Leffler functions and their analogues were given by many researchers and they studied the compositional properties with generalized fractional calculus operators. The publications by [1, 13–16] provide a thorough description of these generalizations and their characteristics.

2. Unified integral representations

We start with recalling the result due to Prudnikov et al. [18] for establishing our main result in the present article.

$$\int_r^s \frac{(x-r)^{\delta_1-1}(s-x)^{\delta_2-1}}{[(s-r)+l_1(x-r)+l_2(s-x)]^{\delta_1+\delta_2}}\, dx = \frac{\Gamma(\delta_1)\Gamma(\delta_2)}{\Gamma(\delta_1+\delta_2)}\frac{(1+l_1)^{-\delta_1}(1+l_2)^{-\delta_2}}{(s-r)},$$

$$(12)$$

provided that $s \neq r, \Re(\delta_1) > 0, \Re(\delta_2) > 0$ and the constants l_1 and l_2 are such that none of the expression $1+l_1, 1+l_2, (s-r)+l_1(x-r)+l_2(s-x)$, where $r \leq x \leq s$ is zero.

Theorem 1. *Let* $s \neq r, \Re(\mu) > 0, \Re(\rho) > 0$ *and the constants* l_1 *and* l_2 *are such that none of the expression* $1+l_1, 1+l_2, (s-r)+l_1(x-r)+l_2(s-x)$, *where* $r \leq x \leq s$ *is zero and let* $\alpha, \beta, \rho, \mu \in \mathbf{C}$ *such that* $\Re(\alpha) > 0, \Re(\beta) > 0, \Re(\rho) > 0, \Re(\mu) > 0$. *Then, the following integral formula holds true:*

$$\int_r^s \frac{(x-r)^{\mu-1}(s-x)^{\rho-1}}{[(s-r)+l_1(x-r)+l_2(s-x)]^{\mu+\rho}}\, E_{\alpha,\mu}^{\beta,\rho}\left(\frac{(x-r)^\alpha}{[(s-r)+l_1(x-r)+l_2(s-x)]^\alpha}\right) dx$$

$$= \frac{(1+l_1)^{-\mu}(1+l_2)^{-\rho}}{(s-r)}\, \Gamma(\rho)\, E_{\alpha,\mu+\rho}^{\beta,\rho}\left((1+l_1)^{-\alpha}\right).$$

$$(13)$$

Proof. Using (7) and reversing the order of integration and summation, we have

$$\int_r^s \frac{(x-r)^{\mu-1}(s-x)^{\rho-1}}{[(s-r)+l_1(x-r)+l_2(s-x)]^{\mu+\rho}}\, E_{\alpha,\mu}^{\beta,\rho}\left(\frac{(x-r)^\alpha}{[(s-r)+l_1(x-r)+l_2(s-x)]^\alpha}\right) dx$$

$$= \sum_{m=0}^\infty \frac{(\beta)_{\rho m}}{\Gamma(\alpha m+\mu)}\frac{1}{m!}\int_r^s \frac{(x-r)^{\alpha m+\mu-1}(s-x)^{\rho-1}}{[(s-r)+l_1(x-r)+l_2(s-x)]^{\alpha m+\mu+\rho}}\, dx.$$

Unified Integral Representations of 4-parameter Mittag-Leffler Functions 295

Using result (12), this yields

$$\int_r^s \frac{(x-r)^{\mu-1}(s-x)^{\rho-1}}{[(s-r)+l_1(x-r)+l_2(s-x)]^{\mu+\rho}}\ E_{\alpha,\mu}^{\beta,\rho}\left(\frac{(x-r)^\alpha}{[(s-r)+l_1(x-r)+l_2(s-x)]^\alpha}\right)dx$$

$$=\sum_{m=0}^\infty \frac{(\beta)_{\rho m}}{\Gamma(\alpha m+\mu)}\frac{1}{m!}\frac{\Gamma(\alpha m+\mu)\Gamma(\rho)}{\Gamma(\alpha m+\mu+\rho)}\frac{(1+l_1)^{-(\alpha m+\mu)}(1+l_2)^{-\rho}}{(s-r)}$$

$$=\frac{(1+l_1)^{-\mu}(1+l_2)^{-\rho}}{(s-r)}\Gamma(\rho)\sum_{m=0}^\infty \frac{(\beta)_{\rho m}}{\Gamma(\alpha m+\mu+\rho)}\frac{(1+l_1)^{-\alpha m}}{m!}$$

$$=\frac{(1+l_1)^{-\mu}(1+l_2)^{-\rho}}{(s-r)}\ \Gamma(\rho)\ E_{\alpha,\mu+\rho}^{\beta,\rho}\left((1+l_1)^{-\alpha}\right).$$

This completes the proof. □

Corollary 1. *Let the conditions stated in Theorem 1 be satisfied and setting $\alpha = \rho = 1$, the following integral representation holds true:*

$$\int_r^s \frac{(x-r)^{\mu-1}(s-x)^{1-1}}{[(s-r)+l_1(x-r)+l_2(s-x)]^{\mu+1}}\ {}_1F_1\left(\beta;\mu;\frac{(x-r)}{[(s-r)+l_1(x-r)+l_2(s-x)]}\right)dx$$

$$=\frac{(1+l_1)^{-\mu}(1+l_2)^{-1}}{\mu(s-r)}\ {}_1F_1\left(\beta;\mu+1;\frac{1}{1+l_1}\right).$$

Theorem 2. *Let $s \neq r, \Re(\mu) > 0, \Re(\rho) > 0$ and the constants l_1 and l_2 are such that none of the expression $1+l_1, 1+l_2, (s-r)+l_1(x-r)+l_2(s-x)$, where $r \leq x \leq s$ is zero and let $\alpha, \beta, \rho, \mu \in \mathbf{C}$ such that $\Re(\alpha) > 0, \Re(\beta) > 0, \Re(\rho) > 0, \Re(\mu) > 0$. Then, the following integral representation holds true:*

$$\int_r^s \frac{(x-r)^{\rho-1}(s-x)^{\mu-1}}{[(s-r)+l_1(x-r)+l_2(s-x)]^{\mu+\rho}}\ E_{\alpha,\mu}^{\beta,\rho}\left(\frac{(s-x)^\alpha}{[(s-r)+l_1(x-r)+l_2(s-x)]^\alpha}\right)dx$$

$$=\frac{(1+l_1)^{-\rho}(1+l_2)^{-\mu}}{(s-r)}\ \Gamma(\rho)\ E_{\alpha,\mu+\rho}^{\beta,\rho}\left((1+l_2)^{-\alpha}\right).$$

$$(14)$$

Proof. Using (7) and reversing the order of integration and summation, we have

$$\int_r^s \frac{(x-r)^{\rho-1}(s-x)^{\mu-1}}{[(s-r)+l_1(x-r)+l_2(s-x)]^{\mu+\rho}}\ E_{\alpha,\mu}^{\beta,\rho}\left(\frac{(s-x)^\alpha}{[(s-r)+l_1(x-r)+l_2(s-x)]^\alpha}\right)dx$$

$$=\sum_{m=0}^\infty \frac{(\beta)_{\rho m}}{\Gamma(\alpha m+\mu)}\frac{1}{m!}\int_r^s \frac{(x-r)^{\rho-1}(s-x)^{\alpha m+\mu-1}}{[(s-r)+l_1(x-r)+l_2(s-x)]^{\alpha m+\mu+\rho}}dx.$$

On applying (12), this immediately yields the right hand side of (14). □

Corollary 2. *Assuming the assumptions stated in Theorem 2 are true and substituting* $\alpha = \rho = 1$, *the following integral representation holds true:*

$$\int_r^s \frac{(x-r)^{1-1}(s-x)^{\mu-1}}{[(s-r)+l_1(x-r)+l_2(s-x)]^{\mu+1}} \, {}_1F_1\left(\beta;\mu;\frac{(s-x)}{[(s-r)+l_1(x-r)+l_2(s-x)]}\right) dx$$

$$= \frac{(1+l_1)^{-1}(1+l_2)^{-\mu}}{\mu(s-r)} \, {}_1F_1\left(\beta;\mu+1;\frac{1}{1+l_2}\right).$$

Theorem 3. *Let* $s \neq r, \Re(\mu) > 0, \Re(\rho) > 0$ *and the constants* l_1 *and* l_2 *are such that none of the expression* $1+l_1, 1+l_2, (s-r)+l_1(x-r)+l_2(s-x)$, *where* $r \leq x \leq s$ *is zero and let* $\alpha, \beta, \rho, \mu \in \mathbf{C}$ *such that* $\Re(\alpha) > 0, \Re(\beta) > 0, \Re(\rho) > 0, \Re(\mu) > 0$. *Then, the following integral representation is valid:*

$$\int_r^s \frac{(x-r)^{\mu-1}(s-x)^{\rho-1}}{[(s-r)+l_1(x-r)+l_2(s-x)]^{\mu+\rho}} \, E_{\alpha,\mu}^{\beta,\rho}\left(\frac{(x-r)^\alpha(s-x)^\alpha}{[(s-r)+l_1(x-r)+l_2(s-x)]^{2\alpha}}\right) dx$$

$$= \frac{(1+l_1)^{-\mu}(1+l_2)^{-\rho}}{(s-r)\,\Gamma(\beta)} \, {}_2\Psi_1\left[\begin{matrix}(\beta,\rho),(\rho,\alpha)\\(\mu+\rho,2\alpha)\end{matrix}\middle|(1+l_1)^{-\alpha}(1+l_2)^{-\alpha}\right].$$

$$(15)$$

Proof. Using (7) and reversing the order of integration and summation, we find that

$$\int_r^s \frac{(x-r)^{\mu-1}(s-x)^{\rho-1}}{[(s-r)+l_1(x-r)+l_2(s-x)]^{\mu+\rho}} \, E_{\alpha,\mu}^{\beta,\rho}\left(\frac{(x-r)^\alpha(s-x)^\alpha}{[(s-r)+l_1(x-r)+l_2(s-x)]^{2\alpha}}\right) dx$$

$$= \sum_{m=0}^{\infty} \frac{(\beta)_{\rho m}}{\Gamma(\alpha m + \mu)} \frac{1}{m!} \int_r^s \frac{(x-r)^{\alpha m+\mu-1}(s-x)^{\alpha m+\rho-1}}{[(s-r)+l_1(x-r)+l_2(s-x)]^{2\alpha m+\mu+\rho}} \, dx.$$

Using result (12), this yields

$$\int_r^s \frac{(x-r)^{\mu-1}(s-x)^{\rho-1}}{[(s-r)+l_1(x-r)+l_2(s-x)]^{\mu+\rho}} \, E_{\alpha,\mu}^{\beta,\rho}\left(\frac{(x-r)^\alpha(s-x)^\alpha}{[(s-r)+l_1(x-r)+l_2(s-x)]^{2\alpha}}\right) dx$$

$$= \sum_{m=0}^{\infty} \frac{(\beta)_{\rho m}}{\Gamma(\alpha m + \mu)} \frac{1}{m!} \frac{\Gamma(\alpha m + \mu)\Gamma(\alpha m + \rho)}{\Gamma(2\alpha m + \mu + \rho)} \frac{(1+l_1)^{-(\alpha m+\mu)}(1+l_2)^{-(\alpha m+\rho)}}{(s-r)}$$

$$= \frac{(1+l_1)^{-\mu}(1+l_2)^{-\rho}}{(s-r)} \sum_{m=0}^{\infty} \frac{(\beta)_{\rho m}\Gamma(\alpha m + \rho)}{\Gamma(2\alpha m + \mu + \rho)} \frac{(1+l_1)^{-\alpha m}(1+l_2)^{-\alpha m}}{m!}.$$

In accordance with (4), this completes the proof. $\square$

Corollary 3. *Under the conditions stated in Theorem 3 and for* $\alpha = \rho = 1$, *the following integral representation holds true:*

$$\int_r^s \frac{(x-r)^{\mu-1}(s-x)^{1-1}}{[(s-r)+l_1(x-r)+l_2(s-x)]^{\mu+1}} \, {}_1F_1\left(\beta;\mu;\frac{(x-r)(s-x)}{[(s-r)+l_1(x-r)+l_2(s-x)]^2}\right) dx$$

$$= \frac{(1+l_1)^{-\mu}(1+l_2)^{-1}}{(s-r)} \frac{\Gamma(\mu)}{\Gamma(\beta)} \, {}_2\Psi_1\left[\begin{matrix}(\beta,1),(1,1)\\(\mu+1,2)\end{matrix}\,\middle|\,(1+l_1)^{-1}(1+l_2)^{-1}\right].$$

Now we recall the Lavoie-Trottier integral formula [8]:

$$\int_0^1 u^{\delta_1-1}(1-u)^{2\delta_2-1}\left(1-\frac{u}{3}\right)^{2\delta_1-1}\left(1-\frac{u}{4}\right)^{\delta_2-1} du = \left(\frac{2}{3}\right)^{2\delta_1}\frac{\Gamma(\delta_1)\Gamma(\delta_2)}{\Gamma(\delta_1+\delta_2)}, \tag{16}$$

where $\Re(\delta_1) > 0, \Re(\delta_2) > 0$.

Theorem 4. *Let* $\alpha, \beta, \mu, \rho \in \mathbb{C}$ *such that* $\Re(\alpha) > 0, \Re(\beta) > 0, \Re(\mu) > 0, \Re(\rho) > 0, \Re(\mu+\rho) > 0,$ *and* $\omega, u > 0.$ *Then, the following integral formula is valid:*

$$\int_0^1 u^{\mu-1}(1-u)^{2\rho-1}\left(1-\frac{u}{3}\right)^{2\mu-1}\left(1-\frac{u}{4}\right)^{\rho-1} E_{\alpha,\mu}^{\beta,\rho}\left[\omega\left(u\left(1-\frac{u}{3}\right)^2\right)^\alpha\right] du$$

$$= \Gamma(\rho)\left(\frac{2}{3}\right)^{2\mu} E_{\alpha,\mu+\rho}^{\beta,\rho}\left(\omega\left(\frac{2}{3}\right)^{2\alpha}\right). \tag{17}$$

Proof. Using (7) and reversing the order of integration and summation, we have

$$\int_0^1 u^{\mu-1}(1-u)^{2\rho-1}\left(1-\frac{u}{3}\right)^{2\mu-1}\left(1-\frac{u}{4}\right)^{\rho-1} E_{\alpha,\mu}^{\beta,\rho}\left[\omega\left(u\left(1-\frac{u}{3}\right)^2\right)^\alpha\right] du$$

$$= \sum_{m=0}^\infty \frac{(\beta)_{\rho m}}{\Gamma(\alpha m+\mu)}\frac{\omega^m}{m!}\int_0^1 u^{(\alpha m+\mu)-1}(1-u)^{2\rho-1}\left(1-\frac{u}{3}\right)^{2(\alpha m+\mu)-1}\left(1-\frac{u}{4}\right)^{\rho-1} du.$$

Applying the result (16), we arrive at

$$\int_0^1 u^{\mu-1}(1-u)^{2\rho-1}\left(1-\frac{u}{3}\right)^{2\mu-1}\left(1-\frac{u}{4}\right)^{\rho-1} E_{\alpha,\mu}^{\beta,\rho}\left[\omega\left(u\left(1-\frac{u}{3}\right)^2\right)^\alpha\right] du$$

$$= \sum_{m=0}^\infty \frac{(\beta)_{\rho m}}{\Gamma(\alpha m+\mu)}\frac{\omega^m}{m!}\left(\frac{2}{3}\right)^{2(\alpha m+\mu)}\frac{\Gamma(\alpha m+\mu)\Gamma(\rho)}{\Gamma(\alpha m+\mu+\rho)}$$

$$= \Gamma(\rho)\sum_{m=0}^\infty \frac{(\beta)_{\rho m}}{\Gamma(\alpha m+\mu+\rho)}\frac{\omega^m}{m!}\left(\frac{2}{3}\right)^{2(\alpha m+\mu)}.$$

Now in the view of (7), this yields the right-hand side of Theorem 4. $\square$

Corollary 4. *If the conditions stated in Theorem 4 are true and $\alpha = \rho = 1$, then following integral representation holds true:*

$$\int_0^1 u^{\mu-1}(1-u)^{2-1}\left(1-\frac{u}{3}\right)^{2\mu-1}\left(1-\frac{u}{4}\right)^{1-1}{}_1F_1\left(\beta;\mu;\omega u\left(1-\frac{u}{3}\right)^2\right)du$$

$$= \frac{1}{\mu}\left(\frac{2}{3}\right)^{2\mu}{}_1F_1\left(\beta;\mu+1;\frac{4}{9}\omega\right).$$

Theorem 5. *Let $\alpha, \beta, \mu, \rho \in \mathbf{C}$ such that $\Re(\alpha) > 0$, $\Re(\beta) > 0$, $\Re(\mu) > 0$, $\Re(\rho) > 0$, $\Re(\mu+\rho) > 0$, and $\omega, u > 0$. Then, the following integral formula holds true:*

$$\int_0^1 u^{\rho-1}(1-u)^{2\mu-1}\left(1-\frac{u}{3}\right)^{2\rho-1}\left(1-\frac{u}{4}\right)^{\mu-1}E_{\alpha,\mu}^{\beta,\rho}\left[\omega\left((1-u)^2\left(1-\frac{u}{4}\right)\right)^{\alpha}\right]du$$

$$= \Gamma(\rho)\left(\frac{2}{3}\right)^{2\rho}E_{\alpha,\mu+\rho}^{\beta,\rho}(\omega).$$

$$(18)$$

Proof. Using (7) and reversing the order of integration and summation, we have

$$\int_0^1 u^{\rho-1}(1-u)^{2\mu-1}\left(1-\frac{u}{3}\right)^{2\rho-1}\left(1-\frac{u}{4}\right)^{\mu-1}E_{\alpha,\mu}^{\beta,\rho}\left[\omega\left((1-u)^2\left(1-\frac{u}{4}\right)\right)^{\alpha}\right]du$$

$$= \sum_{m=0}^{\infty}\frac{(\beta)_{\rho m}}{\Gamma(\alpha m+\mu)}\frac{\omega^m}{m!}\int_0^1 u^{\rho-1}(1-u)^{2(\alpha m+\mu)-1}\left(1-\frac{u}{3}\right)^{2\rho-1}\left(1-\frac{u}{4}\right)^{(\alpha m+\mu)-1}du.$$

Using result (16) and accordance to (7), this yields the right-hand side of Theorem 5. □

Corollary 5. *If the conditions stated in Theorem 5 are true and $\alpha = \rho = 1$, then the following integral representation holds true:*

$$\int_0^1 u^{1-1}(1-u)^{2\mu-1}\left(1-\frac{u}{3}\right)^{2-1}\left(1-\frac{u}{4}\right)^{\mu-1}{}_1F_1\left(\beta;\mu;\omega(1-u)^2\left(1-\frac{u}{4}\right)\right)du$$

$$= \frac{4}{9\mu}{}_1F_1\left(\beta;\mu+1;\omega\right).$$

Theorem 6. *Let $\alpha, \beta, \mu, \rho, \delta_1, \delta_2 \in \mathbf{C}$ such that $\Re(\alpha) > 0, \Re(\beta) > 0, \Re(\mu) > 0, \Re(\rho) > 0, \Re(\delta_1) > 0, \Re(\delta_2) > 0, \Re(\delta_2 + k) > 0$, and $\omega, u > 0$. Then, the following integral formula is valid:*

$$\int_0^1 u^{\delta_1-1}(1-u)^{2(\delta_2+k)-1}\left(1-\frac{u}{3}\right)^{2\delta_1-1}\left(1-\frac{u}{4}\right)^{(\delta_2+k)-1}E_{\alpha,\mu}^{\beta,\rho}\left(wu\left(1-\frac{u}{3}\right)^2\right)du$$

$$= \left(\frac{2}{3}\right)^{2\delta_1}\frac{\Gamma(\delta_2+k)}{\Gamma(\beta)}{}_2\Psi_2\left[\begin{array}{c}(\beta,\rho),(\delta_1,1)\\(\mu,\alpha),(\delta_1+\delta_2+k,1)\end{array}\middle|\frac{4}{9}\omega\right].$$

$$(19)$$

Proof. Using (7) and reversing the order of integration and summation, we find that

$$\int_0^1 u^{\delta_1-1}(1-u)^{2(\delta_2+k)-1}\left(1-\frac{u}{3}\right)^{2\delta_1-1}\left(1-\frac{u}{4}\right)^{(\delta_2+k)-1}E_{\alpha,\mu}^{\beta,\rho}\left(wu\left(1-\frac{u}{3}\right)^2\right)du$$

$$=\sum_{m=0}^{\infty}\frac{(\beta)_{\rho m}}{\Gamma(\alpha m+\mu)}\frac{\omega^m}{m!}\int_0^1 u^{(\delta_1+m)-1}(1-u)^{2(\delta_2+k)-1}\left(1-\frac{u}{3}\right)^{2(\delta_1+m)-1}\left(1-\frac{u}{4}\right)^{(\delta_2+k)-1}du.$$

Using result (16), we arrive at

$$\int_0^1 u^{\delta_1-1}(1-u)^{2(\delta_2\mid k)-1}\left(1-\frac{u}{3}\right)^{2\delta_1-1}\left(1-\frac{u}{4}\right)^{(\delta_2+k)-1}E_{\alpha,\mu}^{\beta,\rho}\left(wu\left(1-\frac{u}{3}\right)^2\right)du$$

$$=\sum_{m=0}^{\infty}\frac{(\beta)_{\rho m}}{\Gamma(\alpha m+\mu)}\frac{\omega^m}{m!}\left(\frac{2}{3}\right)^{2(\delta_1+m)}\frac{\Gamma(\delta_1+m)\Gamma(\delta_2+k)}{\Gamma(\delta_1+\delta_2+m+k)}.$$

Further, using the definition (4), we can easily obtain the desired result (19). $\qquad\square$

Corollary 6. *If the conditions stated in Theorem 6 are true and $\alpha=\rho=1$, then the following integral representation holds true:*

$$\int_0^1 u^{\delta_1-1}(1-u)^{2(\delta_2+k)-1}\left(1-\frac{u}{3}\right)^{2\delta_1-1}\left(1-\frac{u}{4}\right)^{(\delta_2+k)-1}{}_1F_1\left(\beta;\mu;wu\left(1-\frac{u}{3}\right)^2\right)du$$

$$=\frac{\Gamma(\mu)}{\Gamma(\beta)}\left(\frac{2}{3}\right)^{2\delta_1}\Gamma(\delta_2+k)\;{}_2\Psi_2\left[\begin{matrix}(\beta,1),(\delta_1,1)\\(\mu,1),(\delta_1+\delta_2+k,1)\end{matrix}\bigg|\frac{4}{9}\omega\right].$$

Theorem 7. *Let $\alpha,\beta,\mu,\rho,\delta_1,\delta_2\in\mathbf{C}$ such that $\mathfrak{R}(\alpha)>0,\mathfrak{R}(\beta)>0,\mathfrak{R}(\mu)>0,\mathfrak{R}(\rho)>0,\mathfrak{R}(\delta_1)>0,\mathfrak{R}(\delta_2)>0,\mathfrak{R}(\delta_1+k)>0$, and $w,u>0$. Then, the following integral formula holds true:*

$$\int_0^1 u^{(\delta_1+k)-1}(1-u)^{2\delta_2-1}\left(1-\frac{u}{3}\right)^{2(\delta_1+k)-1}\left(1-\frac{u}{4}\right)^{\delta_2-1}E_{\alpha,\mu}^{\beta,\rho}\left(\omega(1-u)^2\left(1-\frac{u}{4}\right)\right)du$$

$$=\left(\frac{2}{3}\right)^{2(\delta_1+k)}\frac{\Gamma(\delta_1+k)}{\Gamma(\beta)}\;{}_2\Psi_2\left[\begin{matrix}(\beta,\rho),(\delta_2,1)\\(\mu,\alpha),(\delta_1+\delta_2+k,1)\end{matrix}\bigg|\omega\right].$$

$$(20)$$

Proof. Using (7), we have

$$\int_0^1 u^{(\delta_1+k)-1}(1-u)^{2\delta_2-1}\left(1-\frac{u}{3}\right)^{2(\delta_1+k)-1}\left(1-\frac{u}{4}\right)^{\delta_2-1}E_{\alpha,\mu}^{\beta,\rho}\left(\omega(1-u)^2\left(1-\frac{u}{4}\right)\right)du$$

$$=\sum_{m=0}^{\infty}\frac{(\beta)_{\rho m}}{\Gamma(\alpha m+\mu)}\frac{\omega^m}{m!}\int_0^1 u^{(\delta_1+k)-1}(1-u)^{2(\delta_2+m)-1}\left(1-\frac{u}{3}\right)^{2(\delta_1+k)-1}\left(1-\frac{u}{4}\right)^{(\delta_2+m)-1}du.$$

Using result (16) and the definition (4), one can easily obtain the right-hand side of Theorem 7. $\qquad\square$

Corollary 7. *If the conditions stated in Theorem 7 are true and $\alpha = \rho = 1$, then the following integral representation holds true:*

$$\int_0^1 u^{(\delta_1+k)-1}(1-u)^{2\delta_2-1}\left(1-\frac{u}{3}\right)^{2(\delta_1+k)-1}\left(1-\frac{u}{4}\right)^{\delta_2-1}{}_1F_1\left(\beta;\mu;\omega(1-u)^2\left(1-\frac{u}{4}\right)\right)du$$

$$=\frac{\Gamma(\mu)}{\Gamma(\beta)}\left(\frac{2}{3}\right)^{2(\delta_1+k)}\Gamma(\delta_1+k)\ {}_2\Psi_2\left[\begin{matrix}(\beta,1),(\delta_2,1)\\(\mu,1),(\delta_1+\delta_2+k,1)\end{matrix}\bigg|\omega\right].$$

Furthermore, we need to recall the following Oberhettinger's integral formula [12]:

$$\int_0^\infty x^{\delta_1-1}\left(x+a+\sqrt{x^2+2ax}\right)^{-\delta_2}dx=2\delta_2\ a^{-\delta_2}\left(\frac{a}{2}\right)^{\delta_1}\frac{\Gamma(2\delta_1)\Gamma(\delta_2-\delta_1)}{\Gamma(1+\delta_1+\delta_2)},\tag{21}$$

provided $\Re(\delta_2) > \Re(\delta_1) > 0$.

Theorem 8. *Let $\delta_1, \delta_2, \alpha, \mu, \beta, \rho \in \mathbf{C}$ such that $\Re(\alpha) > 0, \Re(\beta) > 0, \Re(\mu) > 0, \Re(\rho) > 0$, and $\Re(\delta_2) > \Re(\delta_1) > 0, \Re(\delta_2) > -1$. Then the following integral formula is valid:*

$$\int_0^\infty x^{\delta_1-1}\left(x+a+\sqrt{x^2+2ax}\right)^{-\delta_2}E_{\alpha,\mu}^{\beta,\rho}\left(\frac{\omega}{x+a+\sqrt{x^2+2ax}}\right)dx$$

$$=\frac{a^{\delta_1-\delta_2}}{2^{\delta_1-1}}\frac{\Gamma(2\delta_1)}{\Gamma(\beta)}\times{}_3\Psi_3\left[\begin{matrix}(\beta,\rho),(\delta_2-\delta_1,1),(\delta_2+1,1)\\(\mu,\alpha),(\delta_1+\delta_2+1,1),(\delta_2,1)\end{matrix}\bigg|\frac{\omega}{a}\right],\quad x>0.\tag{22}$$

Proof. Making use of (7) in the integrand of (22) and interchanging the order of integral sign and summation, we find that

$$\int_0^\infty x^{\delta_1-1}\left(x+a+\sqrt{x^2+2ax}\right)^{-\delta_2}E_{\alpha,\mu}^{\beta,\rho}\left(\frac{\omega}{x+a+\sqrt{x^2+2ax}}\right)dx$$

$$=\sum_{m=0}^\infty\frac{(\beta)_{\rho m}}{\Gamma(\alpha m+\mu)}\frac{\omega^m}{m!}\times\int_0^\infty x^{\delta_1-1}\left(x+a+\sqrt{x^2+2ax}\right)^{-\delta_2-m}dx.\tag{23}$$

Using the result (21), we obtain

$$\int_0^\infty x^{\delta_1-1}\left(x+a+\sqrt{x^2+2ax}\right)^{-\delta_2}E_{\alpha,\mu}^{\beta,\rho}\left(\frac{\omega}{x+a+\sqrt{x^2+2ax}}\right)dx$$

$$=\sum_{m=0}^\infty\frac{(\beta)_{\rho m}}{\Gamma(\alpha m+\mu)}\frac{\omega^m}{m!}\times 2(\delta_2+m)\ a^{-\delta_2-m}\left(\frac{a}{2}\right)^{\delta_1}\frac{\Gamma(2\delta_1)\Gamma(\delta_2+m-\delta_1)}{\Gamma(1+\delta_1+\delta_2+m)}.\tag{24}$$

In accordance with result (21), equation (24) yields the proof of Theorem 8. $\qquad\square$

Unified Integral Representations of 4-parameter Mittag-Leffler Functions 301

Corollary 8. *Let the conditions stated in Theorem 8 are valid and* $\alpha = \rho = 1$. *Then the following integral representation holds true:*

$$\int_0^\infty x^{\delta_1-1} \left(x + a + \sqrt{x^2 + 2ax}\right)^{-\delta_2} {}_1F_1\left(\beta; \mu; \frac{\omega}{x + a + \sqrt{x^2 + 2ax}}\right) dx$$

$$= \frac{a^{\delta_1-\delta_2}}{2^{\delta_1-1}} \frac{\Gamma(2\delta_1)\Gamma(\mu)}{\Gamma(\beta)} \times {}_3\Psi_3\left[\begin{matrix}(\beta,1),(\delta_2-\delta_1,1),(\delta_2+1,1) \\ (\mu,1),(\delta_1+\delta_2+1,1),(\delta_2,1)\end{matrix}\middle|\frac{\omega}{a}\right].$$

Theorem 9. *Let* $\delta_1, \delta_2, \alpha, \mu, \beta, \rho \in \mathbf{C}$ *such that* $\Re(\alpha) > 0, \Re(\beta) > 0, \Re(\mu) > 0, \Re(\rho) > 0$, *and* $\Re(\delta_2 - \delta_1) > 0, \Re(\delta_2) > -1, \Re(\delta_1) > 0$. *Then the following integral formula is valid:*

$$\int_0^\infty x^{\delta_1-1} \left(x + a + \sqrt{x^2 + 2ax}\right)^{-\delta_2} E_{\alpha,\mu}^{\beta,\rho}\left(\frac{\omega x}{x + a + \sqrt{x^2 + 2ax}}\right) dx$$

$$= \frac{a^{\delta_1-\delta_2}}{2^{\delta_1-1}} \frac{\Gamma(\delta_2-\delta_1)}{\Gamma(\beta)} \times {}_3\Psi_3\left[\begin{matrix}(\beta,\rho),(2\delta_1,2),(\delta_2+1,1) \\ (\mu,\alpha),(\delta_1+\delta_2+1,2),(\delta_2,1)\end{matrix}\middle|\frac{\omega}{2}\right], \quad x > 0. \tag{25}$$

Proof. Making use of (7) in the integrand of (25) and interchanging the order of integral sign and summation, we find that

$$\int_0^\infty x^{\delta_1-1} \left(x + a + \sqrt{x^2 + 2ax}\right)^{-\delta_2} E_{\alpha,\mu}^{\beta,\rho}\left(\frac{\omega x}{x + a + \sqrt{x^2 + 2ax}}\right) dx$$

$$= \sum_{m=0}^\infty \frac{(\beta)_{\rho m}}{\Gamma(\alpha m + \mu)} \frac{\omega^m}{m!} \times \int_0^\infty x^{\delta_1+m-1} \left(x + a + \sqrt{x^2 + 2ax}\right)^{-\delta_2-m} dx. \tag{26}$$

Using the result (21), we have

$$\int_0^\infty x^{\delta_1-1} \left(x + a + \sqrt{x^2 + 2ax}\right)^{-\delta_2} E_{\alpha,\mu}^{\beta,\rho}\left(\frac{\omega x}{x + a + \sqrt{x^2 + 2ax}}\right) dx$$

$$= \sum_{m=0}^\infty \frac{(\beta)_{\rho m}}{\Gamma(\alpha m + \mu)} \frac{\omega^m}{m!} \times 2(\delta_2+m)\, a^{-\delta_2-m} \left(\frac{a}{2}\right)^{\delta_1+m} \frac{\Gamma(2(\delta_1+m))\Gamma(\delta_2-\delta_1)}{\Gamma(1+\delta_1+\delta_2+2m)}. \tag{27}$$

In accordance with result (21), equation (27) yields the proof of Theorem 9. $\square$

Corollary 9. *Let the conditions stated in Theorem 9 are valid and* $\alpha = \rho = 1$. *Then the following integral representation holds true:*

$$\int_0^\infty x^{\delta_1-1} \left(x + a + \sqrt{x^2 + 2ax}\right)^{-\delta_2} {}_1F_1\left(\beta; \mu; \frac{\omega x}{x + a + \sqrt{x^2 + 2ax}}\right) dx$$

$$= \frac{a^{\delta_1-\delta_2}}{2^{\delta_1-1}} \frac{\Gamma(\delta_2-\delta_1)\Gamma(\mu)}{\Gamma(\beta)} \times {}_3\Psi_3\left[\begin{matrix}(\beta,1),(2\delta_1,2),(\delta_2+1,1) \\ (\mu,1),(\delta_1+\delta_2+1,2),(\delta_2,1)\end{matrix}\middle|\frac{\omega}{2}\right].$$

3. Concluding remarks

The studies described above served as inspiration for the development of new finite integral formulations in this study, which are connected to the four-parameter Mittag-Leffler function and can separate a number of sub-results in terms of generalized special functions from our primary results by selecting different values for the parameters.

Statements & Declarations

Conflicts of interest/Competing interests

The authors declare that they have no competing interests.

Funding

Not applicable.

References

[1] D. Baleanu, P. Agarwal, S.D. Purohit, *Certain fractional integral formulas involving the product of generalized Bessel functions*, Scientific World J. 2013, 1-9.

[2] S. Chandak, S.K.Q. Al-Omari, D.L. Suthar, *Unified integral associated with the generalized V-function*, Adv. Differ. Equ. (2020), 560.

[3] J. Choi, P. Agarwal, *Certain unified integrals involving a product of Bessel functions of the first kind*, Honam Math. J. 35(4) (2013), 667-677.

[4] A. Erdélyi, W. Magnus, F. Oberhettinger, F.G. Tricomi, *Higher transcendental functions*, Vol. I., McGraw-Hill, New York, 1953.

[5] R.K. Jain, A. Bhargava, M. Rizwanullah, *Certain New Integrals Including Generalized Bessel-Maitland Function and M-Series*, Int. J. Appl. Comput. Math 8 (2022), 14.

[6] N. Khan, M.I. Khan, T. Usman, K. Nonlaopon, S. Al-Omari, *Unified integrals of generalized Mittag–Leffler functions and their graphical numerical investigation*, Symmetry 14 (2022), 869.

[7] A.A. Kilbas, M. Saigo, R.K. Saxena, *Generalized Mittag-Leffler function and generalized fractional calculus operators*, Integral Transforms Spec. Funct. 17 (2004), 31-49.

[8] J.L. Lavoie, G. Trottier, *On the sum of certain Appell's series*, Ganita 20(1) (1969), 31-32.

[9] G.M. Mittag-Leffler, *Sur la nouvelle fonction $E_\alpha(x)$*, C. R. Acad. Sci. Paris 137 (1903), 554-558.

[10] K.S. Nisar, R.K. Parmar, A.H. Abusufian, *Certain new unified integrals with the generalized k-Bessel function*, Far East J. Math. Sci. 100 (2016), 1533-1544.

[11] K.S. Nisar, D.L. Suthar, S.D. Purohit, M. Aldhaifallah, *Some unified integral associated with the generalized Struve function*, Proc. Jangjeon Math. Soc. 20(2) (2017), 261-267.

[12] F. Oberhettinger, *Tables of Mellin Transforms*, Springer-Verlag, New York, 1974.

[13] A. Pal, *Some finite integrals involving Mittag-Leffler confluent hypergeometric function*, Analysis, 44(1) (2024), 17-24.

[14] A. Pal, R.K. Jana, A.K. Shukla, *Some integral representations of the $_pR_q(\alpha, \beta; z)$ function*, Int. J. Appl. Comput. Math 6 (2020), 72.

[15] A. Pal, R.K. Jana, A.K. Shukla, *Generalized fractional calculus operators and the $_pR_q(\lambda, \eta; z)$ function*, Iran. J. Sci. Technol. Trans. Sci. 44 (2020), 1815-1825.

[16] A. Pal, R.K. Jana, G.S. Khammash, A.K. Shukla, *The incomplete exponential $_pR_q(\alpha, \beta; z)$ function with applications*, Georgian Math. J., vol. 29(1) (2022), 95-107.

[17] T.R. Prabhakar, *A singular integral equation with a generalized Mittag-Leffler function in the kernel*, Yokohama Math. J. 19 (1971), 7-15.

[18] A.P. Prudnikov, Y.A. Brychkov, O.I. Marichev, *Integral and Series V.1. More Special Function*; Gordon and Breach: New York, NY, USA; London, UK, 1992.

[19] E.D. Rainville, *Special functions*, The Macmillan Company, New York, 1960.

[20] P. Singh, S. Jain, C. Cattani, *Some unified integrals for generalized Mittag-Leffler functions*, Axioms 10 (2021), 261.

[21] H.M. Srivastava, Ž. Tomovski, *Fractional calculus with an integral operator containing a generalized Mittag-Leffler function in the kernel*, Appl. Math. Comput., 211 (2009), 198-210.

[22] A. Wiman, *Über den fundamentalsatz in der Theorie der Funktionen $E_\alpha(x)$*, Acta Math. 29 (1905), 191-201.

[23] E.M. Wright, *On the coefficients of power series having exponential singularities*, J. Lond. Math. Soc. 8 (1933), 71-79.

[24] E.M. Wright, *The asymptotic expansion of the generalized hypergeometric functions*, J. London Math. Soc. 10 (1935), 286-293.

[25] E.M. Wright, *The asymptotic expansion of integral functions defined by Taylor series*, Philos. Trans. Roy. Soc. London A 238 (1940), 423-451.

© 2025 World Scientific Publishing Company
https://doi.org/10.1142/9789819812202_bmatter

Author Index

Avalos-Ramos, C., 221

Bharali, H., 123
Bhatnagar, S., 25
Bhatter, S., 185

Croitoru, A., 59, 95, 143

Gavriluţ, A., 59

Iosif, A., 59

Kainth, S. P. S., 1
Kalita, H., 123, 143
Koley, S., 205
Kumari, K., 291

Mahanta, S., 175

Nishant, 185

Pal, A., 291
Paul, S., 205
Purohit, S. D., 185

Robdera, M. A., 263

Sambucini, A. R., 59
Sekhose, V., 123
Singh, N., 1
Skvortsov, V., 43
Stamate, C., 95

Talvila, E., 157

www.ingramcontent.com/pod-product-compliance
Lightning Source LLC
Chambersburg PA
CBHW050353090625
27790CB00004B/21